高职高专“十二五”规划教材

机电设备安装与维修专业英语

● 李文波 主编 ● 华 杰 主审

English for Electromechanical Equipment Installation and Maintenance

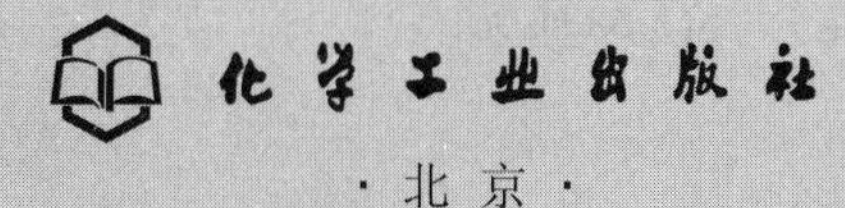

· 北京 ·

本书包含三个模块，共 75 篇课文。第一个模块内容为机械设备，分为两个部分：机械基础知识与工艺设备，介绍了英文识图、工程材料、机械零件、热处理、机加工方法及机床、CAD/CAM以及液压气压传动的理论、钳工铆工基本操作、管道、压力容器、热交换器、反应器、泵、压缩机、球罐、设备安装记录表以及工程安全和求职应聘所需的知识，共 27 篇课文。第二模块为焊接模块，分为三个部分：焊接基础知识、焊接方法和焊接检验，介绍了焊接的分类、位置、焊接接头及焊缝、焊接电流与极性、焊接工艺规程、焊接机器人、焊条电弧焊等焊接方法、焊接缺陷以及射线探伤等焊接检验方面的知识，共 23 篇课文。第三模块为电气仪表，较为全面地介绍了电工学、电磁学、电气测量、二极管、晶体管、集成电路、PLC、逻辑门、电机、变压器的基本理论以及电气系统和仪表系统安装的知识，共 25 篇课文。

本教材适用于高职高专机电设备安装及维修、焊接和电气仪表等专业的学生使用。

图书在版编目（CIP）数据

机电设备安装与维修专业英语/李文波主编. —北京：化学工业出版社，2015.7（2023.2重印）

高职高专“十二五”规划教材

ISBN 978-7-122-24189-4

Ⅰ.①机… Ⅱ.①李… Ⅲ.①机电工程-英语-高等职业教育-教材 Ⅳ.①H31

中国版本图书馆 CIP 数据核字（2015）第 119757 号

责任编辑：高 钰　　装帧设计：刘丽华

责任校对：边 涛

出版发行：化学工业出版社（北京市东城区青年湖南街 13 号 邮政编码 100011）

印　　刷：三河市航远印刷有限公司

装　　订：三河市宇新装订厂

787mm×1092mm 1/16 印张 18½ 字数 462 千字 2023 年 2 月北京第 1 版第 8 次印刷

购书咨询：010-64518888　　售后服务：010-64518899

网　　址：http://www.cip.com.cn

凡购买本书，如有缺损质量问题，本社销售中心负责调换。

定　　价：58.00 元

为了进一步贯彻［2014］19号文《国务院关于加快发展现代职业教育的决定》的文件精神，满足企业对技术工人专业英语水平的要求，使专业英语能成为学生顺利择业、就业的工具之一，编者在结合机电设备安装及维修专业一体化课程所涵盖内容的基础上，编写了本书。

本书采用模块式结构，由三个专业模块（机械设备模块、焊接模块和电气仪表模块）组成。其中，机械设备模块分为机械基础知识和工艺设备两部分，27篇课文；焊接模块分为焊接基础知识、焊接方法和焊接检验三个部分，23篇课文；电气仪表模块中，包含有25篇课文。教师可在指导学生学完本专业模块的基础上，适当选用其他模块中为本专业所需的知识内容进行教授，同时学生自学时，也可选学自己感兴趣的部分。

书中课文由属于课前预习范畴的“看和学或译”、课文正文、知识拓展以及课后练习组成。其中，课文篇幅尽可能短小精悍，图文并茂；练习多以连线、填空和判断为主，旨在让学生消除畏学烦学专业英语的心理，增强学习信心提高学习兴趣。

本书由湖南省工业技师学院李文波主编，湖南理工学院教授华杰主审。陈梅春、杨育红、贾华川、吴艺、张宏武、李春晓、易灿以及梁建民参加了编写和搜集资料，在此一并表示感谢。

为了方便教师教学，本书还配有部分译文、电子教案和习题答案，并将免费提供给采用本书作为教材的院校使用。如有需要，请发电子邮件至 cipedu@163.com 获取，或登录 www.cipedu.com.cn 免费下载。

由于编者水平有限，本书还有待进一步完善，书中错误和不当之处，敬请读者和同行们批评指正。

编　者

2015年4月

Contents
目录

Mechanical Equipments Part

Part Ⅰ Basic Knowledge of Mechanical Engineering

Lesson 1 Drawings (Ⅰ)

Look and select

Look at the pictures and select the correct terms from the box.

designer	angle	drawing	view

1. ____________________

2. ____________________

3. ____________________

4. ____________________

Text

1. Coordinate system

The basic of all input AutoCAD is the Cartesian coordinate system. Fig.1.1 (a) illustrates the axis for two-dimension (2D) and Fig.1.1 (b) for three-dimension (3D).

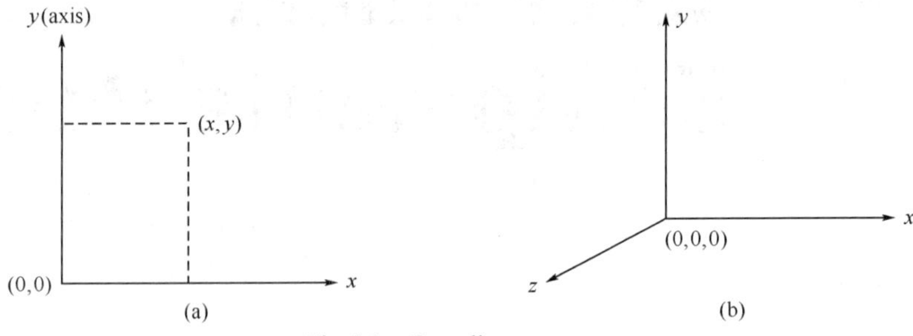

Fig.1.1 Coordinate system

2. Types of views

There are many views as follows:

(1) The six basic views An orthographic projection of an object can be seen from the front, top, bottom, rear, right and left sides: the front view, top view, bottom view, rear view, right side view and left side view. They are shown in Fig.1.2.

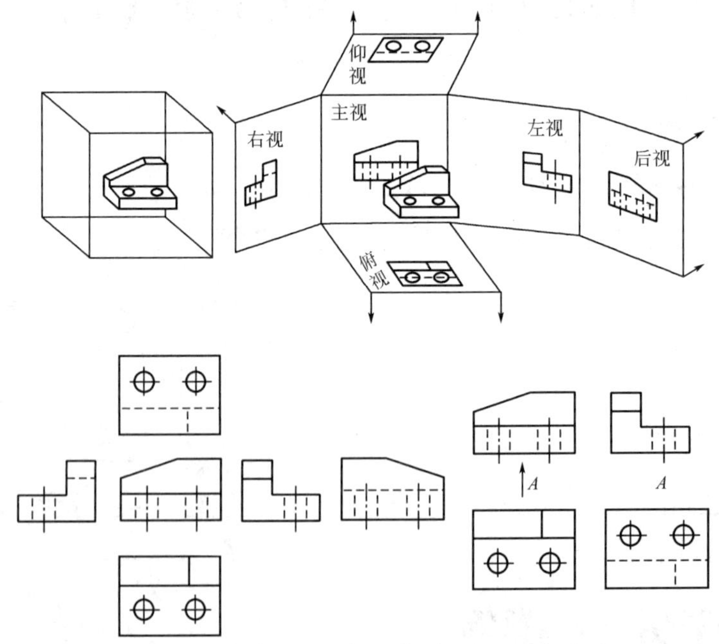

Fig.1.2 Six basic views

(2) Auxiliary view Any view created by projecting 90° to an inclined surface, datum plane, or along an axis.

a. Partial view When a symmetrical objects is drafted, two views are sufficient to represent it (typically, one view is omitted). *A* partial view (see Fig.1.3) can be used to substitute one of the two views. Sectional and auxiliary views are also commonly used to represent part detail.

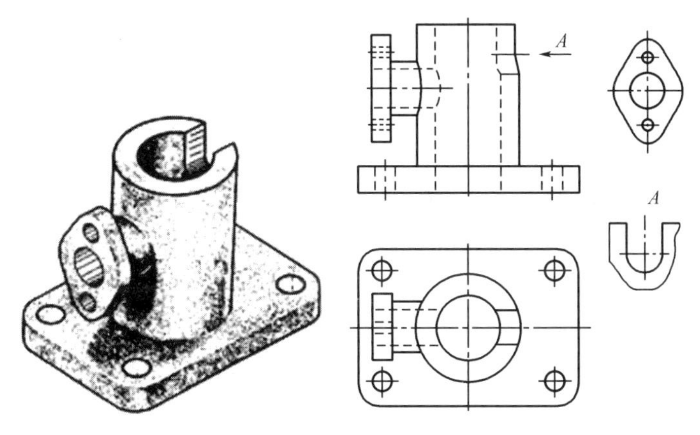

Fig.1.3 Partial view

b. Inclined view Any view created by projecting an object onto an plane which is not parallel to any basic projecting plane, which is illustrated in Fig.1.4.

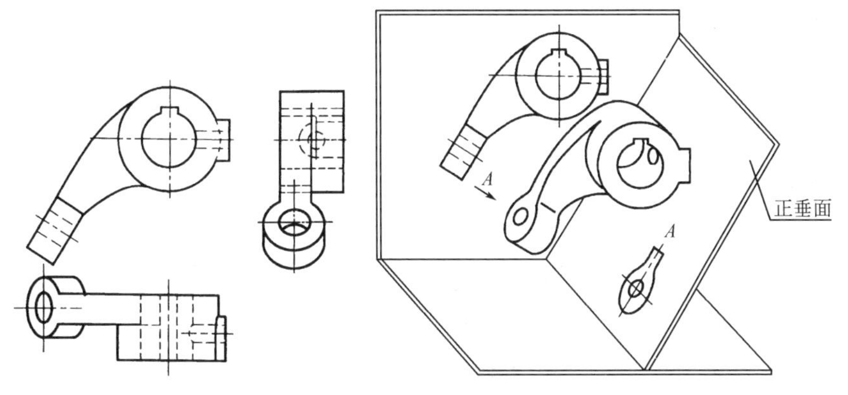

Fig.1.4 Inclined view

c. Revolved view *A* planar, area cross-section was revolved 90° degrees about the cutting plane line and offset alone its length as shown in Fig.1.5.

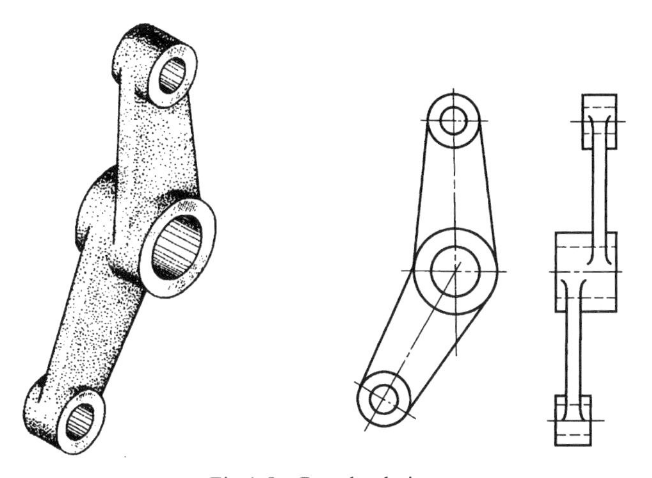

Fig.1.5 Revolved view

(3) Section view illustrated in Fig.1.6, displays a cross-section for a particular view.

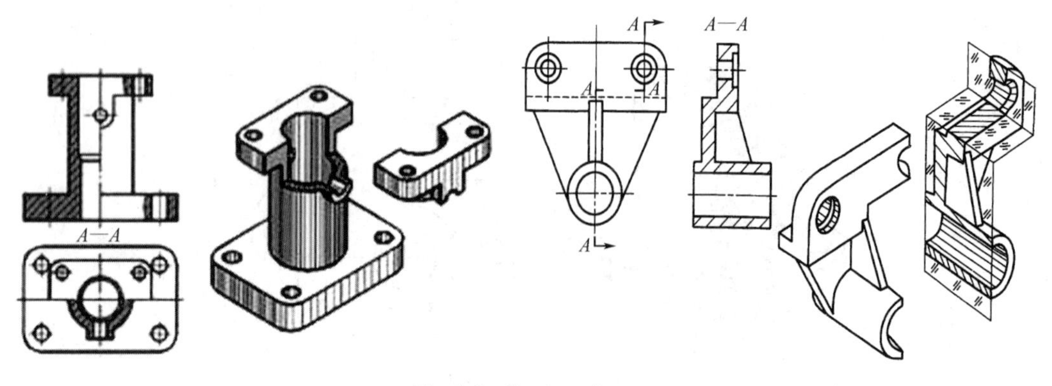

Fig.1.6 Section view

(4) Broken View Used on large objects to remove a section between two points and move the remaining sections close together as per Fig.1.7.

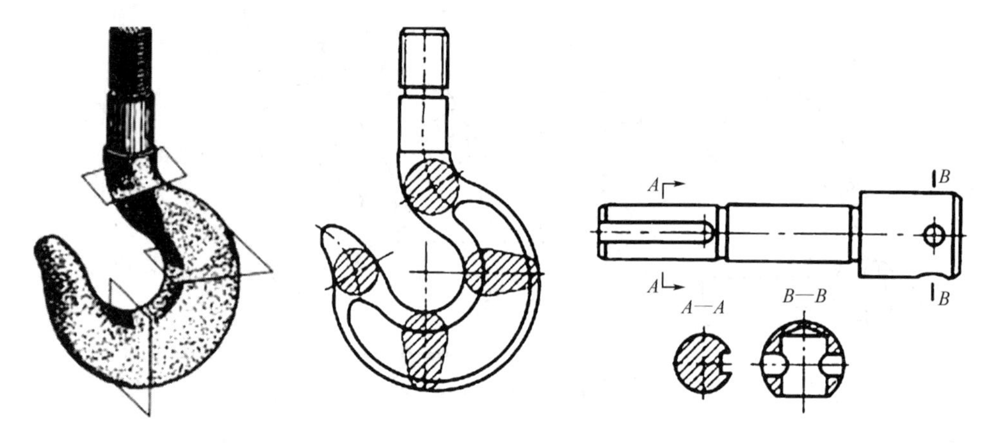

Fig.1.7 Broken View

Knowledge extention

My name is Liu Xiaobing. I'm 18 years old. I am a student in Hunan Indust rial Technician collage (Central South Industrial School). My major is Electromechanical Equipment Installation and Maintenance. I like my major very much. I want to become a good skilled worker in the future.

Exercises

1. Answer the following questions according to the hints given above.

(1) What's your name?

(2) What's your major?

(3) What's your job?

(4) Try to make a simple self-introduction.

2. Match the English with the Chinese (Table 1.1).

Table 1.1 Six basic views

1	the front view	a	主视图
2	the top view	b	仰视图
3	the bottom view	c	俯视图
4	the rear view	d	左视图
5	the right side view	e	后视图
6	the left side view	f	右视图

1. ________ 2. ________ 3. ________ 4. ________ 5. ________ 6. ________

3. Match the English with the Chinese. Draw lines.

A.

(1) auxiliary view a. 斜视图
(2) partial view b. 旋转视图
(3) inclined view c. 辅助视图
(4) revolved view d. 剖视图
(5) section view e. 断面图
(6) broken View f. 局部视图

B.

(1) 2D a. 三维
(2) 3D b. 国际标准组织
(3) ISO c. 二维
(4) HSE d. 健康，安全，环境
(5) SS e. 碳钢
(6) CS f. 不锈钢
(7) AS g. 合金钢

Words and phrases

drawing ['drɔ:ɪŋ] 图纸
designer [dɪ'zaɪnə(r)] 设计师
angle ['æŋgl] 角
assembly[ə'semblɪ] 装配
view 视图
coordinate [kəʊ'ɔ:dɪneɪt] 坐标系
axis ['æksɪs] 坐标轴
orthographic [ˌɔ:θə'græfɪk] 正交的
auxiliary view 辅助视图
inclined view 斜视图
section view 剖视图
electromechanical [ɪ'lektrəʊmɪ'kænɪkəl] 机电的
major ['meɪdʒə(r)] 专业
equipment [ɪ'kwɪpmənt] 设备
installation [ˌɪnstə'leɪʃn] 安装
maintenance ['meɪntənəns] 维修保养
skilled 有技术的
system ['sɪstəm] 系统
industrial [ɪn'dʌstriəl] 工业的
projection [prə'dʒekʃn] 投影
partial view 局部视图
revolved view 旋转视图
broken view 断面图

Lesson 2 Drawings（Ⅱ）

Look and learn

Assembly drawing

a. Dimensions — Overall dimensions of an assembled object are usually indicated as shown in Fig.2.1.

b. Internal parts — If there are internal assemblies, sectional view should be used.

c. Parts list — An example of parts list is shown in Table 2.1.

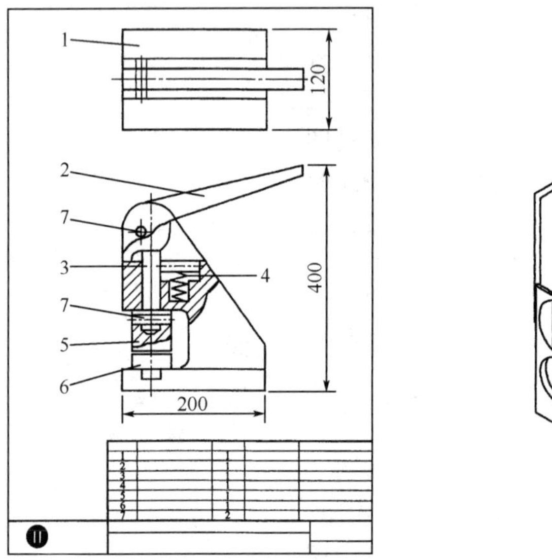

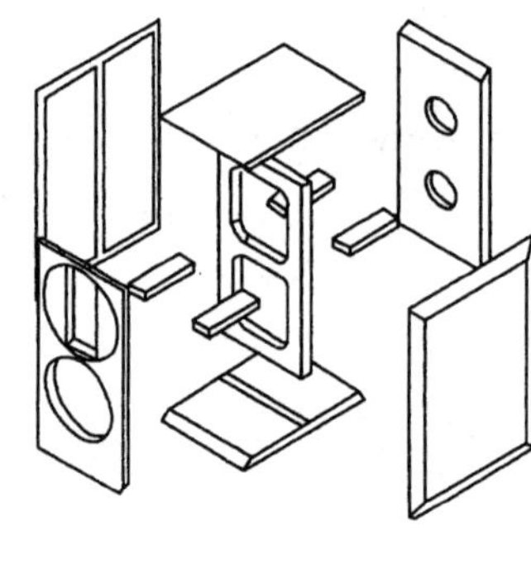

Fig.2.1 Assembly drawing

Example of parts list (Table 2.1).

Table 2.1 Parts list

Item No.	Description	Qty	Material	Remarks

Text

A complete drawing (Fig.2.2) should include at least the following information.

1. Sufficient orthographic **views** of the part concerned.
2. **Dimensions** and instructional notes.
3. **Scale** used.

4. **Projection** used, for example, first or third angle.
5. Drafting **standard** reference, for example, AS 1100.
6. **Name** or title of drawing.
7. Dimensional **unit**s which apply.
8. **Tolerance**s where necessary.
9. **Surface finish** requirements.
10. **Special treatment**s needed.
11. Type of **material** used: AS, SS, CS (HCS, MCS, LCS).
12. **Drawn** by, **checked** by, **appr**oved by , **issued** by .
13. Relevant **date**s of action by those concerned.
14. **Zone** reference system when necessary.
15. Drawing sheet **size**.
16. Name of **company** (CO. LTD.) or **department (** DEPT. **)** as applicable.
17. Drawing sheet reference, for example, **sheet 1 of 2**.

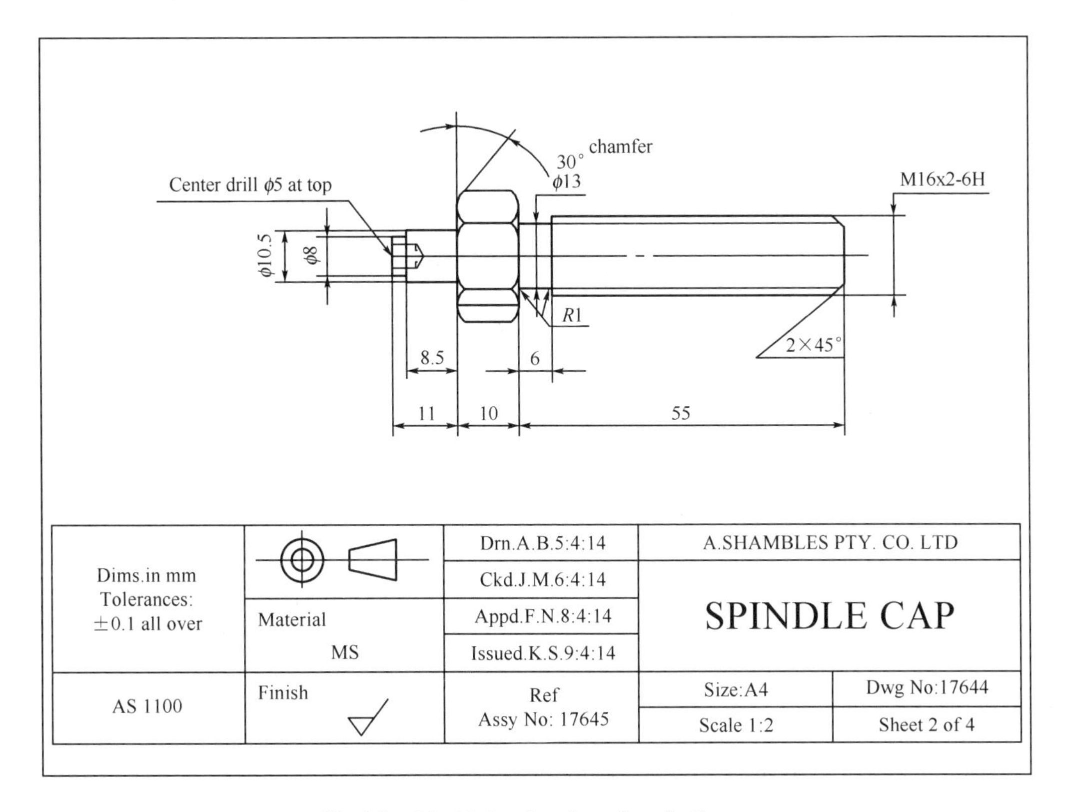

Fig.2.2 Machining drawing of a spindle cap

Knowledge extention

Understand the engineering drawing (Fig.2.3) (stock size: ϕ1.5×0.9).

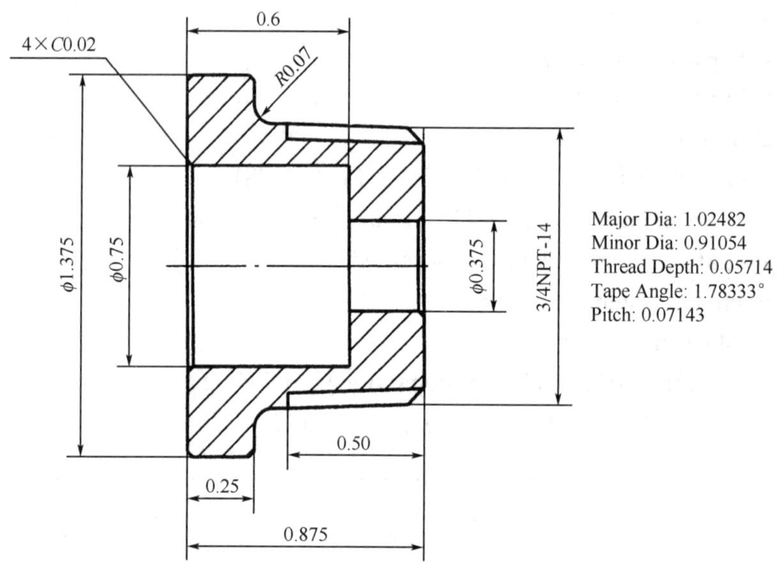

Fig.2.3 Part drawing

Translate the title bars (Table 2.2).

Table 2.2 Title bars

	Name	Date		
Drawn by	Jack	12: 10: 14	TUBE UNION	
Checked by	Jim	13: 10: 14		
Eng. Dept. Appr.				
Mfg. Appr.				
Unless otherwise specified, dimensions are in inches. Angle: 0.1° xx: 0.01 xxx: 0.005			Size A	Dwg. No. 001
			Material: 20 steel	
			Scale: 1:1	Weight: 200g · Sheet 1of 1

Exercises

1. Match the English with the Chinese. Draw lines.

A.

(1) item No.	a. 数量
(2) description	b. 零件位号
(3) quantity	c. 备注
(4) material	d. 内容
(5) remark	e. 材料

B.

(1) engineering	a. 部门
(2) department	b. 工程
(3) manufacture	c. 明确，说明
(4) otherwise	d. 制造
(5) specify	e. 英寸
(6) inch	f. 管接头
(7) weight	g. 否则，另外
(8) tube union	h. 重量

2. Translate the English into Chinese (Table 2.3):

Table 2.3 Title bars of drawing about spindle cap

English	Chinese
Dims. in mm	
Tolerance ± 0.1 all over	
Material MS	
Finish	
Drn.	
Ckd.	
Appd.	
Issued	
SPINDLE CAP 指	
Size A4	
Scale 1:2	
Dwg No 17644	
Sheet 2 of 4	

Words and phrases

view [vjuː] 视图
dimension [dɪˈmɛnʃən, daɪ-] 尺寸
scale [skeɪl] 比例
draftsman [ˈdrɑːftsmən] 制图人
standard [ˈstændəd] 标准
AS=American standard 美国标准
name 姓名
unit 单位
tolerance [ˈtɒlərəns] 公差，允许度
sheet size 纸张尺寸
department [dɪˈpɑːtmənt] 部门
surface finish 表面粗糙度
special treatment 特殊要求
material [məˈtɪəriəl] 材料
carbon steel 碳钢
check 检查人
approve 审批人
issue 分发
date 日期
zone [zəʊn] 区域
company [ˈkʌmpənɪ] 公司
sheet 1 of 2 共 2 页，第 1 页

Lesson 3 Mechanical Parts

Look and select

Look at the pictures and select the correct terms from the box.

pins	splines	keys	parts

1. ____________

2. ____________

3. ____________

4. ____________

Text

1. Shafts

As a machine component, a shaft is commonly a cylindrical bar that supports and rotates with devices for receiving and delivering rotary motion and torque (See Fig.3.1).

Fig.3.1 Shafts

2. Shaft Accessories

Keys, Splines, and Pins The primary purpose of keys, splines, and pins is to prevent relative rotary movement.

3. Clutch

A clutch is a device for quickly and easily connecting or disconnecting a rotary shaft with a rotating coaxial shaft (Fig.3.2).

4. Screws

Screws have been used as fastener for a long time. A screw consists of a circular cylinder (or truncated cone) with a helical groove in it (Fig.3.3).

5. Springs

A spring is a load-sensitive, energy-storing device, the chief characteristics of which are an ability to tolerate large deflections without failure and to recover its initial size and shape when loads are removed (Fig.3.4).

Fig.3.2 Clutch

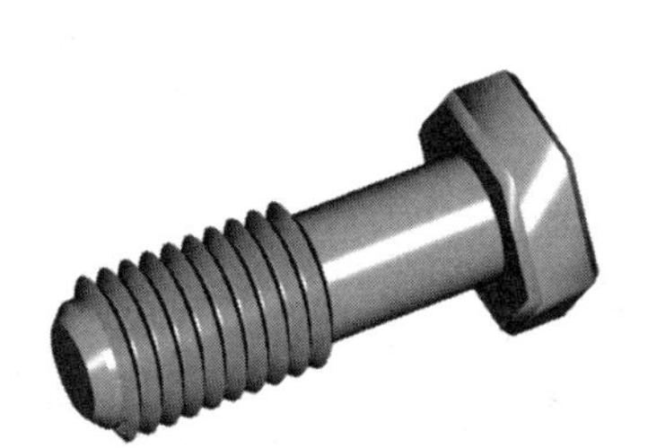

Fig.3.3 Screw

Fig.3.4 Springs

6. Ball Bearings

Ball bearings are used in almost every kind of machine and device with rotating parts. The bearing must be provided with adequate mounting, lubrication, and sealing (Fig.3.5).

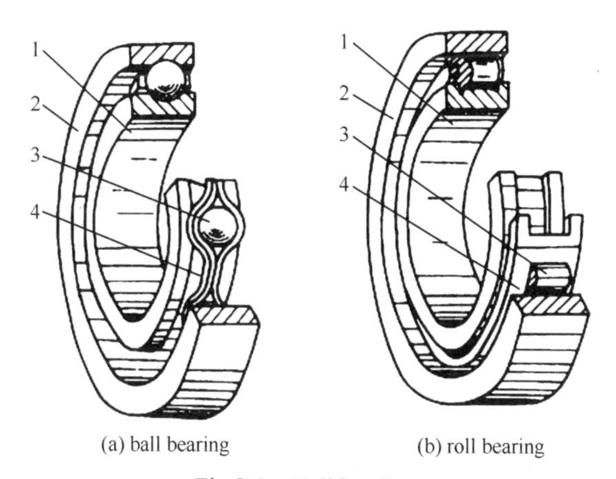

(a) ball bearing (b) roll bearing

Fig.3.5 Ball bearings

1—inner ring; 2—outer ring; 3—ball/roll; 4—cage/separator

7. Cam

A cam is a machine member that drives a follower through a specified motion.

8. Gears

Basic gear forms: Spur gear and pinion, bevel gears and worm gear and worm (Fig.3.6).

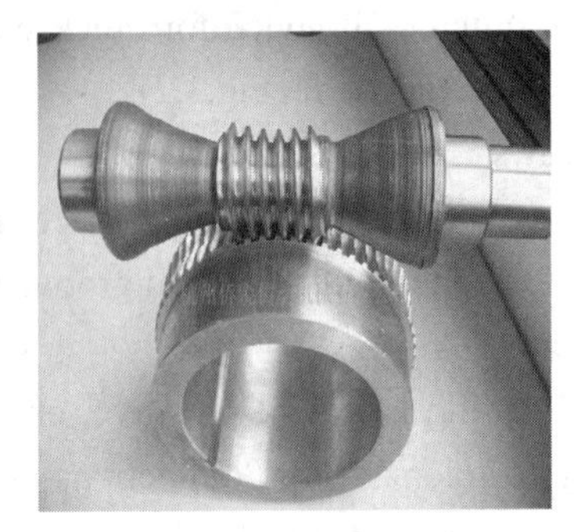

Fig.3.6 Gears

9. Couplings

A coupling is a device for connecting the ends of adjacent shafts. There are several types of shaft coupling: rigid couplings and flexible coupling (Fig.3.7).

Fig.3.7 Couplings

10. Brakes

Brakes include the external embrace block type, internal expansion type and band brake (Fig.3.8).

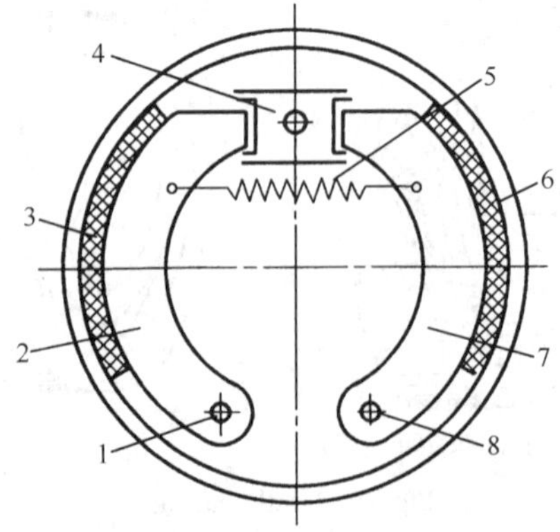

Fig. 3.8 Internal expansion type

1, 8—pivot; 2, 7—brake shoes; 3—friction disc; 4—pump; 5—spring; 6—brake wheel

Knowledge extention

1. Fit

The fit between two mating parts is the relationship which results from the clearance or interference obtained. There are three classes of fit (Fig.3.9).

(a) Clearance fit: the shaft is always smaller than the hole.

(b) Transition fit: the limits are such that the condition may be of clearance or interference fit.

(c) Interference fit: the shaft is always larger than the hole.

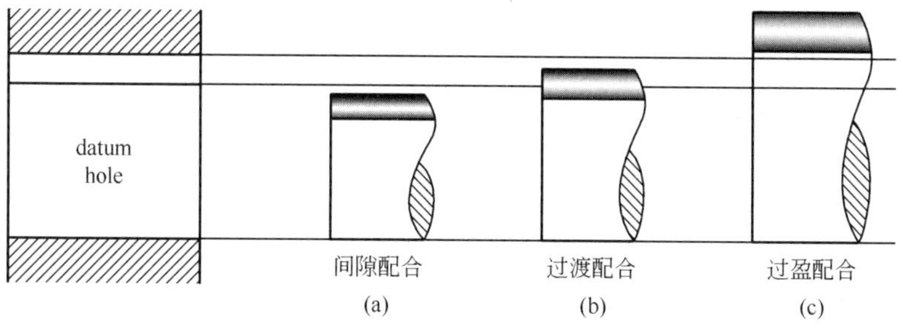

Fig.3.9 Fits

2. Limits of Size and Tolerance

In deciding the limits necessary for a particular dimension, there are three considerations: functional importance, interchangeability and economics (Fig.3.10).

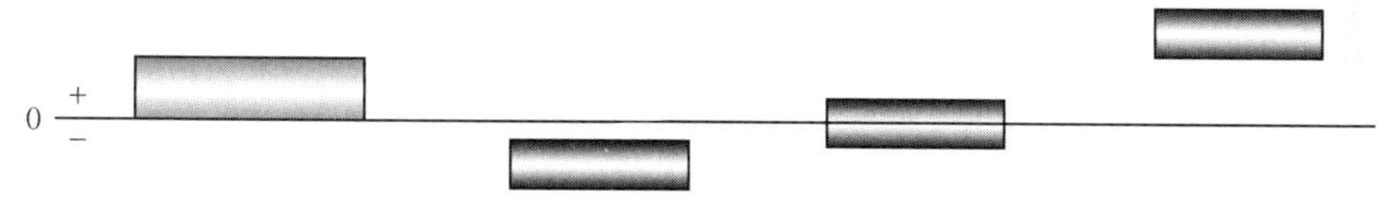

Fig.3.10 Tolerance zones

Exercises

1. Answer the following questions according to the text.

(1) What is the definition of shaft?

(2) What do the shaft accessories include? What are they used for?

(3) What are the springs?

(4) What are the gears classified into?

2. Match the English with the Chinese. Draw lines.

(1) shaft	a. 离合器
(2) shaft accessories	b. 弹簧
(3) clutch	c. 轴
(4) screw	d. 球轴承
(5) spring	e. 轴附件
(6) ball bearing	f. 凸轮
(7) cam	g. 联轴器

(8) gear h. 螺钉
(9) coupling i. 制动器
(10) brake j. 齿轮

3. Fill in the blanks with the proper words in the text.

(1) A ball bearing usually consists of ______ parts: an inner ring, ______, the ball and the ______ or the ______.

(2) Brakes include the external __________ type, internal ______ type and __________ .

(3) There are three classes of fit: __________ fit, __________ fit, and interference fit.

(4) In deciding the limits necessary for a particular dimension, there are three considerations: __________, interchangeability and __________.

Words and phrases

mechanical [mɪ'kænɪkəl]机械的
pin [pɪn] 销
key [ki:] 钥匙，键
accessory [ək'sesəri] 附件
screw [skru:] 螺钉
bearing ['beərɪŋ] 轴承
gear [gɪə(r)] 齿轮
coupling ['kʌplɪŋ] 联轴器
tolerance ['tɑ:lərəns] 公差
part [pɑ:t]部件，零件
spline [splaɪn] 花键
shaft [ʃɑ:ft] 轴
clutch [klʌtʃ] 离合器
spring [sprɪŋ]弹簧
cam [kæm] 凸轮
brake [breɪk] 制动器
fit [fɪt] 配合

Lesson 4 Materials

Look and select

Look at the pictures and select the words from the box.

carbon steel	copper	aluminum	ceramic

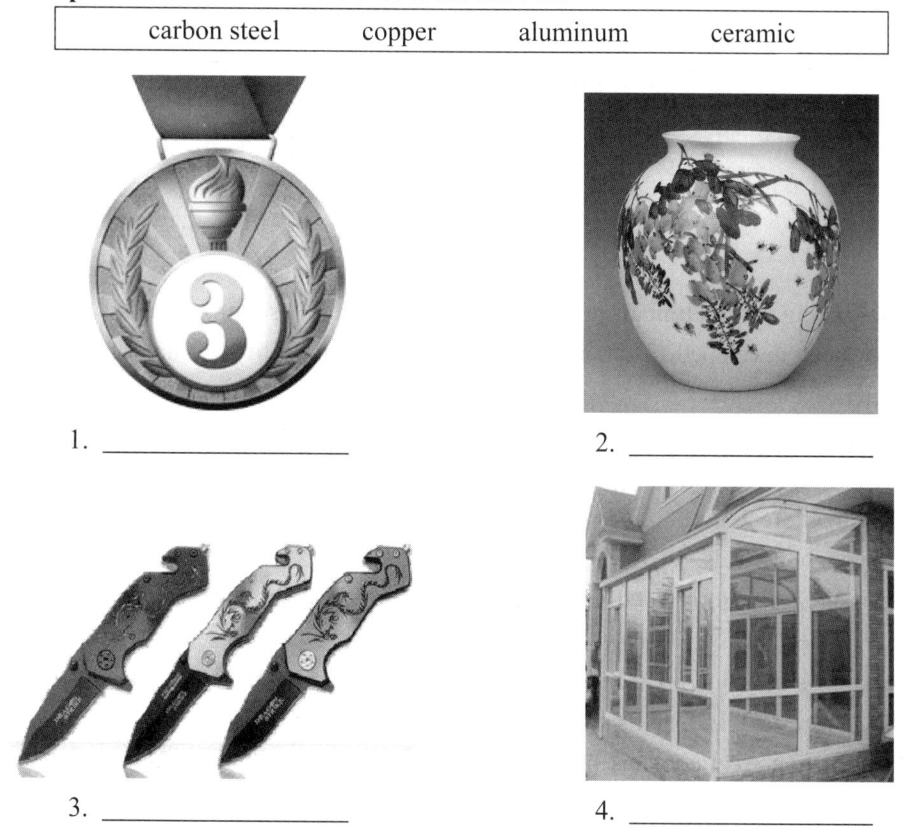

1. ________ 2. ________

3. ________ 4. ________

Text

Material properties

The properties of materials may be divided into the following four groups:

Physical properties include color, density, melting point, freezing point, specific heat, heat of fusion, thermal conductivity, thermal expansion, electrical conductivity, magnetic properties, and so on.

Chemical properties include anti-oxidation ability and anti-corrosion ability, which can also be important during the manufacturing processes because, they can influence the formation of surface films, affecting friction and lubrication, and thermal and electrical conductivity.

Mechanical properties, illustrated in Fig.4.1, are the characteristic responses of a material to applied forces. These properties fall into four broad categories: strength, hardness, plasticity and toughness.

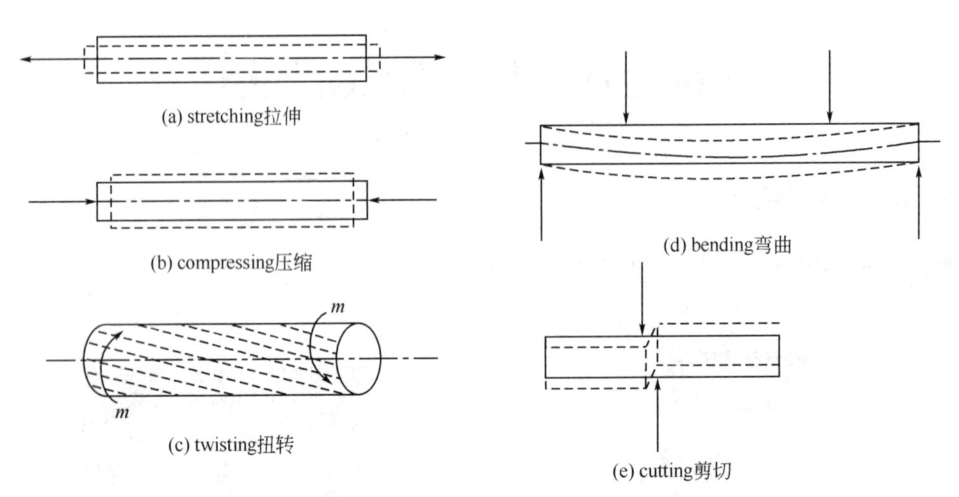

Fig.4.1 Mechanical properties

Manufacturing or technical properties: To evaluate these properties, various testing methods have been developed to describe the castability, forgeability, weldability, machinability, and formability of a material.

Knowledge extention

1. Classification of steel

(1) Plain carbon steels

Plain carbon steels are divided into three categories: low-carbon steel, medium-carbon steel, and high-carbon steel.

Low-carbon steel contains from 0.02 to 0.30 percent carbons by mass. Items such as bolts, nuts, washers, sheet steel, and shafts are made of low-carbon steel.

Medium-carbon steel contains from 0.30 to 0.60 percent carbons and is used where greater tensile strength is required.

High-carbon steel, also know as tool steel, contains over 0.60 percent carbon and may range as high as 1.7 percent.

(2) Alloy steels

Alloy steel may be defined as steel containing other elements, in addition to carbon, that produce the desired qualities in the steel.

2. Categories of tool materials

Ferrous materials: tool steel, alloy steel, carbon steel, and cast iron.

Nonferrous materials: aluminum, magnesium, zinc, lead, bismuth, copper, and a variety of alloys.

Nonmetallic materials: woods, plastic, rubbers, epoxy resins, ceramics, and diamonds etc.

3. Cutting tools

The various materials from which most metal cutting tools are made can be classified into: carbon tool steel, high-speed steel, cast alloys, cemented carbides, ceramics and diamonds.

Exercises

1. How many groups can the properties of materials be divided into? What are they?

2. Fill in the blanks with the detailed properties we learned from the text (Table 4.1).

Table 4.1 Properties of materials

Properties of materials 材料的性能	Description 具体内容
Physical properties 物理性能	
Chemical properties 化学性能	
Mechanical properties 力学性能	
Manufacturing or technical properties 工艺性能	

3. Match the English with the Chinese. Draw lines.

(1) mechanical properties　　a. 扭转
(2) stretching　　b. 弯曲
(3) compressing　　c. 机械性能
(4) twisting　　d. 压缩
(5) bending　　e. 剪切
(6) cutting　　f. 拉伸

4. Fill in the blanks with the proper words in the text.

(1) Plain carbon steels are divided into three categories: ________ steel, medium-carbon steel, and ________ steel.

(2) Categories of tool materials include ________, nonferrous materials and ________.

(3) The various materials from which most metal cutting tools are made can be classified into: ________, high-speed steel, ________, cemented carbides, ________ and ________.

Words and phrases

property ['prɒpəti] 性能
mechanical [mɪ'kænɪkəl] 机械的
castability [kɑːstə'bɪlɪtɪ] 可铸造性
density ['densəti] 密度
fusion ['fjuːʒn] 熔合
thermal 热的，热量的
conductivity 传导性，传导率
expansion 膨胀
elasticity 弹力，弹性
ductility 展延性
cast iron 铸铁
magnesium 镁
lead 铅
metal 金属
machinability 机械加工性，切削性
formability [fɔːmə'bɪlɪtɪ] 可成形性
classification [ˌklæsɪfɪ'keɪʃn] 分类
plain carbon steel 普通碳钢
low-carbon steel 低碳钢
medium-carbon steel 中碳钢
high-carbon steel 高碳钢
alloy steel 合金钢
tensile strength 抗拉强度
toughness 韧性，刚性
aluminum 铝
zinc 锌
bismuth 铋
plastic 塑料

Lesson 5 Heat Treatment

Look and translate

Look at the pictures, understand and translate the words of them into Chinese:

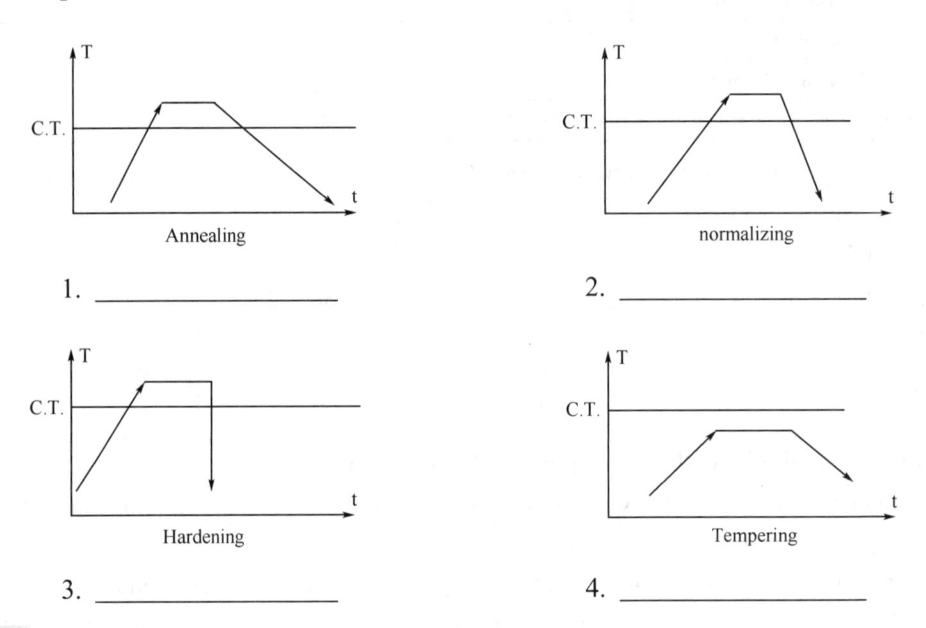

1. ________________　2. ________________

3. ________________　4. ________________

Text

Heat treatment is to produce predictable changes in the internal structure of metals by controlling the time and temperature.

Types of heat treating operations

Four operations are detailed in this lesson as the basis of heat treatment. Explanations of these operations are shown as the following.

Annealing is the process of softening steel by a heating and cooling cycle, so that it may be bent or cut easily. In annealing, steel is heated above a transformation temperature and cooled very slowly after it has reached a suitable temperature. The distinguishing characteristics of full annealing are: (a) temperature above the critical temperature and (b) very slow cooling, usually in the furnace.

Normalizing is identical with annealing, except that the steel is air cooled; this is much faster than cooling in a furnace. Steel is normalized to refine grain size, make its structure more uniform, or to improve machinability.

Hardening is carried out by quenching a steel, that is, cooling it rapidly from a temperature above the transformation temperature. Steel is quenched in water or brine for the most rapid cooling, in oil for some alloy steels, and in air for certain higher alloy steels. After steel is quenched, it is usually very hard and brittle; it may even crack if dropped. To make the steel more ductile, it must be tempered.

Tempering consists of reheating a quenched steel to a suitable temperature below the transformation temperature for an appropriate time and cooling back to room temperature. This process can make steel tough.

Knowledge extention

Stress relieving is the heating of steel to a temperature below the transformation temperature, as in tempering, but is done primarily to relieve internal stress and thus prevent distortion or cracking during machining (Fig.5.1). This is sometimes called process annealing.

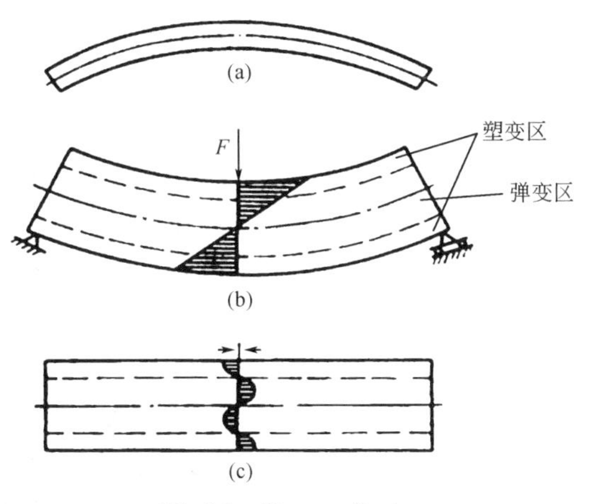

Fig.5.1 Stress relieving

Exercises

1. Fill in the blanks with the proper words in the text.

(1) Heat treatment is to produce predictable __________ in the internal structure of metals by controlling the __________ and __________.

(2) Four operations detailed in this lesson are ______________, normalizing, hardening and __________.

2. Match the English with the Chinese. Draw lines.

(1) critical temperature	a. 特性
(2) cooling	b. 之上
(3) medium	c. 临界温度
(4) characteristic	d. 之下
(5) above	e. 介质
(6) below	f. 冷却

3. Complete the four processes of heat treatment in English (Table 5.1).

Table 5.1 Four processes of heat treatment

项目	临界温度	冷却速度	冷却介质	材料特性
Annealing			Cool in the furnace	
Normalizing	Above	Air cooled faster than annealing		refine grain size, make its structure more uniform, or to improve machinability
Hardening			Cool in water or brine for CS Cool in oil for AS Cool in air for HAS	
Tempering		Room cool	Cool back to room temp.	

Words and phrases

heat [hi:t] 热量，加热
annealing [ə'ni:lɪŋ] 退火
hardening ['hɑ:dnɪŋ] 淬火
comparison [kəm'pærɪsən]比较
above ……之上
slowly 慢慢地
soften ['sɔfən, 'sɑfən] 使……变软
bend [bend] 弯曲
easily 容易地
refine [rɪ'faɪn] 细化
size 规格，尺寸
uniform ['ju:nɪfɔ:m] 均匀的，统一的
machinability [məʃi:nə'bɪlɪtɪ] 切削加工性
brine [braɪn] 盐水
hard [hɑ:d] 硬的，坚硬的
crack[kræk]裂纹
below ……之下
temperature ['temprətʃə(r)] 温度
treatment ['tri:tmənt]处理，对待
normalizing ['nɔ:məlaɪzɪŋ] 正火
tempering ['tempərɪŋ] 回火
process [prə'ses] 工艺，过程
cool [ku:l] 冷却，凉快的，酷的
furnace ['fɜ:nɪs] 炉子
steel [sti:l] 钢
cut [kʌt] 切割
air [eə(r)]空气
grain [greɪn] 颗粒，晶粒，谷粒
structure ['strʌktʃɚ] 结构
improve [ɪm'pru:v] 改善，提高
rapidly ['ræpɪdlɪ] 急速地
oil [ɔɪl] 油
brittle ['brɪtl] 脆，淬
even 甚至
back 返回，回到
tough [tʌf] 韧性地

Lesson 6 Basic Machining Methods

Look and select

Look at the pictures and select the correct terms from the box.

engineer	drawing	machining	robot

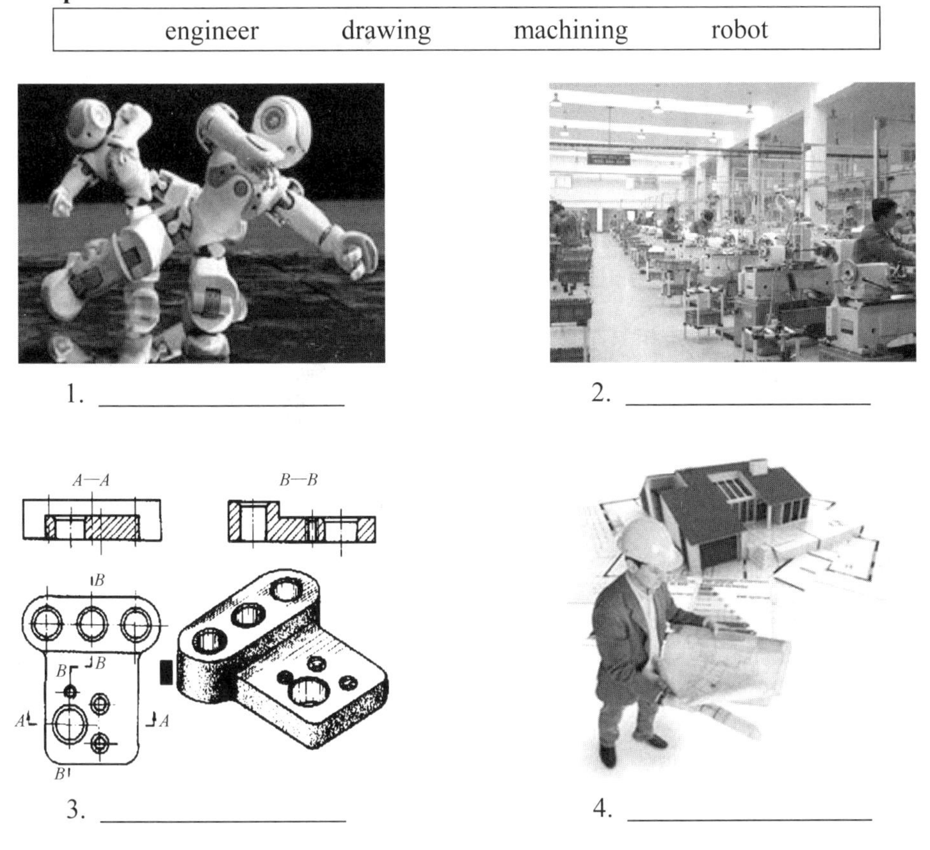

1. ______________ 2. ______________

3. ______________ 4. ______________

Text

1. Drilling consists of cutting a round hole by means of a rotating drill (Fig.6.1).

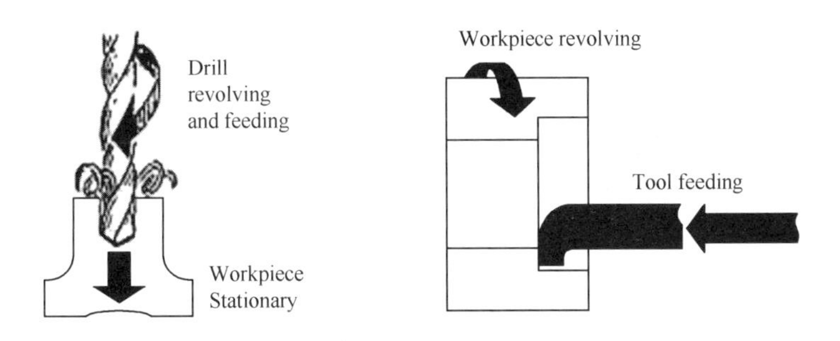

Fig.6.1 Drilling and boring

2. The **lathe,** as the **turning** machine is commonly called, is the father of all machine tools (Fig.6.2).

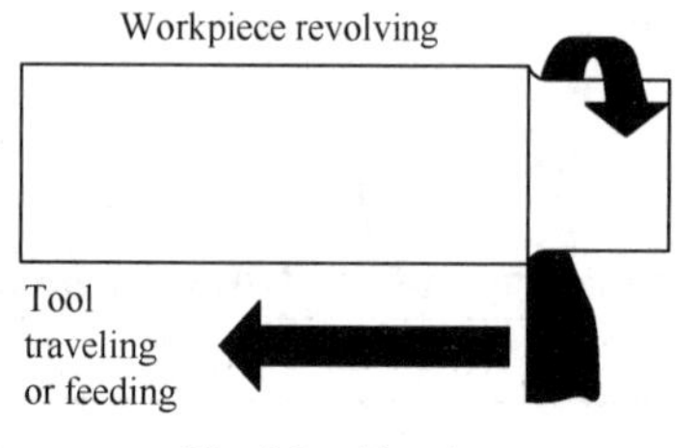

Fig.6.2 Turning

3. Planing and **shaping** metal with a machine tool is a process similar to planning wood with a hand plane (Fig.6.3).

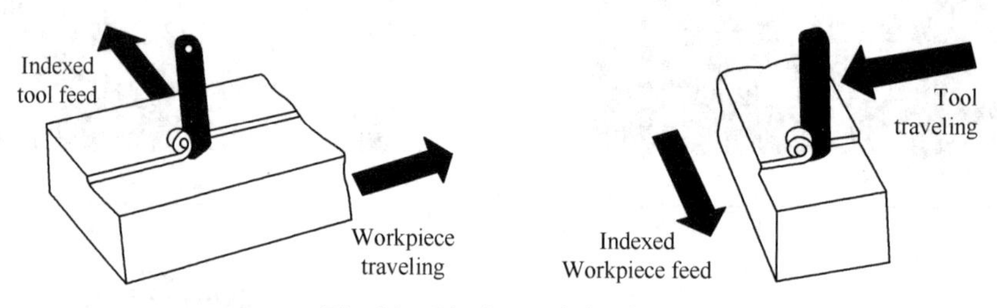

Fig.6.3 Planing and shaping

4. Milling consists of machining a piece of metal by bringing it into contact with a rotating cutting tool which has multiple cutting edges (Fig.6.4).

5. Grinding consists of shaping a piece of work by bringing it into contact with a rotating abrasive wheel (Fig.6.5).

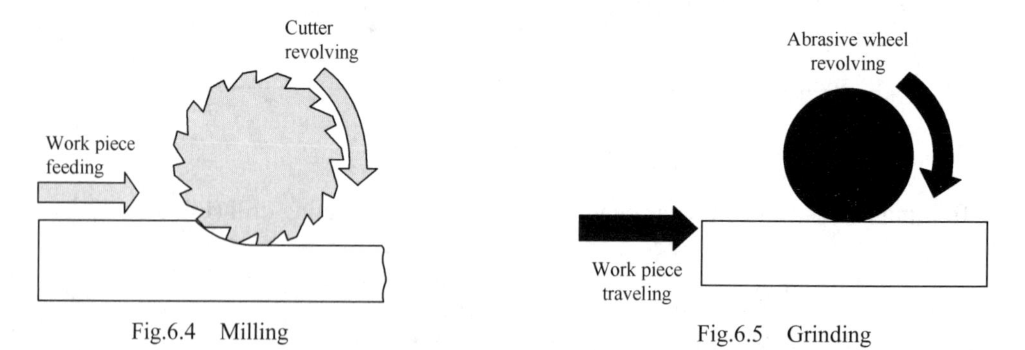

Fig.6.4 Milling

Fig.6.5 Grinding

Knowledge extention

Freedoms of robots

In order to achieve flexibility of motion in a three-dimensional space, a robotic manipulator needs to be to move in at least three dimensions. Fig.6.6 describes a linear movement and a rotational movement.

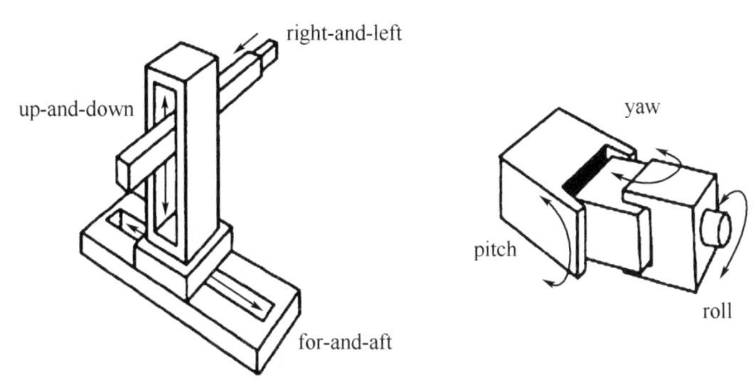

Fig.6.6 Freedoms of robots

Exercises

1. Fill in the blanks with the proper words in the text.

(1) We have learned five basic machining methods in the text: ________, turning, planing and ________ , milling and ________.

(2) In order to achieve flexibility of ________ in a ________ space, a robotic manipulator needs to be to move in at least three dimensions: ________, up and down and ________. Fig.6.6 describes a ________ movement and a ________ movement.

2. Match the English with the Chinese. Draw lines.

(1) revolve a. 刀具
(2) feed b. 工件
(3) workpiece c. 移动
(4) tool d. 静止的
(5) travel e. 进给
(6) stationary f. 旋转

Words and phrases

basic ['beɪsɪk] 基本的，基础的
method ['meθəd] 方法
robot ['robət] 机器人
revolve [rɪ'vɒlvɪŋ] 旋转
workpiece ['wəkpi:s] 工件
tool [tu:l] 工具，刀具
turning ['tə:nɪŋ] 车削
planing [pleɪnɪŋ] 龙门刨
milling ['mɪlɪŋ] 铣削
abrasive [ə'bresɪv] 磨砂的
machining [mə'ʃi:nɪŋ] 加工，切削
engineer [ˌendʒɪ'nɪə(r)] 工程师
drilling ['drɪlɪŋ] 钻削
feed ['fi:dɪŋ] 进给，喂养
stationary ['steɪʃənrɪ] 静止的
lathe [leɪð] 车床
travel ['trævəlɪŋ] 移动
shaping ['ʃeɪpɪŋ] 牛头刨
cutter ['kʌtə(r)] 切割器，铣刀

Lesson 7 Lathes

Look and select

Look at the pictures and select the correct terms from the box.

NC lathe	turret lathe	vertical lathe	bench lathe

1. ____________________

2. ____________________

3. ____________________

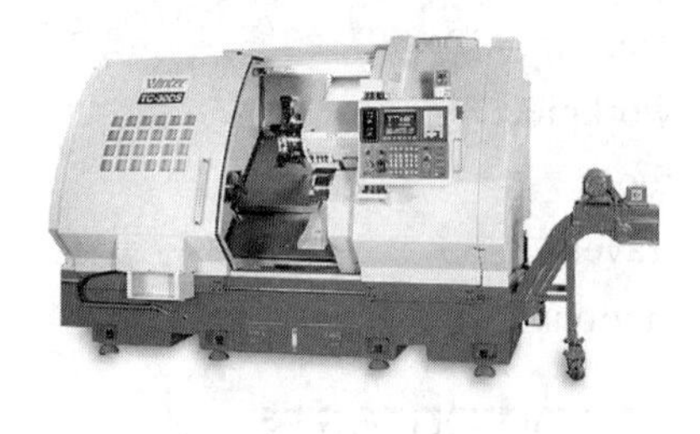

4. ____________________

Text

1. General conception

Engine lathe, as shown in Fig.7.1, is a machine that removes material by rotating the work against a cutter. It is the oldest of all machine tools as well as the important machine in modern production.

2. The main parts of lathe

(1) The **bed** of engine lathe is the foundation unit supporting the other unit and parts. It must be rigid in construction; and the V ways and flat ways of the bed must be precisely machined, so that they may ensure the proper fitting and alignment of all working parts which are mounted on the bed.

(2) Headstock contains the main spindle, a series of different sized gears, an electric motor and chucks etc.

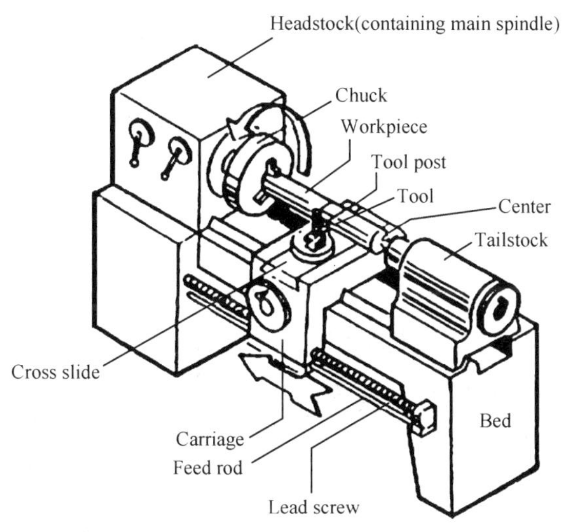

Fig.7.1 Lathe

(3) Tailstock is properly camped at the right end of the bed. Usually it is used to machine long workpiece.

(4) Carriage is a movable part. It carries the compound rest and slides on ways of the bed.

(5) Feed gearbox contains gears and other parts which are necessary in transmitting motion from the main spindle to the feed rod or lead screw.

Knowledge extention

Other types of lathes

Bench lathes, often found in watch repair shops, are employed to process small machine parts.

Turret lathes are equipped with a turret for mounting a number of different cutting tools used in sequence.

Vertical lathes, including vertical turret lathes and vertical boring mill, are characterized by the vertical spindle. They are suitable for turning and boring.

Other various types of lathes are on the market: for example the single spindle or multispindle screw machines, automatic lathes, and NC lathes etc.

Exercises

1. Fill in the blanks with the proper words in the text.

(1) Engine lathe is a machine that removes material by __________ the work against a _________. It is the _________ of all machine tools as well as the _________ machine in modern production.

(2) The kinds of lathe we learned are ________, vertical lathe, ________, turret lathe etc.

(3) Headstock contains the ____________, a series of different sized ____________, an electric __________ and __________ etc.

2. Match the English with the Chinese. Draw lines.

(1) tool post　　a. 溜板，滑架

(2) carriage　　b. 丝杠

(3) feed rod c. 刀架，刀座
(4) lead screw d. 进给杆
(5) cross slide e. 横向滑板/拖板

3. Translate the words in the picture (Fig.7.2) into English.

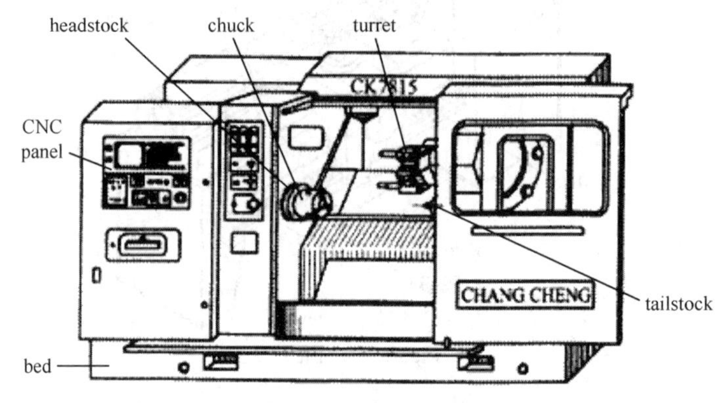

Fig.7.2 Structure of NC lathe

headstock ________________________; chuck ________________________;
turret ________________________; tailstock ________________________;
CNC panel ________________________; bed ________________________.

Words and phrases

engine lathe ['endʒɪn] [leɪð] 普通车床
remove [rɪ'mu:v] 去除
as well as 以及
main part 主要部件
headstock ['hedstɒk] 床头箱，主轴箱
gear [gɪə(r)] 齿轮
chuck [tʃʌk] 卡盘，夹头
workpiece ['wɜ:kˌpi:s] 工件
carriage ['kærɪdʒ] 溜板，滑架
gearbox ['giəbɔks] 齿轮箱，变速箱
feed rod 进给杆
turret lathe ['tɜ:rɪt, 'tʌr-] 六角头车床
cross slide [krɒs] [slaɪd] 横向滑板/拖板
tool post 刀架，刀座
automatic lathe [ˌɔ:tə'mætɪk] 自动车床
multispindle ['mʌltɪzpɪndl] 多轴
NC lathe 数控车床
machine [mə'ʃi:n] 机器
material [mə'tɪəriəl] 材料
important machine 重要机器/机床
bed 床身
spindle ['spɪndl] 主轴
motor ['məʊtə(r)] 电机，马达
tailstock ['teɪlstɒk] 尾座
center ['sentə] 中心
feed [fi:d] 进给
contain [kən'teɪn] 包含，包括
leadscrew 导杆，丝杠
bench lathe [bentʃ] 台式车床
vertical lathe ['vɜ:tɪkl] 立式车床
bed way 床身导轨
single spindle 单轴
screw [skru:] 螺纹，螺钉
etc. 等等
rotating the work against a cutter 工件绕着刀具旋转
modern production ['mɔdən] [prə'dʌkʃn] 现代生产

Lesson 8 Milling Machines

Look and select

Look at the pictures and select the correct terms from the box.

NC miller	cutter	column-and-knee miller	planer miller

1. ________________

2. ________________

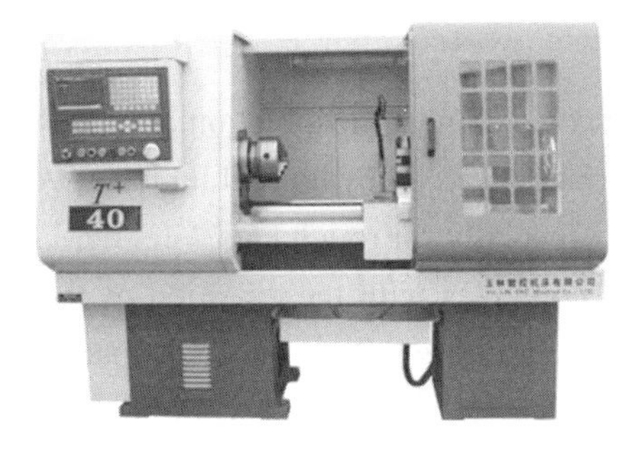

3. ________________

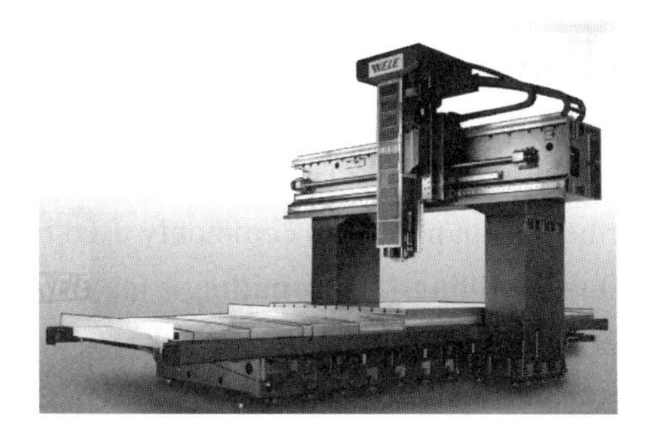

4. ________________

Text

1. General conception

The milling machine is a machine that removes metal from the work with a revolving milling cutters as the work is fed against it.

2. The main parts of milling machine

Parts of the milling machine: (1) column; (2) spindle; (3) arbor; (4) overarm; (5) table; (6) saddle; (7) knee; (8) base; (9) motor.

A wide variety of milling machines are available: for example, the plain column-and-knee type, universal column-and-knee type (Fig.8.1), bed-type, and planer type.

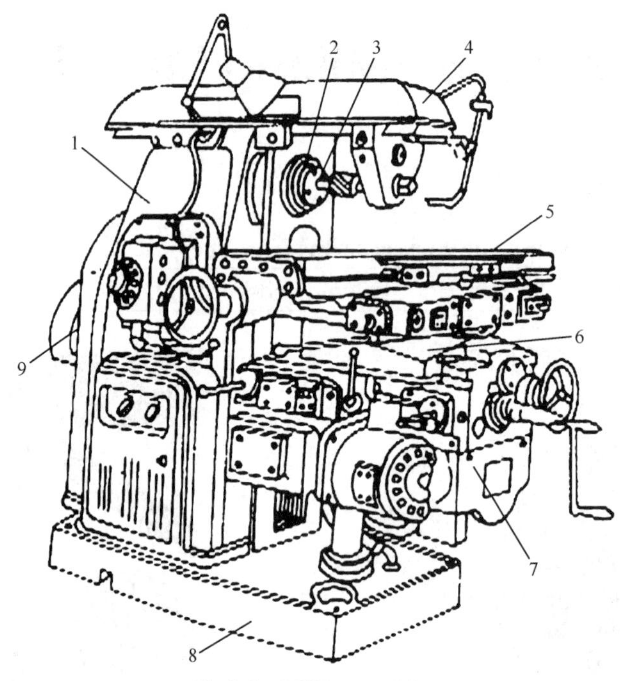

Fig.8.1 Milling machine

Knowledge extention

Milling methods:

Up milling (conventional milling): the workpiece is fed against the direction of cutter rotation, as shown in Fig.8.2 (a). The quality of the machined surface obtained by up milling is not very high. Nevertheless, up milling is commonly used in industry, especially for rough cuts.

Down milling (climb milling): the cutter rotation coincides with the direction of feed at the contact point between the tool and the workpiece, as shown in Fig.8.2 (b). The advantages of this method include higher quality of the machines surface and easier clamping of workpieces, since the cutting forces act downword.

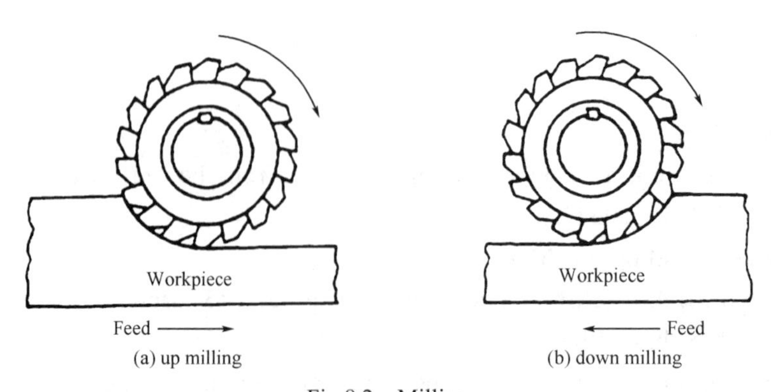

Fig.8.2 Milling

Exercises

1. Fill in the blanks with the proper words in the text.

(1) The milling machine is a machine that __________ metal from the work with a revolving __________ as the work is ________ against it.

(2) A wide variety of milling machines are ________: for example, the plain column-and-knee type, ________ type, ________ type, and ________ type.

2. Match the English with the Chinese. Draw lines.

(1) column　　a. 刀杆

(2) spindle　　b. 底座

(3) arbor　　c. 床身，支柱

(4) overarm　　d. 工作台

(5) table　　e. 升降台

(6) saddle　　f. 横梁

(7) knee　　g. 主轴

(8) base　　h. 电动机

(9) motor　　i. 床鞍

3. Translate the words in the picture (Fig.8.3) into English.

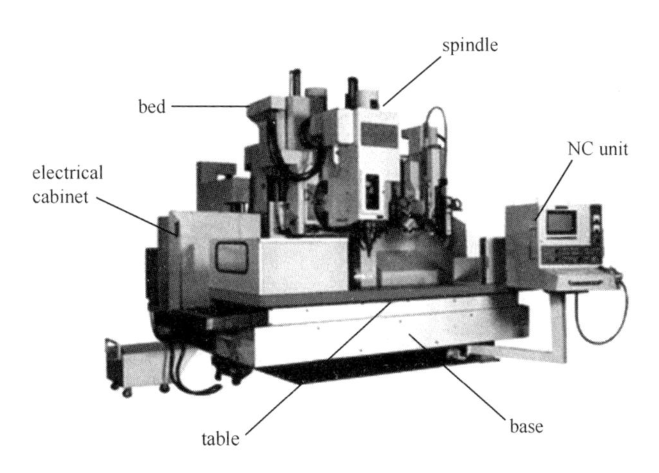

Fig.8.3　Structure of NC milling machine

electrical cabinet ____________________________; bed ____________________________;

spindle ____________________________; NC unit ____________________________;

table ____________________________; base ____________________________.

4. Answer the following questions.

(1) What is up milling? How about the quality of up milling?

(2) What is down milling? What about its advantages?

Words and phrases

milling machine ['mɪlɪŋ] [mə'ʃi:n] 铣床
metal ['metl] 金属
revolving [rɪ'vɒlvɪŋ] 旋转的
be fed against ……倚靠着……
column ['kɔləm] 支柱
arbor ['ɑ:bə] 刀杆
table ['teɪbl] 工作台
knee [ni:] 升降台
motor ['məʊtə(r)] 电动机
bed-type 卧型
NC miller 数控铣床
work 工件
cutter ['kʌtə(r)] 铣刀
part 部分，部件
spindle ['spɪndl] 主轴
overarm ['ɑ:bə] 横梁
saddle ['sædl] 床鞍
base [beɪs] 底座
down milling 顺铣
up milling 逆铣
universal column-and-knee type [ˌju:nɪ'vɜ:sl] 万能支柱升降台型
column-and-knee miller 普通支柱升降台型
planing miller ['pleɪnə(r)] 龙门型铣床

Lesson 9 Drilling Machines

Look and select

Look at the pictures and select the correct terms from the box.

drill	bench-type	upright-type	multispindle

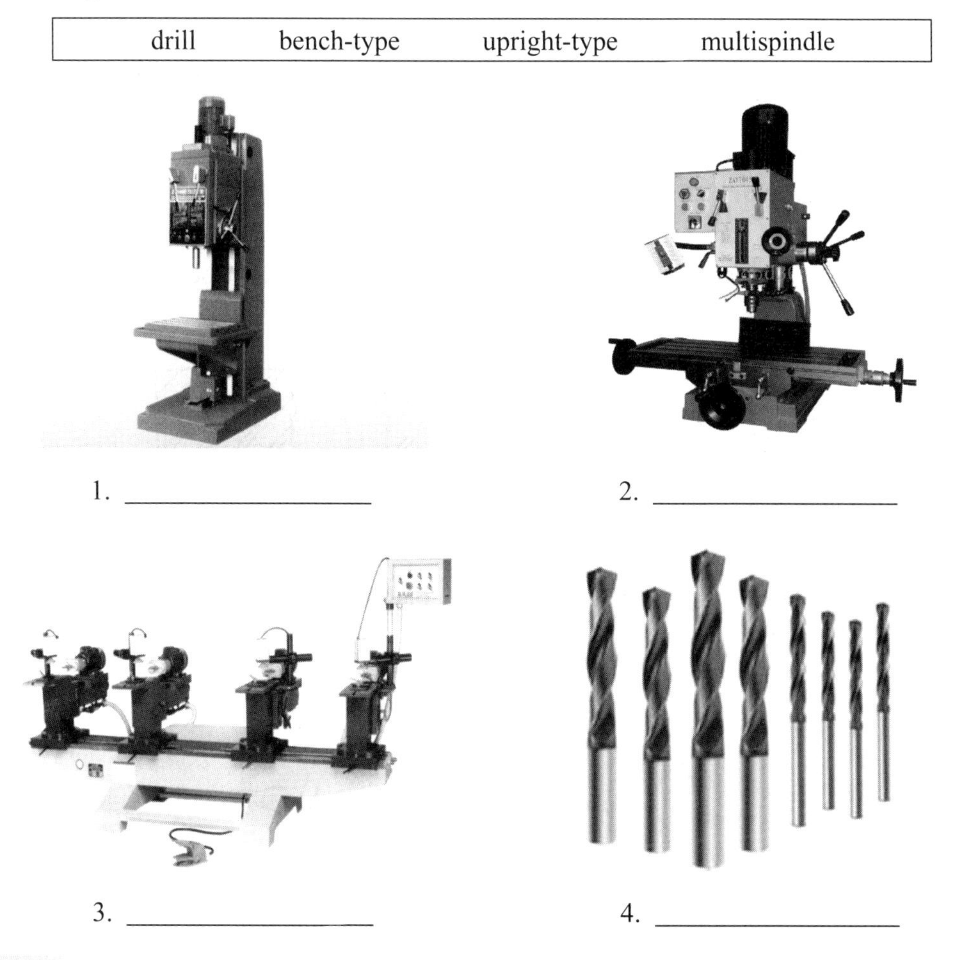

1. ________________ 2. ________________

3. ________________ 4. ________________

Text

1. General conception

A drilling machine is a machine which holds and turns a drill to cut holes in metal. It is one of the most common and useful machines.

2. Types of drilling machines:

Bench-type drilling machines are general-purpose, small machine tools that are usually placed on benches.

Upright drilling machine: Depending upon the size, upright drilling machine tools can be used for light, medium, and even relatively heavy jobs.

Multispindle drilling machines have sturdy construction and require high power; each is capable of drilling many holes simultaneously. This type of drilling machine is used mainly for mass production in jobs having many holes, such as cylinder blocks.

Knowledge extention

Other types of drilling operations

In addition to conventional drilling, there are other operations that are involved in the production of holes in the industrial practice.

Counterboring As a result of counterboring, only one end of a drilled hole is enlarged. This is shown in Fig.9.1 (a).

Spot facing operation is performed to finish off a small surface area around the opening of a hole as shown in Fig.9.1 (b).

Countersinking As shown in Fig.9.1 (c), countersinking is done to enable accommodating the conical seat of a flathead screw so that the screw does not appear above the surface of the part.

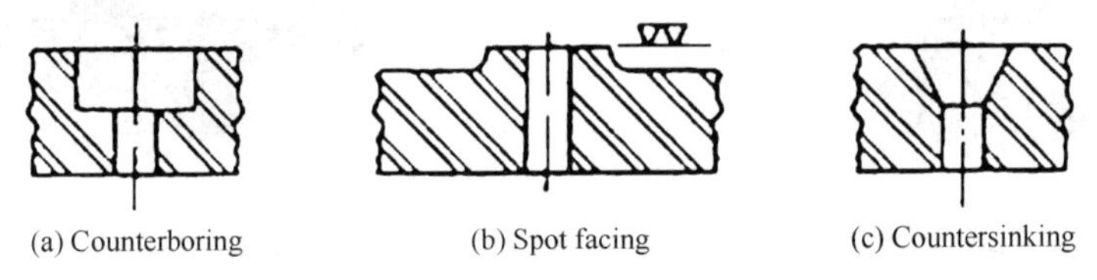

(a) Counterboring (b) Spot facing (c) Countersinking

Fig.9.1 Other types of drilling operations

Reaming is actually a sizing process, by which an already drilled hole is slightly enlarged to the desired size (Fig.9.2).

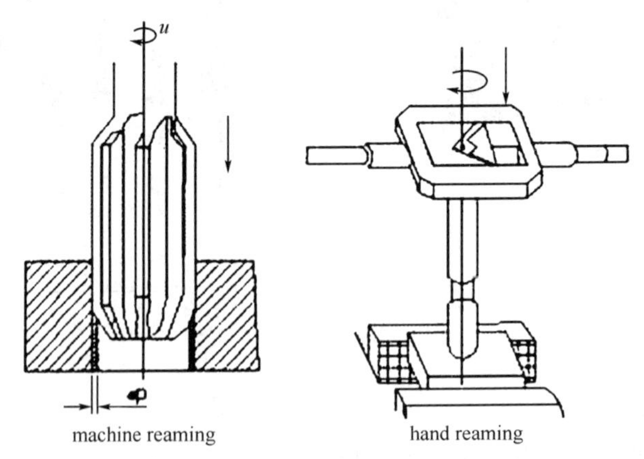

machine reaming hand reaming

Fig.9.2 Reaming

Exercises

1. Fill in the blanks with the proper words in the text.

(1) A drilling machine is a machine which holds and turns a _______ to cut _______ in metal. It is one of the most _______ and _______ machines.

(2) In addition to conventional drilling, there are other ___________ that are involved in the _______ of holes in the industrial _______: _______, spot facing, _______ and _______.

2. Match the English with the Chinese. Draw lines.

(1) drill	a. 台子
(2) bench	b. 钻头
(3) upright	c. 孔
(4) multispindle	d. 机器
(5) hole	e. 手
(6) machine	f. 竖立的
(7) hand	g. 多轴

3. Describe the Multispindle drilling machine.

4. Translate the words in the picture (Fig.9.3) into English.

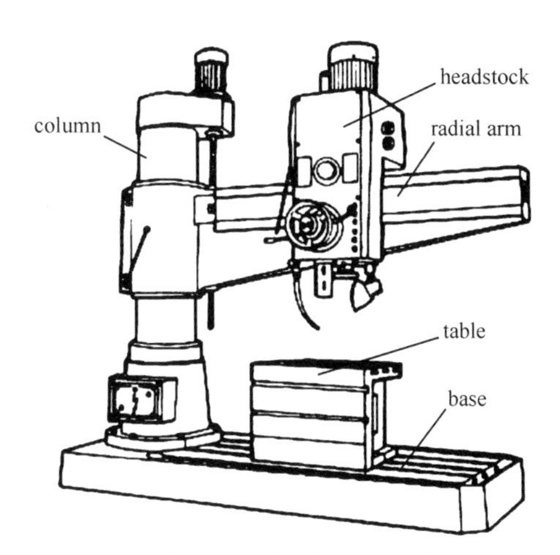

Fig.9.3 Radial drilling machine

column ______________________________; headstock ______________________________;
table ______________________________; base ______________________________;
radial arm ______________________________.

Words and phrases

drill [drɪl] 钻，钻头
bench [bentʃ] 台子
hold and turn a drill 夹住并旋转钻头
place on bench 放置在工作台上
depend upon the size 依据/根据尺寸
simultaneously [saɪməl'tenɪəslɪ] 同时地
counterboring ['kaʊntəbərɪŋ]（平底）锪孔
spot facing [spɒt] ['feɪsɪŋ] 锪端面
countersinking ['kaʊntəˌsɪŋk] 锪锥孔
drilling machine 钻床
upright ['ʌpraɪt] 竖立的，直立的
multispindle ['mʌltɪzpɪndl] 多轴
require high power 要求大功率
enlarge [ɪn'lɑːdʒ] 扩大，放大
be capable of ['keɪpəbl] 能够
cylinder block 气缸柱/座/体
reaming ['riːmɪŋ] 铰孔
boring ['bɔːrɪŋ] 扩孔

bench-type drilling machine 台式钻床
upright drilling machine 立式钻床
multispindle drilling machine 多轴钻床
be used mainly for 主要用于……
mass production [mæs] [prə'dʌkʃn] 大批量/大规模生产
sturdy construction ['stɜ:dɪ] [kən'strʌkʃn] 坚固结实的构造
general-purpose ['dʒɛnərəl] ['pɜ:pəs] 通用型
cut holes in metal [həʊl] ['metl] 在金属中钻孔
most common and useful machine ['kɒmən] 最普遍最有用的机床
in jobs having many holes 有许多孔的工件上
be used for light, medium, and even relatively heavy jobs 适用于轻巧，中等，相对重的工件

Lesson 10 Grinding Machines

Look and select

Look at the pictures and select the correct terms from the box.

NC grinder	external grinder	grinding wheel	universal tool grinder

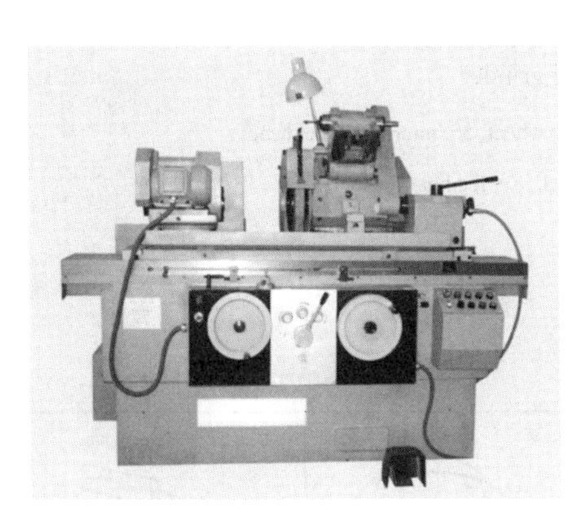

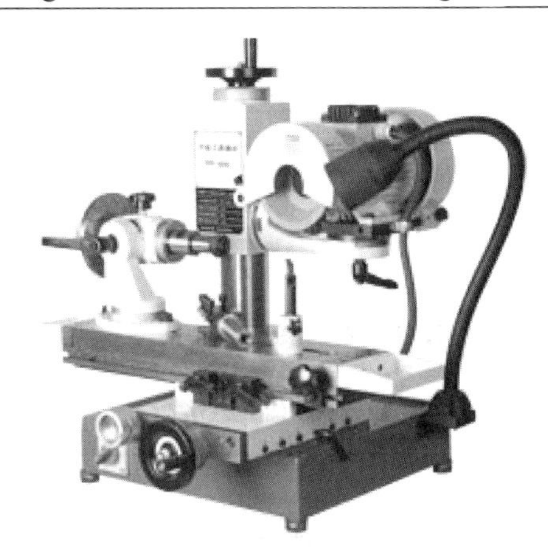

1. 

2. 

3. ____________________

4. ____________________

Text

A grinding machine is a machine which employs a grinding wheel for producing cylindrical, conical or plane surfaces accurately and economically and to the proper shape, size, and finish. The surplus stock is removed by feeding the work against the revolving wheel or by forcing the revolving wheel against the workpiece.

Main parts of a grinding machine contain base, table, headstock, and wheel head.

Classes of grinding machines include internal grinders, centerless grinders, external grinders (Fig.10.1), surface grinders, and tool and cutter grinders.

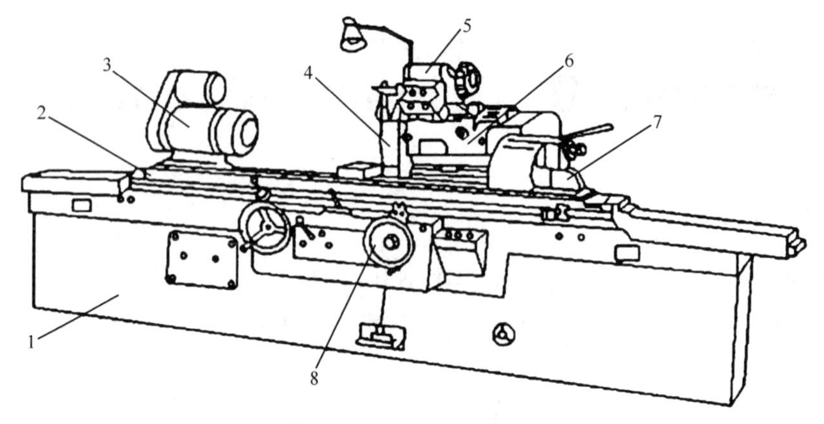

Fig.10.1 External grinder

1—bed; 2—table; 3—headstock; 4—grinding wheel; 5—internal wheel head;

6—grinding wheel rack; 7—tail stock; 8—feed handwheel

Knowledge extention

Primary processes (Fig.10.2)

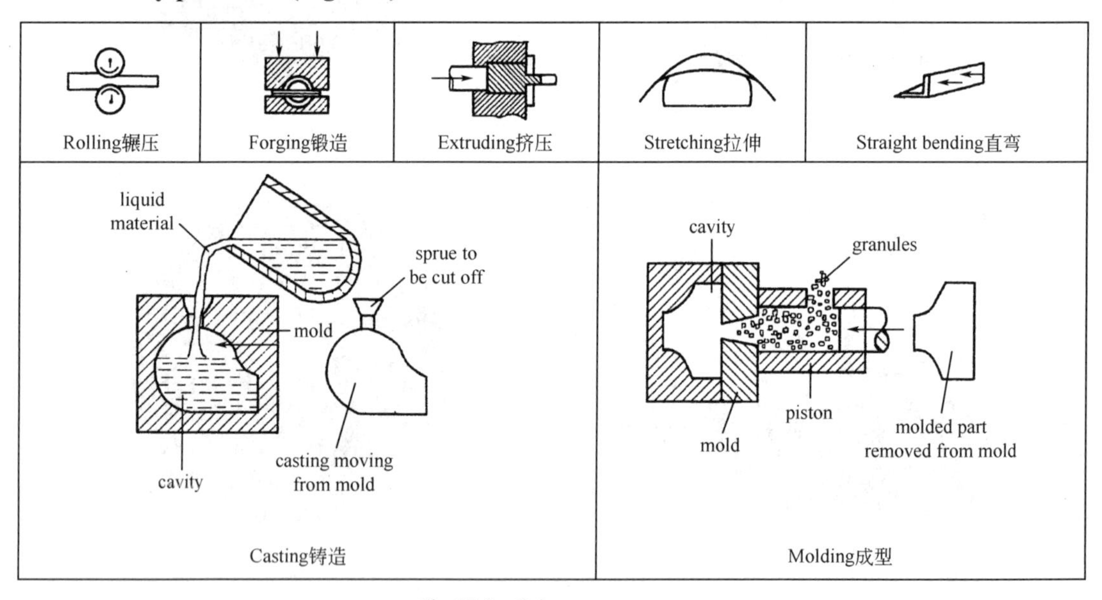

Fig.10.2 Primary processes

Exercises

1. Fill in the blanks with the proper words in the text.

(1) A grinding machine is a machine which employs a ________ for producing ________, conical or ________ surfaces accurately and economically and to the proper ________, ________, and finish.

(2) The surplus stock is removed by ________ the work against the ________ wheel or by forcing the revolving ________ against the ________ .

2. Match the English with the Chinese. Draw lines.

(1) bed
(2) table
(3) headstock
(4) grinding wheel
(5) internal wheel head
(6) grinding wheel rack
(7) tail stock
(8) feed handwheel

a. 头架
b. 内圆磨头
c. 砂轮
d. 砂轮架
e. 尾座
f. 工作台
g. 进给手轮
h. 床身

3. Select the proper words from the box (Fig.10.3):

table	guide way	workpiece	tailstock
spindle	bed	grinding wheel	headstock

Fig.10.3 Structure of grinding machine

1. ______________; 2. ________________; 3. ________________; 4. ________________;
5. ______________; 6. ________________; 7. ________________; 8. ________________.

4. Answer the following questions:

(1) Brief the classes of grinding machines.

(2) What are the primary processes?

Words and phrases

grinding machine ['graɪndɪŋ] [mə'ʃi:n] 磨床
produce [prə'dju:s]产生，加工
plane surface [pleɪn] ['sɜ:fɪs] 平面
size 尺寸，规格
main part [meɪn] [pa:ts] 主要部件
table ['teɪbl] 工作台
wheel head [wi:l] [hed] 磨头
include [□n'klu:d] 包含
internal grinder [in'tə:nəl] 内圆磨床
surface grinder 平面磨床
wheel [wi:l] 轮，砂轮
conical ['kɒnɪkl] 圆锥形的
proper shape ['prɒpə(r)] [ʃeɪp] 适当的形状
finish ['fɪnɪʃ] 光洁度
base [beɪs] 基座
headstock ['hedstɒk] 尾座
class [klɑ:sɪz] 类别，种类
centerless grinder ['sentələs] 无芯磨床
external grinder [eks'tə:nl] 外圆磨床
tool and cutter grinder 工具磨床

Lesson 11 Shaper and Planer

Look and select

Look at the pictures and select the correct terms from the box.

shaper	planer	shaping	planing

1.  ________________ 2. ________________

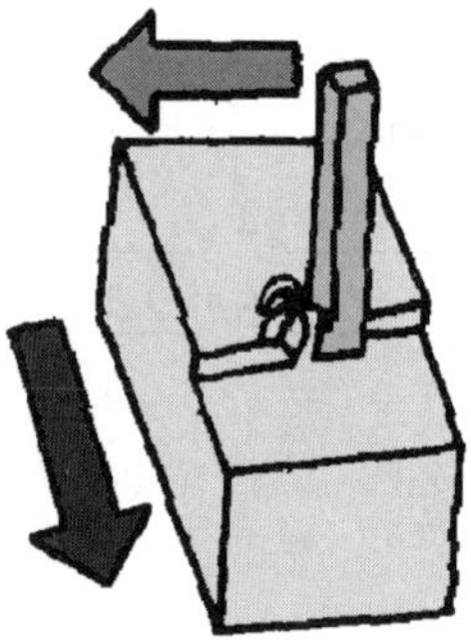

3. 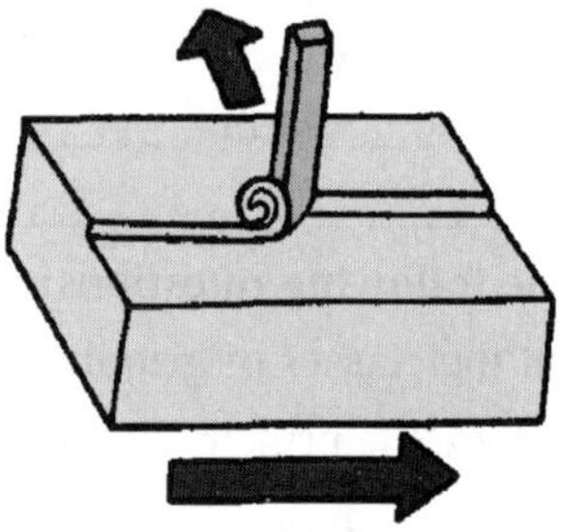 ________________ 4. ________________

Text

The machines are used for machining flat surface, which is performed by a cutter that peel the chip from the work. The main motion is reciprocating and the feed is perpendicular to the direction of the main motion.

The planer (Fig.11.1) is the largest reciprocating machine tools. Since it is larger than the shaper, it can accomplish what the latter cannot do. The main difference between them lies in the function of the tool head and the table, for the planer the table has a reciprocating motion past the tool head, while for the shaper (Fig.11.2) the table is relatively stationary and the tool has a reciprocating motion.

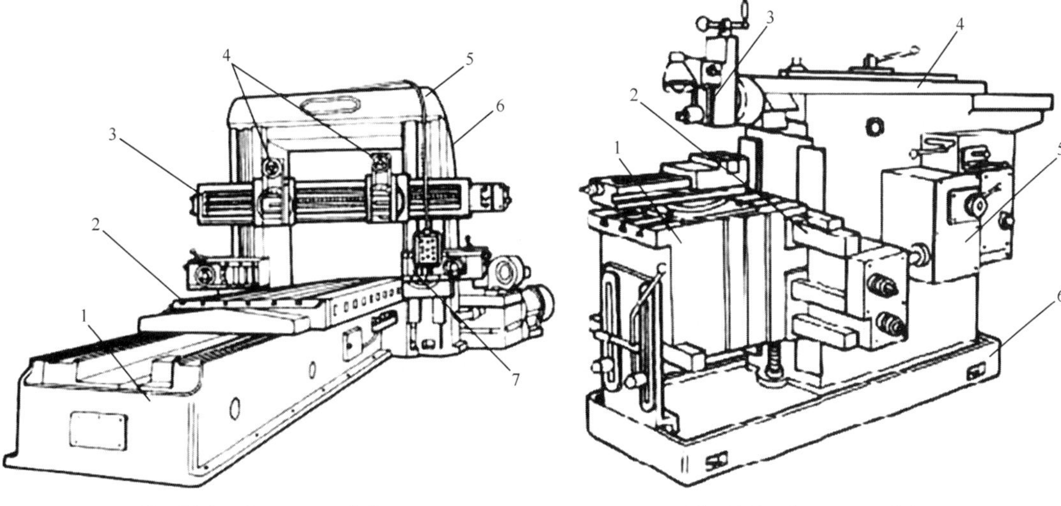

Fig.11.1 Structure of planer

1—bed; 2—table; 3—crossbeam; 4—vertical tool post; 5—top beam; 6—column; 7—side tool post

Fig.11.2 Structure of shaper

1—table; 2—carriage; 3—tool post; 4—ram; 5—bed; 6—base

Knowledge extention

Flat surface machining (Fig.11.3).

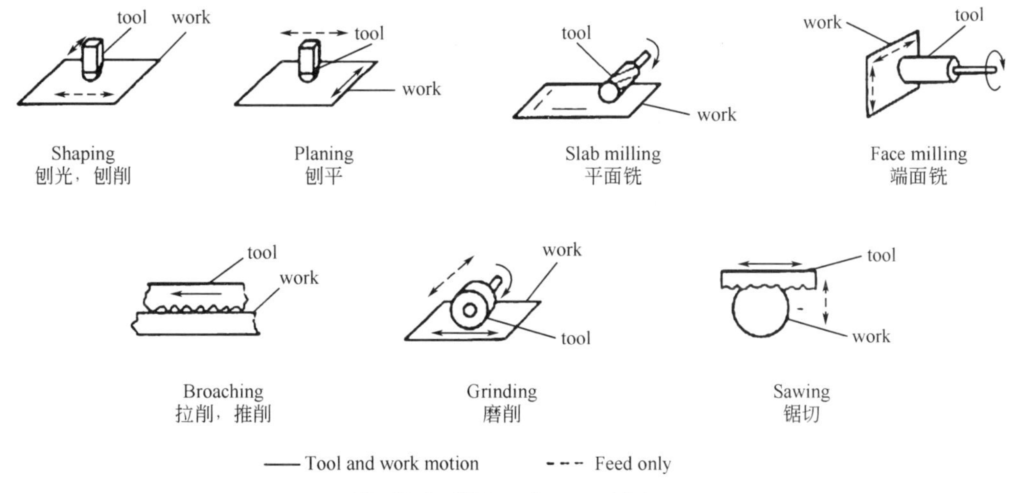

Fig.11.3 Flat surface machining

Exercises

1. Fill in the blanks with the proper words in the text.

The machines are used for machining ________, which is performed by a cutter that peel the ________ from the work. The main motion is ________ and the feed is ________ to the direction of the main motion.

2. Match the English with the Chinese. Draw lines.

(1) planer　　a. 立柱
(2) shaper　　b. 横梁
(3) crossbeam　　c. 龙门刨床
(4) tool post　　d. 牛头刨床
(5) column　　e. 刀架
(6) carriage　　f. 滑枕
(7) ram　　g. 滑板

3. Answer the following questions.

(1) What are the differences between the shaper and planer?
(2) What does the flat surface machining include?

Words and phrases

shaper ['ʃeɪpə] 牛头刨床
planer ['pleɪnə(r)] 龙门刨床
machining flat surface 加工平面
be performed by 凭借……完成/实施
cutter ['kʌtə(r)] 刀具
peel the chip from 从……剥离/削皮
main motion [meɪn] ['məʊʃn] 主运动
feed [fiːd] 进给
be perpendicular to [ˌpɜːpən'dɪkjələ(r)] 与/相对……垂直
direction 方向
be larger than [lɑːdʒə] 比……大一些
accomplish [ə'kʌmplɪʃ] 完成
main difference between 两者间主要区别在于
lie in [laɪ] 在于
tool head [tuːl] [hed] 刀具主轴箱
function ['fʌŋkʃn] 功能
stationary ['steɪʃənrɪ] 静止的

For the planer the table has a reciprocating motion past the tool head, while for the shaper the table is relatively stationary and the tool has a reciprocating motion.

对于龙门刨床，工作台经过刀具主轴箱做往复式运动。而对牛头刨床来说，工作台相对是静止的，刀具做往复式运动。

Lesson 12 Cranes

Look and select

Look at the pictures and select the correct terms from the box.

jib crane	truck crane	bridge crane	crawler crane

1. ________________

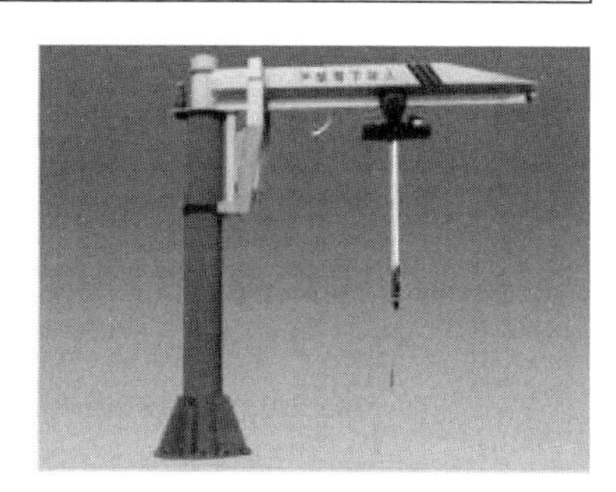

2. ________________

3. ________________

4. ________________

Text

1. Bridge crane

Modern cranes include two main types——bridge cranes (Fig.12.1) and jib cranes. Both use a windlass with steel wire rope wrapped around a powered drum. A bridge crane has a box-girder beam (called a gantry) running on long elevated tracks at each of its ends. The gantry can move backward and forward along the tracks. The hoisting system is carried in a trolley, which moves along the gantry beam. Bridge cranes are commonly set up above a working area to handle such loads as tree trunks and steel beams.

2. Jib crane

A jib crane has a long boom that can swing horizontally to move the load sideways. Many such cranes can also "luff" to control the reach of the crane by angling the boom more or less to the horizontal position. A cantilever crane is a typical example. Cantilever or tower crane is used in constructing high-rise buildings. Anchored to the ground or to the building, the crane is extended upward as the work proceeds.

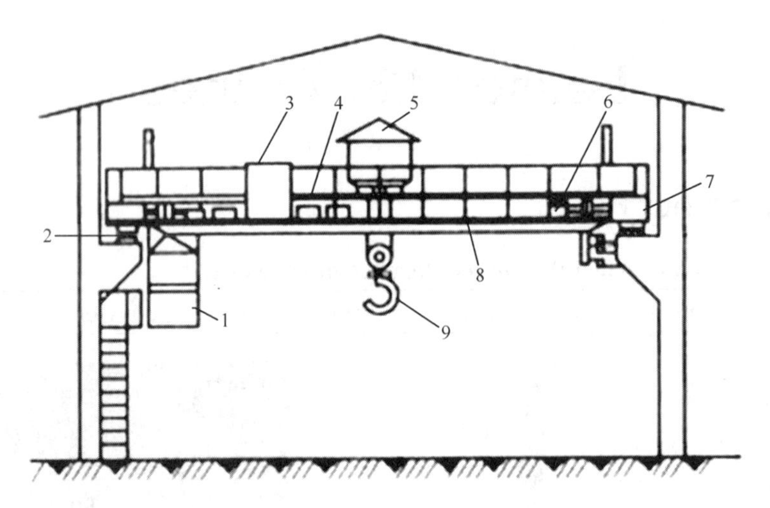

Fig.12.1 The structure of bridge crane

1—driver's cab; 2—track; 3—AC magnetic control panel; 4—resistance box; 5—trolley; 6—motor and transmission mechanism; 7—end beam; 8—gantry beam; 9—lifting hook

Knowledge extention

There are different kinds of engineering trucks (Fig.12.2) used on the building site to fulfill a variety of tasks. Look at the following pictures and study their names in English and Chinese. Discuss their usage in groups.

bull dozer 推土机	dump truck 倾卸卡车	caterpillar crane 履带起重机	road roller 压路机

Fig.12.2 Engineering trucks

Exercises

1. Fill in the blanks with the proper words in the text.

(1) A bridge crane has a ________ (called a gantry) running on long elevated ________ at each of its ends. The gantry can move ________ and forward along the tracks. The ________ is carried in a ________, which moves along the gantry beam.

(2) A jib crane has a long ________ that can ________ horizontally to move the load sideways. Many such cranes can also "luff" to ________ the reach of the crane by ________ the boom more or less to the ________ position.

2. Describe the engineering trucks mentioned in this lesson.

3. Match the English with the Chinese. Draw lines.

A.

(1) jib crane	a. 桥式起重机
(2) truck crane	b. 履带式起重机
(3) bridge crane	c. 悬臂式起重机
(4) crawler crane	d. 卡车式起重机

B.

(1) windlass	a. 横梁
(2) steel wire rope	b. 起重小车
(3) beam	c. 卷扬机
(4) driver's cab	d. 钢丝绳
(5) trolley	e. 轨道
(6) lifting hook	f. 负载
(7) track	g. 吊钩
(8) load	h. 驾驶室

4. Put the words into the proper boxes (Fig.12.3).

a. track　　b. windlass　　c. lifting hook　　d. gantry beam　　e. trolley

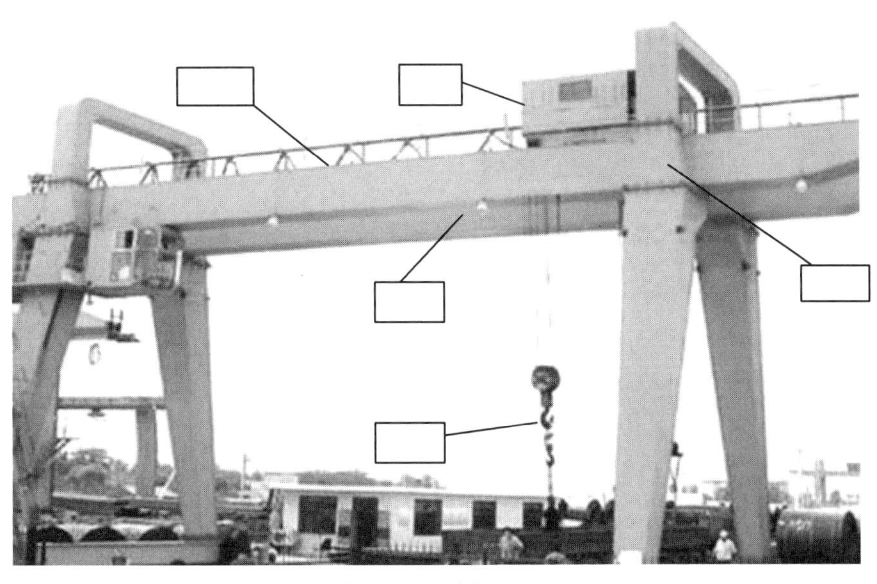

Fig.12.3　Bridge crane

Words and phrases

jib crane [dʒɪb] [kren] 悬臂式起重机

truck crane [trʌk] 卡车起重机

bridge crane 桥式起重机

crawler crane ['krɔːlə(r)] 履带式起重机

windlass['wɪndləs] 卷扬机，绞盘

steel 钢的

wire rope 绳

wrap 缠绕

powered drum 有动力装置的鼓
box-girder beam 箱形梁
run 运行
end 端部
backward 向后地
hoist 起重，升起
be carried in 装在……
move 移动
set up 安装，固定
tree trunk [trʌŋk] 木料堆
lifting hook 吊钩
swing [swɪŋ] 摆动
load [ləʊd] 负载，重物
luff [lʌf] 转舵
angling the boom 调节吊臂的角度
position 位置
typical example 典型例子
constructing high-rise building 建筑高层楼房
around 在……周围
gantry ['gæntri] 行车梁
elevated track 高架轨道
move 移动，运动
forward 向前地
system 系统
trolley 滑车
gantry beam [biːm] 行车梁
working area 工作区域
steel beam 钢梁
long boom [buːm] 长臂
horizontally [ˌharɪ'zantəlɪ] 水平地
sideways 斜向一边
control [kən'trol] 控制
more or less 或多或少
cantilever 悬臂
tower crane 塔吊
Anchored to the ground or to the building is, the crane extended upward as the work proceeds. 固定在地面或建筑物上并随着建筑物的升高而升高

Lesson 13 CAD/CAM

Look and select

Look at the pictures and select the correct terms from the box.

2D	3D	modeling	interface

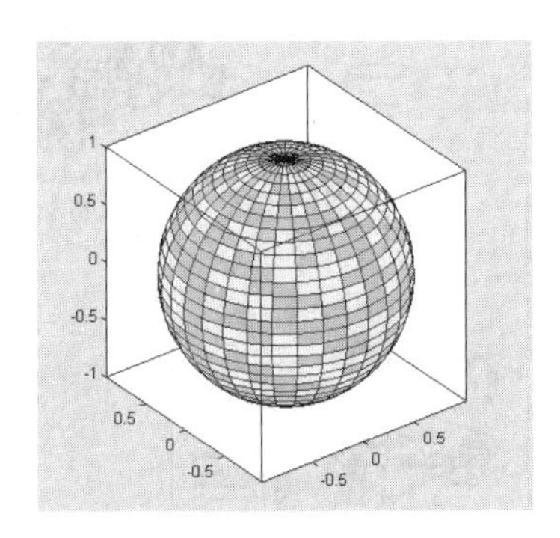

1. ____________

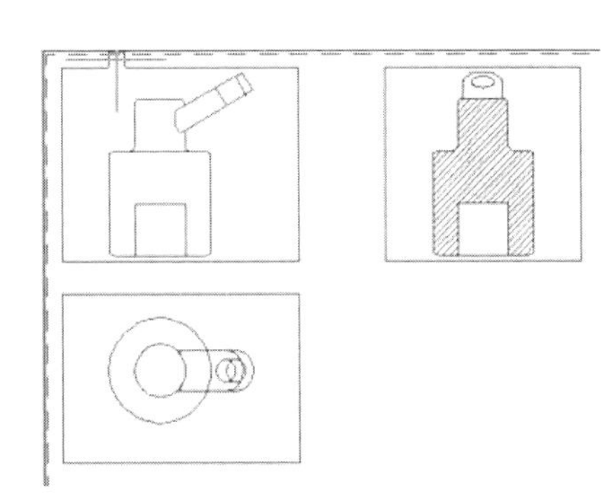

2. ____________

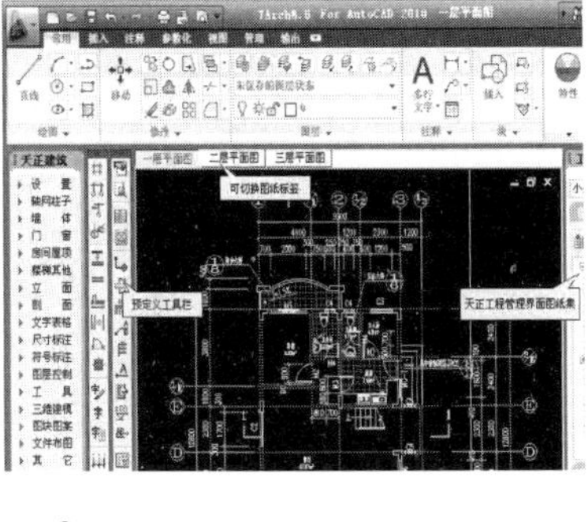

3. ____________

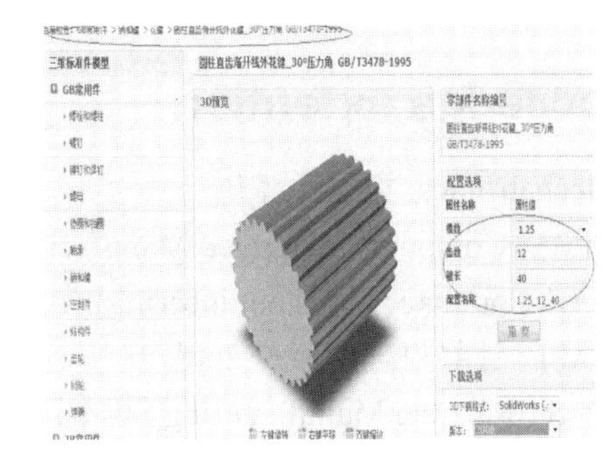

4. ____________

Text

1. CAD

CAD, is the short form for computer–aided design (Fig.13.1). When you design on a computer with CAD tools, you can get a three-dimensional view quickly. You don't have to imagine how a component will look like from two-dimensional drawings. You can put your thoughts into the solid without having to go via paper.

2. CAM

CAM, short for computer–aided manufacture (Fig.13.1), is used computer system to plan, manage, and control the operations of a manufacturing plant through either direct or indirect computer interface with the plant's production resources.

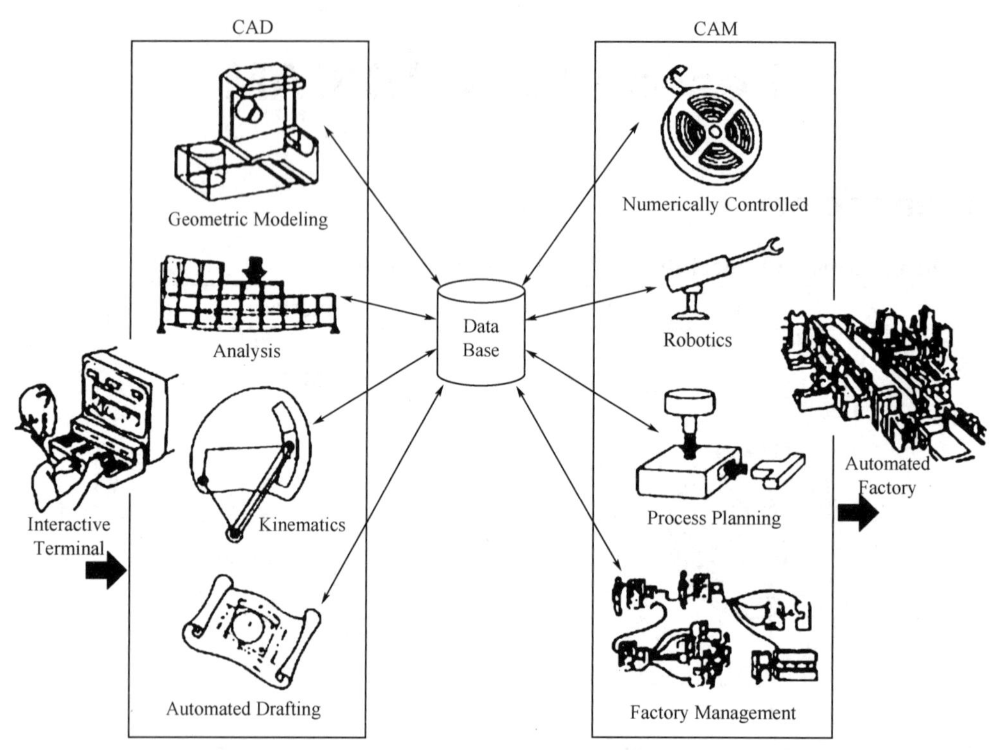

Fig.13.1 Conceptual CAD/CAM system

Knowledge extention

Mechatronics 机电一体化

CIMS(Computer Integrated Manufacturing System) 计算机集成制造系统

CAE(Computer-aided engineering) 计算机辅助工程

CAPE 计算机辅助生产工程

FMS (Flexible Manufacturing System) 柔性制造系统

digital factory 数字化工厂

automation and robots 自动化与机器人

Exercises

1. Fill in the blanks with the proper words in the text.

(1) CAD, is the short for ____________. When you design on a computer with CAD tools, you can get a ____________ view quickly.

(2) CAM, short for __________, is used computer system to _____________, manage, and ______________________ the operations of a manufacturing plant through either direct or indirect _____________ with the plant's production resources.

2. Match the English with the Chinese. Draw lines.

(1) geometric modeling　　　　a. 动态

(2) analysis　　　　b. 几何建模

(3) kinematics　　c. 分析
(4) automated drafting　　d. 自动成图
(5) numerically controlled　　e. 机器人技术
(6) robotics　　f. 工厂管理
(7) process planning　　g. 数字控制
(8) factory management　　h. 工艺规划

3. Describe the CAD/CAM system.

Words and phrases

CAD computer – aided design ['eɪdɪd] [dɪ'zaɪn] 计算机辅助设计
CAM computer – aided manufacture [ˌmænjʊ'fæktʃə(r)] 计算机辅助制造
interface ['ɪntəfeɪs] 计算机接口/界面
modeling ['mɒdlɪŋ] 建模
three-dimensional view 3D 视图
short for……的缩写
component[kəm'pəʊnənt]构建，元件，零件
imagine [ɪ'mædʒɪn] 想象
put your thoughts into 把你的想法付诸……
manage ['mænɪdʒ] 管理
without having to go via paper 无需借助纸张绘出
solid ['sɒlɪd] 实体，实物
computer system ['sɪstəm] 计算机系统
plan [plæn] 计划
two-dimensional drawing 2D 图纸
control [kən'trəʊl] 控制
operation [ˌɒpə'reɪʃn] 运作，操作
manufacturing plant 制造工厂
direct [də'rekt] 直接
indirect [ˌɪndə'rekt] 间接
analysis [ə'næləsɪs] 分析
kinematics [ˌkɪnə'mætɪks] 动态
automated drafting 自动成图
data base 数据库
numerically controlled [njuː'merɪklɪ] 数字控制
automated factory 自动化工厂
robotics [rəʊ'bɒtɪks] 机器人技术
process planning 工艺规划，生产计划
factory management 工厂管理
interactive terminal [ˌɪntər'æktɪv] ['tɜːmɪnl] 交互终端
geometric modeling [ˌdʒiːə'metrɪk] ['mɒdlɪŋ] 几何建模
with the plant's production resources [prə'dʌkʃn] [rɪ'sɔːsɪz] 利用生产资源

Lesson 14 Hydraulic and Pneumatic Transmission

Look and select

Look at the pictures and select the words from the box.

shearing machine	hydraulic press	bulldozer	pneumatic driller

1. ________________

2. ________________

3. ________________

4. ________________

Text

The common machine transmission has the following forms: mechanical transmission, electrical transmission, hydraulic transmission and pneumatic transmission. The hydraulic and pneumatic transmission are mainly introduced in this text.

Hydraulic transmission is a kind of transmission in the form of liquid as medium to transmit energy. The most common machines include the hydraulic press, shearing machine (Fig.14.1), hydraulic jack (Fig.14.2) and pump (Fig.14.3), and bulldozer etc.

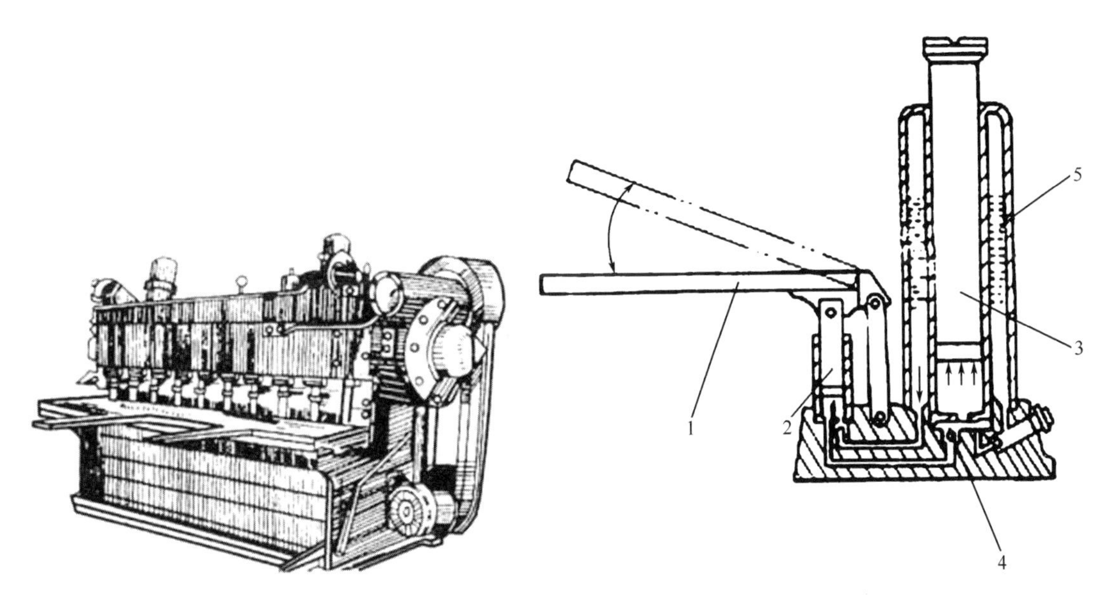

Fig.14.1 Shearing machine

Fig.14.2 Working principle of hydraulic jack

1—lever; 2—small piston; 3—large piston;

4—one-way valve; 5—oil tank

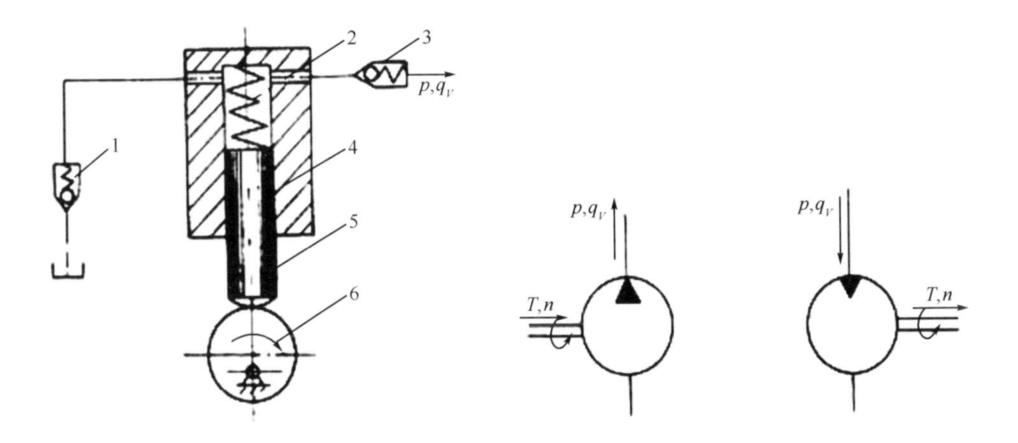

Fig.14.3 Working principle and symbols of hydraulic pump

1, 3—one-way valve; 2—spring; 4—pump body; 5—plunger; 6—eccentric wheel

Pneumatic transmission is a kind of transmission in the form of compressed air as the working medium for the transmission of energy, such as drilling device and nailing gun etc. shown in (Fig.14.4).

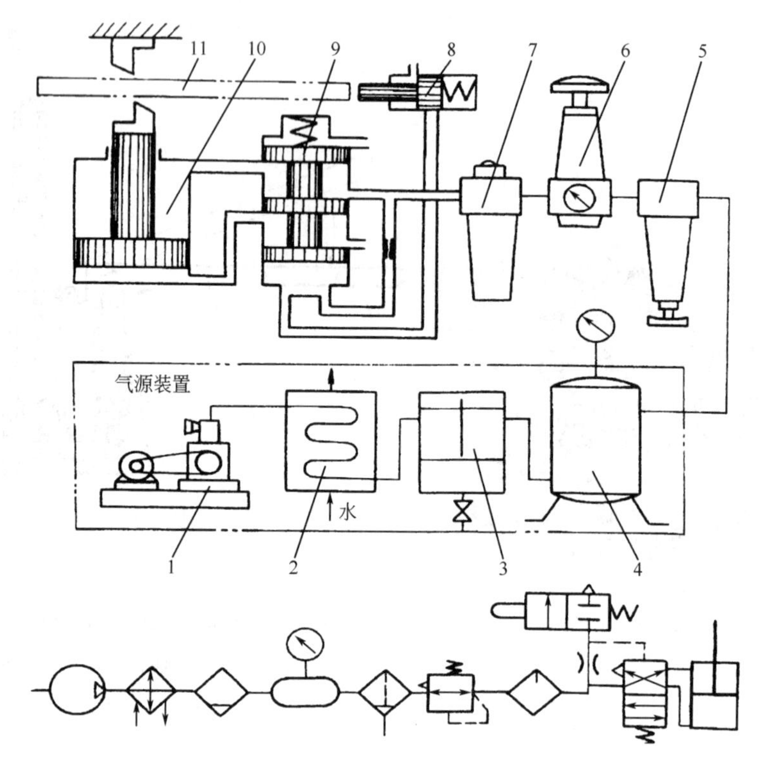

Fig.14.4 Working principle and symbols of pneumatic shearing machine

1—air compressor; 2—cooler; 3—oil-water separator; 4—air tank; 5—water-separating gas filter; 6—relief valve; 7—oil mist set; 8—motion valve; 9—directional valve; 10—air cylinder; 11—material

Knowledge extention

Hydraulic and pneumatic transmission system consist of the following four basic parts:

a. The power unit: hydraulic pump or air compressor

b. The actuating components, including a variety of cylinders and motors

c. Controlling and regulating components, such as pressure valves, flow valves and directional valves etc.

d. Auxiliary components, such as filters, oil tubing, pressure gauges, flow meters, oil tanks, oil mists and silencers etc.

Exercises

1. Fill in the blanks with the proper words in the text.

(1) Hydraulic transmission is a kind of transmission in the form of ________ as medium to transmit ________.

(2) Pneumatic transmission is a kind of transmission in the form of ______________ as the working ________ for the transmission of ________.

2. Match the English with the Chinese. Draw lines.

A.

(1) transmission	a. 机械的
(2) hydraulic	b. 传动，传输
(3) pneumatic	c. 电的
(4) mechanical	d. 液压的
(5) electrical	e. 气动的

B.

(1) shearing	a. 千斤顶
(2) motor	b. 剪切
(3) jack	c. 空压机
(4) valve	d. 电机
(5) air compressor	e. 阀门

3. Answer the following questions.

(1) What are the common forms of machine transmission?

(2) What are the four basic parts of hydraulic and pneumatic system?

Words and phrases

hydraulic [haɪ'drɔːlɪk] 液压/水压的
transmission [træns'mɪʃn] 传输，传送
jack [dʒæk] 千斤顶
bulldozer ['bʊldəʊzə(r)] 推土机
pump [pʌmp] 泵
driller ['drɪlə] 钻子
nailing gun ['neɪlɪŋ] 射钉枪
power unit 动力装置
controlling and regulating 控制调节
flow valve 流量阀
auxiliary [ɔːg'zɪliəri] 辅助
oil tubing 油管
flow meter 流量计
oil mist 油雾器
pneumatic [njuː'mætɪk]气动的
medium ['miːdiəm] 媒介，介质
shearing machine ['ʃɪərɪŋ] 剪切机
principle ['prɪnsəpl] 原理
hydraulic press 液压机
piston ['pɪstən] 活塞
valve [vælv] 阀
air compressor 空气压缩机
pressure valve 压力阀
directional valve 方向阀
filter ['fɪltə(r)] 过滤器
pressure gauge [geɪdʒ] 压力计
oil tank 油箱
silencer ['saɪlənsə(r)] 消声器
actuating component ['æktʃuːˌeɪtɪŋ] [kəm'pəʊnənt] 执行元件
cylinder and motor ['sɪlɪndə(r)] ['məʊtə(r)] 缸与马达

Part II
Process Equipment

Lesson 15 Basic Operations for Fitter

Look and select

Look at the pictures and select the correct terms from the box.

fitter	welder	piper	riveter

1. ________________

2. ________________

3. ________________

4. ________________

Text

During the mechanical machining, fitter shall carry out the following operations shown in Table 15.1.

Table 15.1 Basic operations for fitter

Description	Basic operations	
lining	(a) plane lining	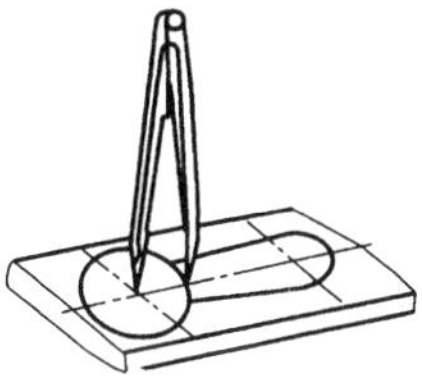
	(b) tridimensional lining	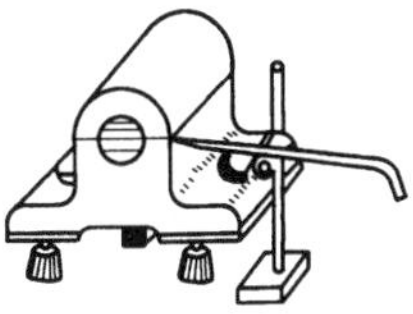
chiseling		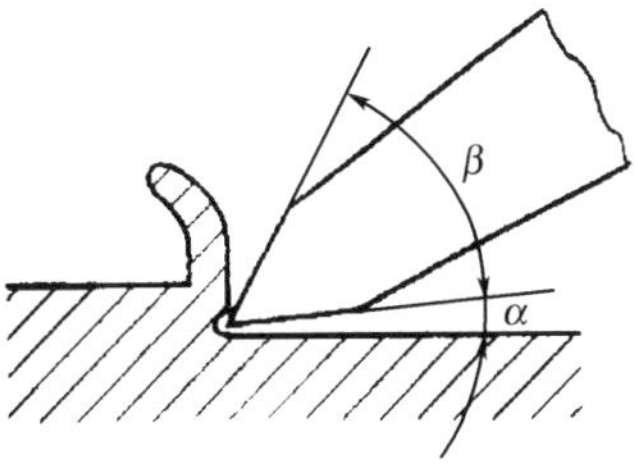
sawing	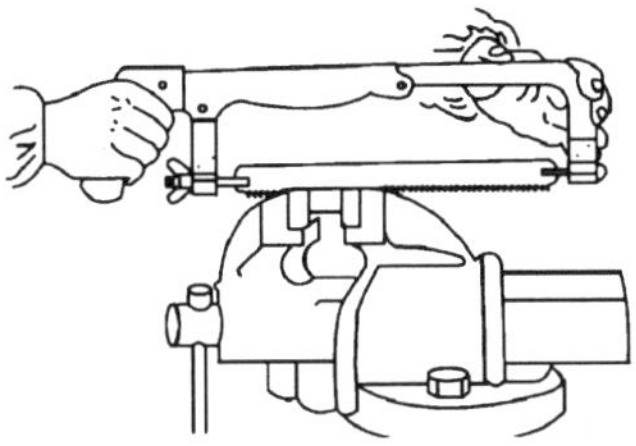	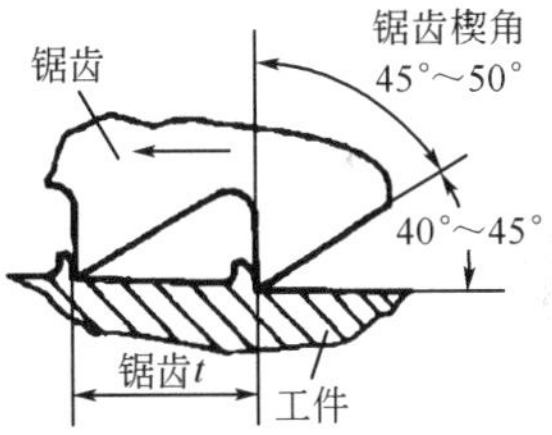
filing	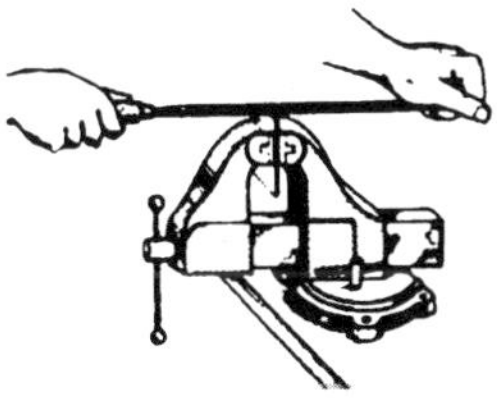	
drilling	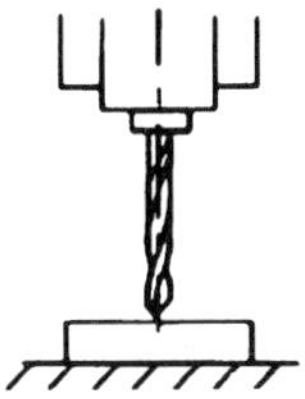	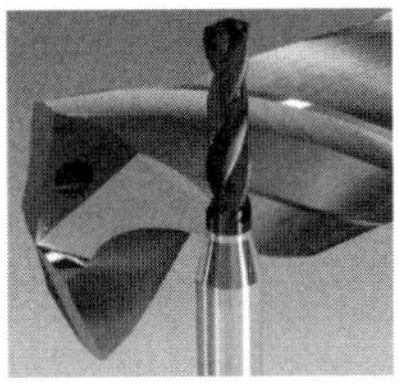
counterboring (a) countersinking(b) spot facing(c)	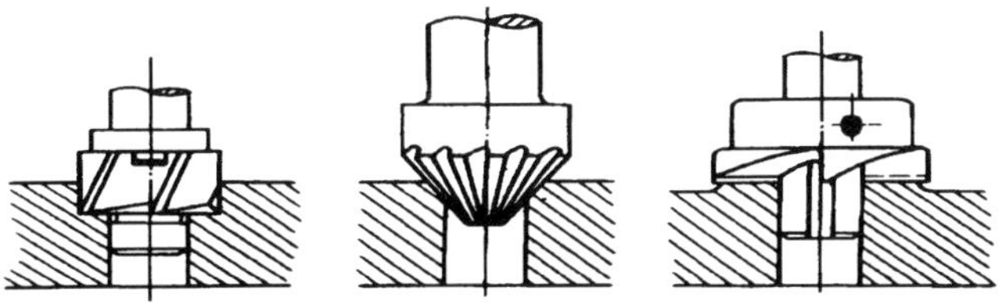 (a) (b) (c)	

续表

Description	Basic operations	
reaming	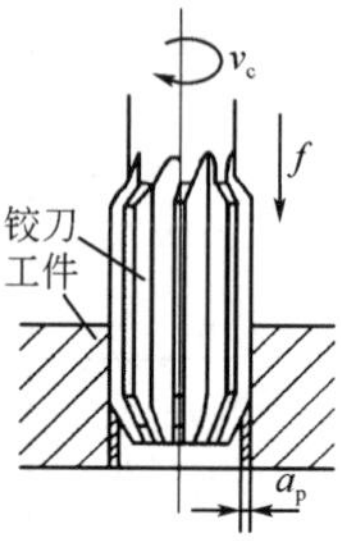 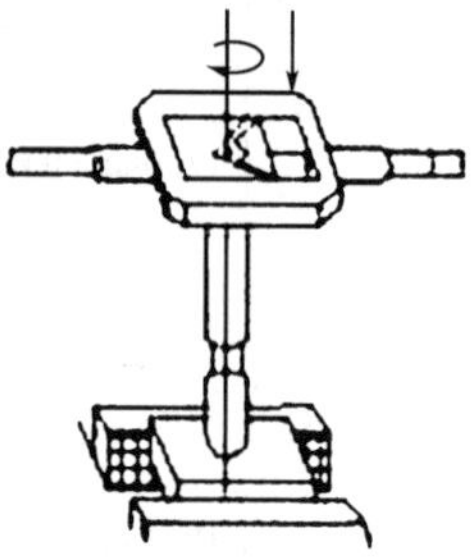	
thread machining	female/internal threading	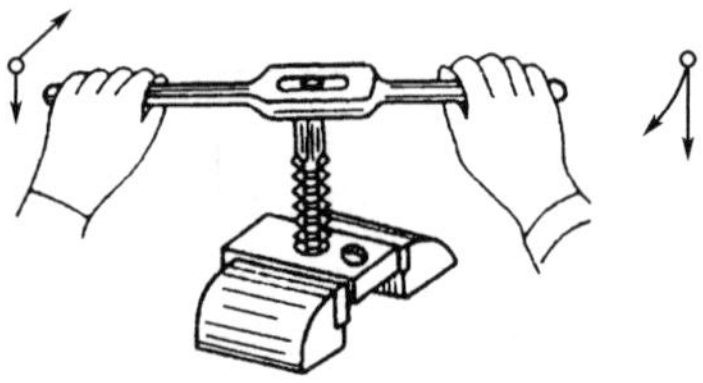
	male/external threading	
scraping	 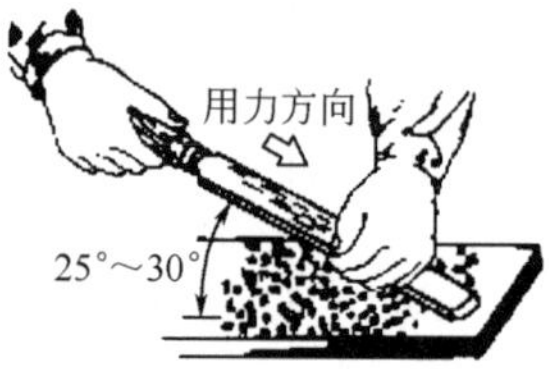	
rubbing	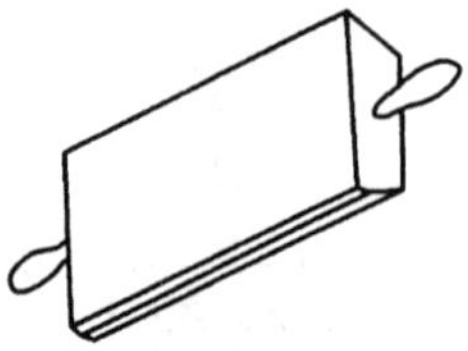 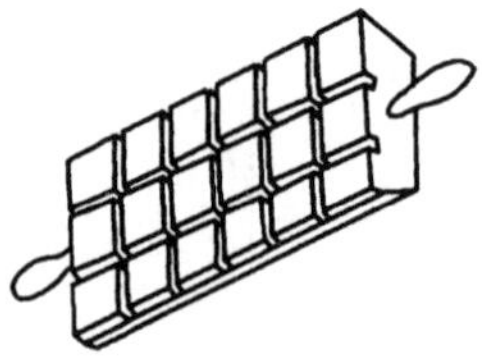	

Knowledge extention

Fitter is one job aggregate of bench work. Fitters' primary tools are vices (Fig.15.1), files (Fig.15.2) and hammers (Fig.15.3).

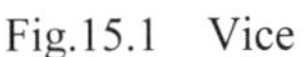

Fig.15.1 Vice

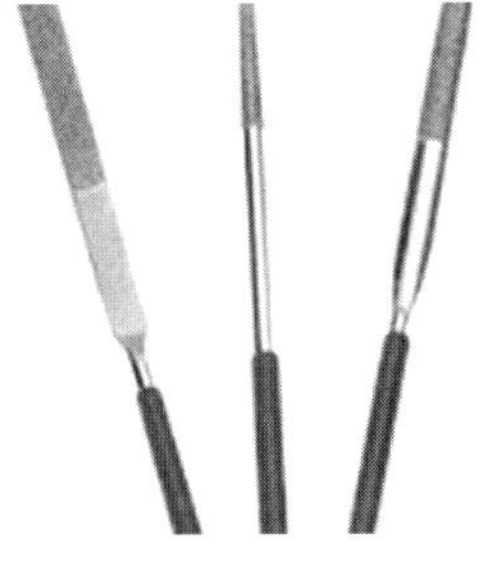

Fig.15.2 Files

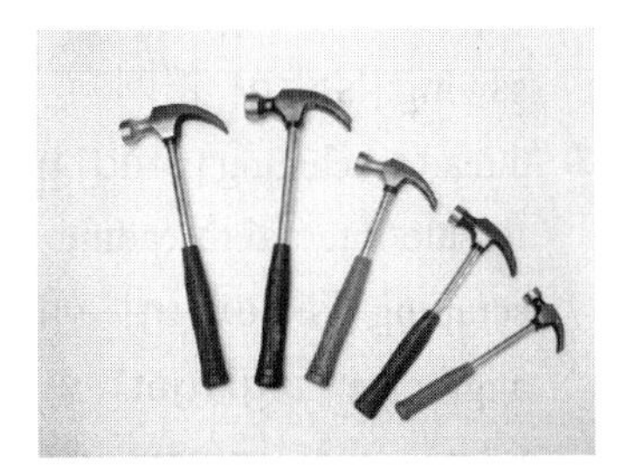

Fig.15.3 Hammers

Exercises

1. Fill in the blanks with the proper words in the text.

Fitter is one job ______ of bench work. Fitters' primary tools are ______, files and ______.

2. Match the English with the Chinese. Draw lines.

A.

(1) fitter	a. 焊工
(2) welder	b. 管工
(3) riveter	c. 钳工
(4) piper	d. 铆工

B.

(1) lining	a. 锯削
(2) chiseling	b. 划线
(3) sawing	c. 锉削
(4) filing	d. 錾削

C.

(1) drilling	a. 铰孔
(2) reaming	b. 钻削
(3) thread machining	c. 研磨
(4) scraping	d. 螺纹加工
(5) rubbing	e. 刮削

D.

(1) counterboring	a. 锪孔口平面
(2) countersinking	b. 锪沉孔
(3) spot facing	c. 锪锥孔

Words and phrases

fitter ['fɪtə(r)] 钳工

welder ['weldə(r)] 焊工

piper ['paɪpə(r)] 管工

riveter ['rɪvɪtə] 铆工

basic ['beɪsɪk] 基本的

operation [ˌɒpə'reɪʃn] 操作

lining ['laɪnɪŋ] 划线

plane [pleɪn] 平面

tridimensional [ˌtraɪdɪ'menʃənəl] 立体的
chiseling [t'ʃɪzlɪŋ] 錾削
sawing ['sɔːɪŋ] 锯削
filing ['faɪlɪŋ] 锉削
thread machining [θred] [mə'ʃiːnɪŋ] 螺纹加工
reaming ['riːmɪŋ] 铰孔
female/internal threading 攻螺纹
male/external threading 套螺纹
scraping ['skreɪpɪŋ] 刮削
rubbing ['rʌbɪŋ] 研磨
aggregate ['ægrɪgət] 总计，组合
bench work [bentʃ] 台面工作
primary tool ['praɪməri] 主要工具
vice [vaɪs] 虎钳
file [faɪl] 锉刀
hammer ['hæmə(r)] 锤子

Lesson 16 Basic Operations for Riveter

Look and select

A. Look at the pictures and select the correct terms from the box.

locating by sample plate	positioning by jig frame
locating by marking	positioning by setting element

smaple plate

1. ____________________ 2. ____________________

$B+\Delta b$

3. ____________________ 4. ____________________

B. Understand the measuring process of level gauge in Chinese (Fig.16.1).

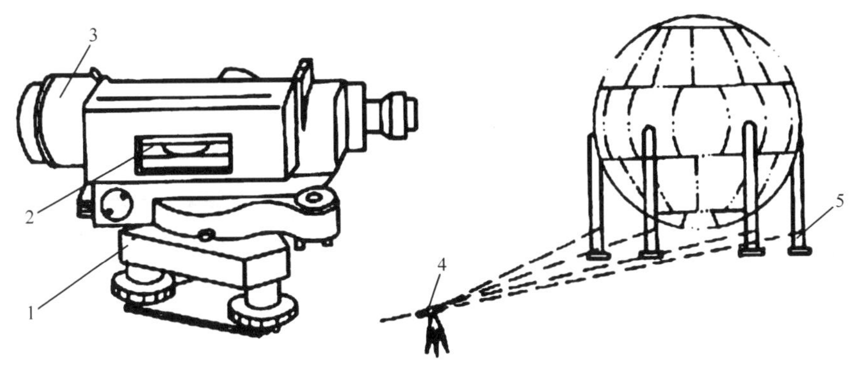

Fig.16.1 The measuring process of level gauge

1. base ______; 2. levelling instrument ______; 3. telescope ______; 4. level gauge ______; 5. datum mark ______

Text

Riveter shall perform the followings according to the chart (Fig.16.2).

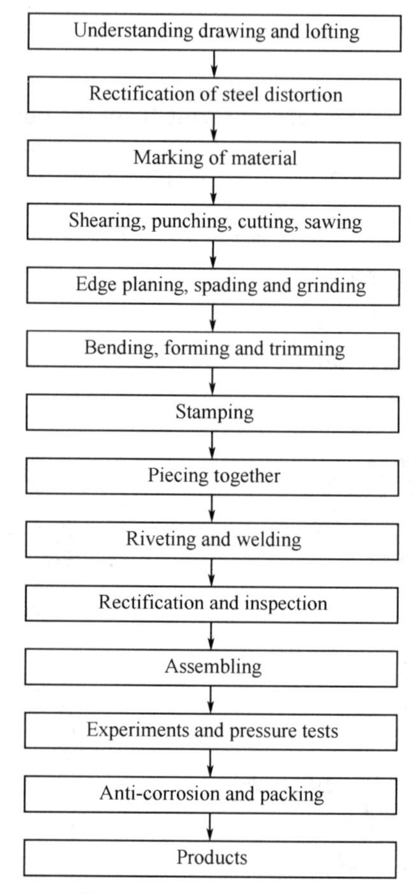

Fig.16.2 Flow Chart of Operations for Riveter

Knowledge extention

During the assembling construction for riveters, commonly used tools are mainly hammer, chisel, hand abrasive wheel, crowbar, wrench, and other various tools for lining as shown in Fig.16.3.

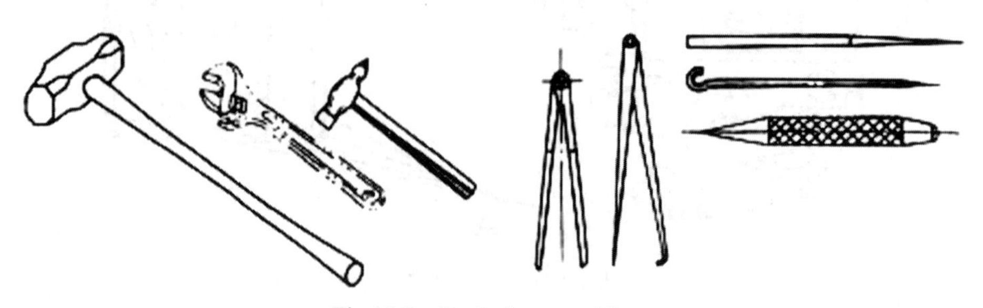

Fig.16.3 Tools for assembling

For measuring instruments, the followings shown in Fig.16.4 are commonly used in the construction: 90° angle square, steel straight ruler, level ruler, steel tape, theodolite and level gauge etc.

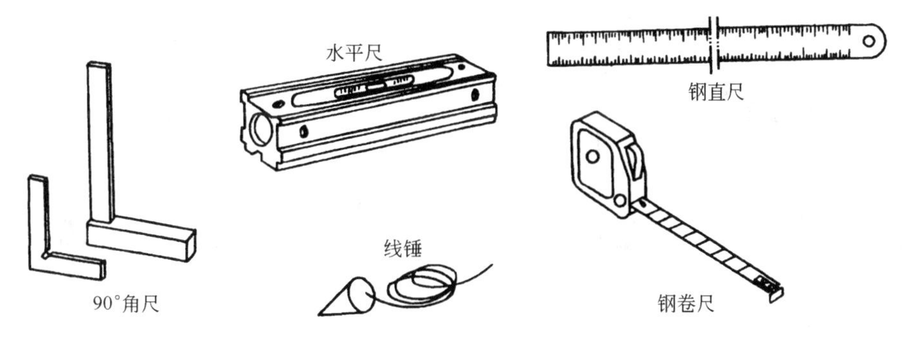

Fig.16.4 Meauring instruments for assembling

Exercises

1. Fill in the blanks with the proper words in the text.

(1) During the assembling construction for riveters, commonly used tools are ________, chisel, ________, crowbar, ________, and various tools for lining.

(2) For measuring instruments, the followings are commonly used in the construction: 90° ________, steel straight ruler, ________, steel tape, ________ and level gauge etc.

(3) The measuring process of level gauge include ________, ________ instrument, telescope, ________ and datum mark.

2. Match the English with the Chinese. Draw lines.

A.

(1) lofting　　a. 矫正
(2) rectification　　b. 放样
(3) steel distortion　　c. 冲切
(4) material marking　　d. 钢材变形
(5) punching　　e. 铲边
(6) spading　　f. 号料

B.

(1) bending　　a. 成形
(2) forming　　b. 弯曲
(3) trimming　　c. 拼装
(4) stamping　　d. 组装
(5) piecing together　　e. 修边
(6) assembling　　f. 冲压

3. Answer the following question.

What are the steps which should be performed after assembling in the operations for riveter?

4. Understand the structure and application of theodolite (Fig.16.5).

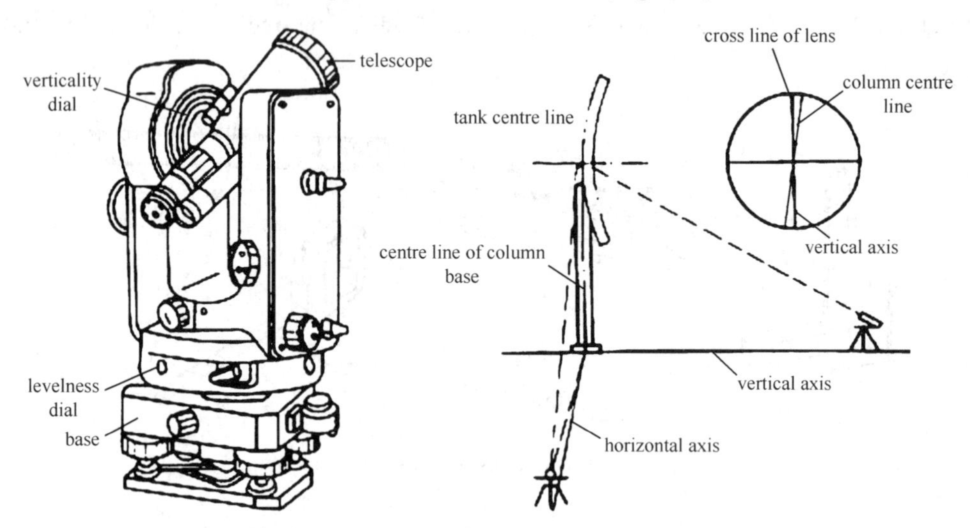

Fig.16.5 Structure and application of theodolite

Words and phrases

locating by sample plate [ləʊ'kəɪtɪŋ] ['sɑ:mpl] [pleɪt] 用样板定位
locating by marking ['mɑ:kɪŋ] 标记定位
positioning by jig frame [dʒɪg] [freɪm] 框/胎架定位
positioning by setting element ['setɪŋ] ['elɪmənt] 设置元件定位
measure ['meʒə(r)] 测量
process ['prəʊses] 过程
level gauge ['levl] [geɪdʒ] 水准仪
rectification [ˌrektɪfɪ'keɪʃn] 矫正，修正
steel distortion [dɪ'stɔ:ʃn] 钢材变形
marking of material ['mɑ:kɪŋ] 号料
understanding drawing and lofting 识图和放样
punch [pʌntʃ] 冲切
edge planing, spading and grinding 刨边，铲边，修磨
trim [trɪm] 修边
stamping [stæmpɪŋ] 冲压
piecing together [pɪsɪŋ] 拼装
assembling [ə'semblɪŋ] 组装
anti-corrosion and packing ['ænti:kər'əʊʒn] ['pækɪŋ] 防腐包装
pressure test 试压
product ['prɒdʌkt] 成品
crowbar ['krəʊbɑ:(r)] 撬棍
wrench [rentʃ] 扳手
level ruler 水平尺
steel tape 钢卷尺
theodolite [θi'ɒdəlaɪt] 经纬仪
angle square ['æŋgl] [skweə(r)] 角尺

Lesson 17 Valves

Look and select

Look at the pictures and select the correct terms from the box.

safety valve	butterfly valve	check valve	gate valve

1. ________________

2. ________________

3. ________________

4. ________________

Text

In the pipeline engineering, commonly used valves are gate valve, globe valve, ball valve, check valve, safety valve, tap and cock etc.

The structure of **gate valves** are shown in Fig.17.1: valve body, hand wheel, valve stem, gland, valve plate and flange.

Globe valves shown in Fig.17.2 are widely used in plumbing and industrial pipelines. The structure is as follows: valve body, hand wheel, valve stem, gland, valve plate, sealing ring and flange.

Check valves shown in Fig.17.3 are to rely on the flow of the medium itself to open and close the discs automatically. They are applied to prevent the medium backflowing.

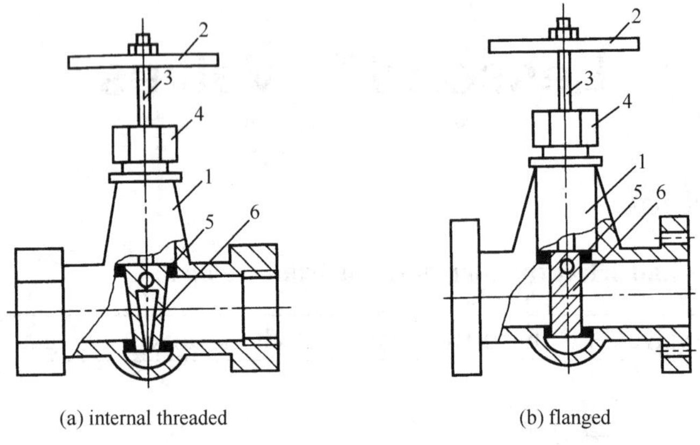

(a) internal threaded (b) flanged

Fig.17.1 Structure of gate valves

1—valvebody; 2—hand wheel; 3—valve stem; 4—gland; 5—valve plate; 6—flange

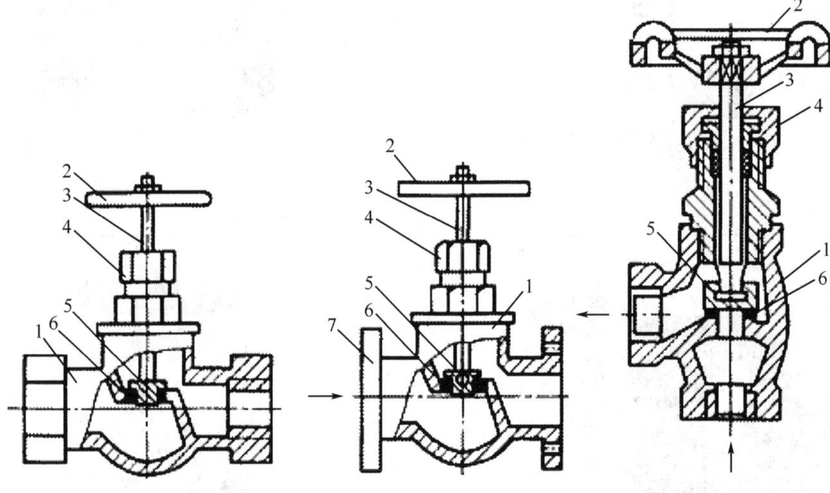

Fig.17.2 Globe valves

1—valve body; 2—hand wheel; 3—valve stem; 4—gland; 5—valve plate; 6—sealing ring; 7—flange

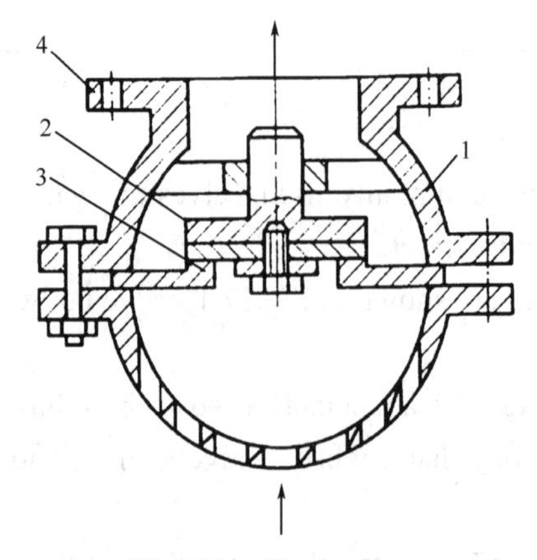

Fig.17.3 Check valves

1—valve body; 2—valve plate; 3—sealing ring; 4—flange

Knowledge extention

The **safety valves**, shown in Fig.17.4, in the system play the role of security protection. When the system pressure exceeds the setting value, the safety valves open automatically to let the partial gas or fluid in the system into the atmosphere or pipeline for ensuring that the pressure in the system does not exceed the allowance.

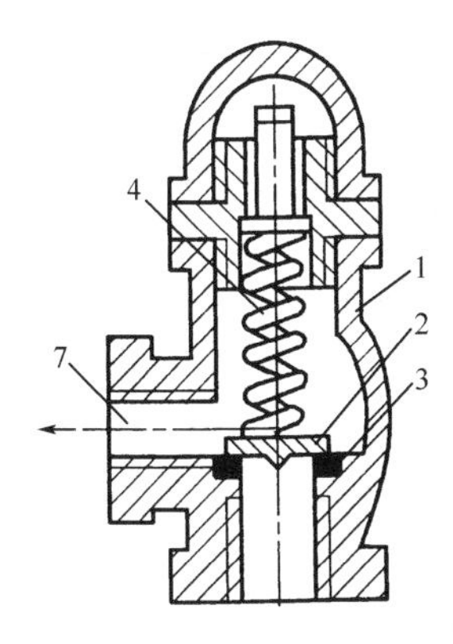

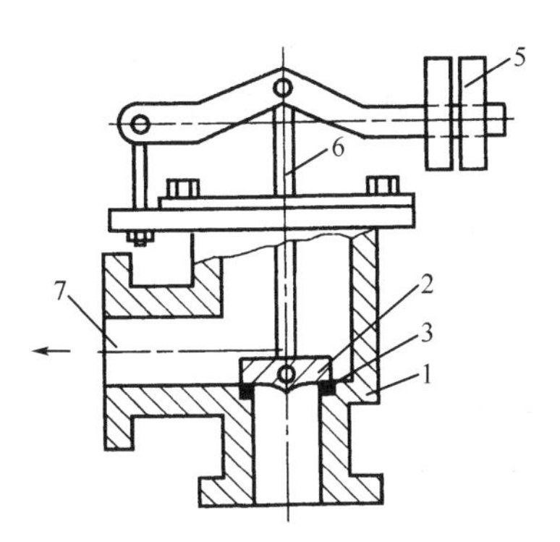

Fig.17.4 Safety valves

1—valve body; 2—valve plate; 3—sealing ring; 4—spring; 5—heavy bob; 6—lever; 7—medium outlet

The main advantages and applications of **cocks** are similar with ball valves (Fig.17.5). The internal thread cock (Fig.17.6) driven by the handle is as follows:

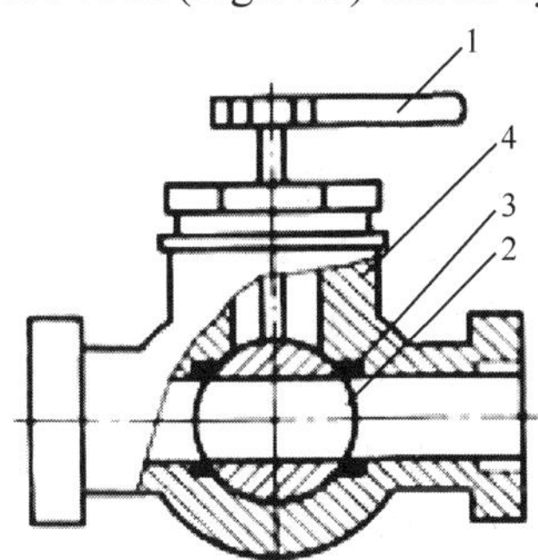

Fig.17.5 Internal thread ball valve

1—handle; 2—ball; 3—sealing ring; 4—valve body

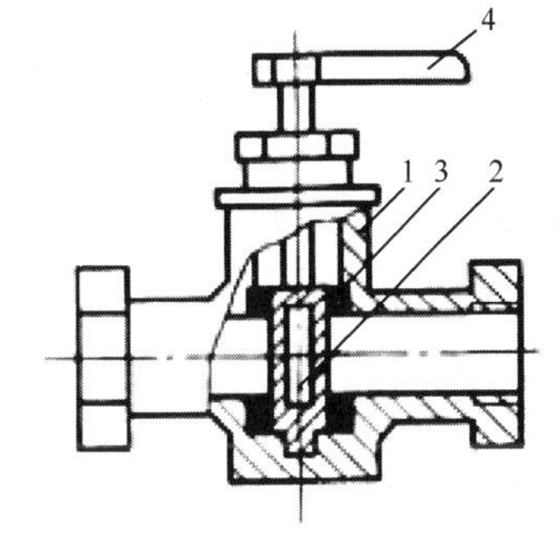

Fig.17.6 Internal thread cock

1—valve body; 2—cylinder; 3—sealing ring; 4—handle

The common **tap** is shown in Fig.17.7:

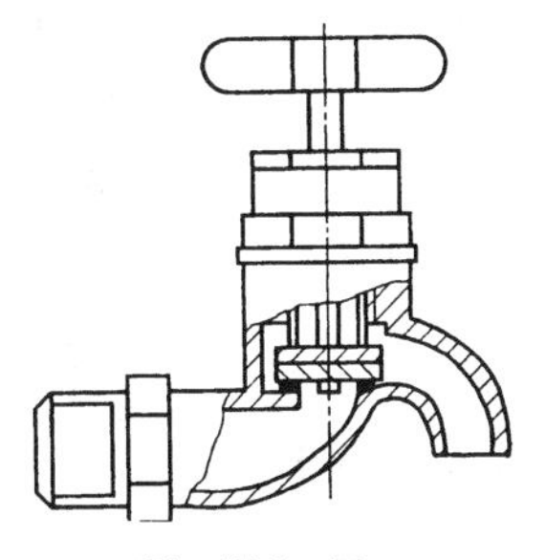

Fig.17.7 Tap

Exercises

1. Fill in the blanks with the proper words in the text.

(1) Globe valves are widely used in ________ and industrial ________, the structure is as follows: valve body, ________, valve stem, ________, valve plate, ________ and flange.

(2) Check valves are to rely on the flow of the ________ itself to ________ and close the ________ automatically. They are applied to prevent the medium ________.

(3) The safety valves in the system play the role of ________.

2. Match the English with the Chinese. Draw lines.

(1) gate valve a. 球阀
(2) globe valve b. 止回阀
(3) ball valve c. 闸阀
(4) check valve d. 安全阀
(5) safety valve e. 截止阀
(6) butterfly valve f. 蝶阀
(7) tap g. 旋塞
(8) cock h. 水龙头

3. Translate the following Chinese words into English.

阀体　阀杆　密封圈　法兰　压盖　手轮　弹簧

Words and phrases

safety valve ['seɪfti] [vælv] 安全阀
check valve [tʃek] 止回阀
engineering [ˌendʒɪ'nɪərɪŋ] 工程
globe valve [gləʊb] 截止阀
tap [tæp] 水龙头
valve body 阀体
valve stem [stem] 阀杆
valve 阀板
plumbing ['plʌmɪŋ] 水暖管道
sealing ring ['si:lɪŋ] [rɪŋ] 密封圈
automatically [ˌɔ:tə'mætɪklɪ] 自动地
prevent [prɪ'vent] 阻止
play the role of 起……作用
atmosphere ['ætməsfɪə(r)] 大气
medium outlet ['mi:diəm] ['aʊtlet] 介质排出口
application [ˌæplɪ'keɪʃn] 适用，应用
internal thread cock 内螺纹式旋塞
butterfly valve ['bʌtəflaɪ] 蝶阀
gate valve [geɪt] 闸阀
pipeline ['paɪplaɪn] 管道
ball valve 球阀
cock [kɒk] 旋塞
hand wheel 手轮
gland [glænd] 压盖
flange [flændʒ] 法兰
as follow 如下
be applied to 应用于
rely on 依靠……
backflow ['bækfləʊ] 回流
handle 手柄
allowance [ə'laʊəns] 允许度
heavy bob ['hevi] [bɒb] 重锤
advantage [əd'vɑ:ntɪdʒ] 优点
be similar with 与……相似
security protection [sɪ'kjʊərəti] [prə'tekʃn] 安全保护
exceed the setting value [ɪk'si:d] ['setɪŋ] ['vælju:] 超出规定值

Lesson 18 Piping

Look and select

Look at the pictures and select the correct terms from the box.

flange	teepipe	ell	reducer

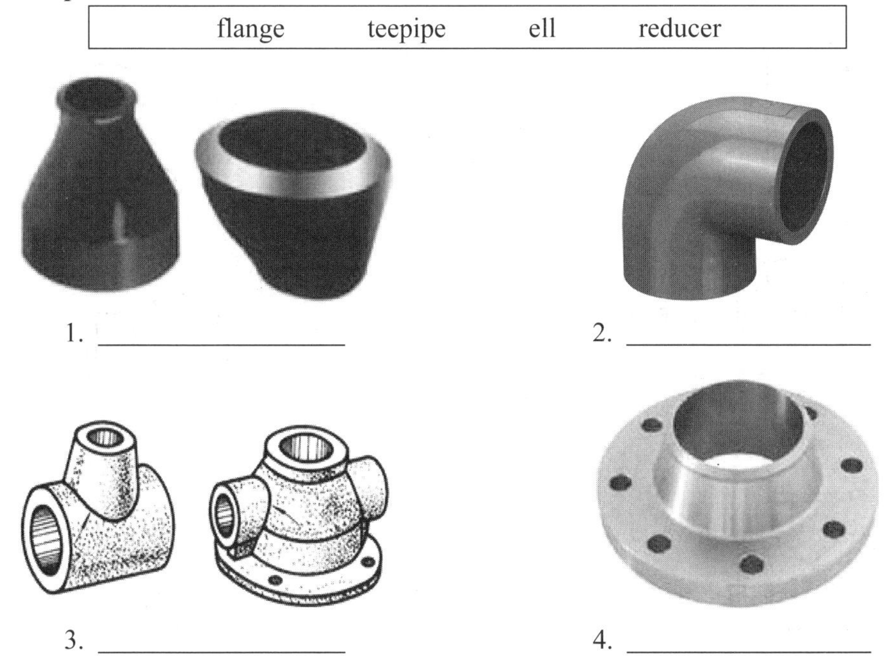

Pipe

1. Home piping and the pipe in process plants

Most **home piping** (Fig.18.1) is of small size and is put together by screws. Screwing sections of pipe together is easy to do with small-sized pipe because it can be easily handled. But **the pipe used in process plants** (Fig.18.1) generally ranges in size from two to twelve inches in diameter and a section of such pipe may weigh many hundreds of pound or kilograms.

Fig.18.1 Home piping and the pipe in process plants

2. Pipe fittings

There are a number of pipe fittings used with all piping systems (see Fig.18.2). When a pipe must change 90 degrees in direction, a pipe **elbow**, or 90-degree **ell**, is used; **45-degree ells** change the direction by 45 degrees. When one pipe joints another at right angles, the juncture made with a **teepipe**. A **pipe cross** joins four pieces of pipe. Pipe of two different sizes can be connected by **reducers**, or pipe reducing fittings. When pipes must be joined at a sharp angle, a **pipe lateral fitting** is used.

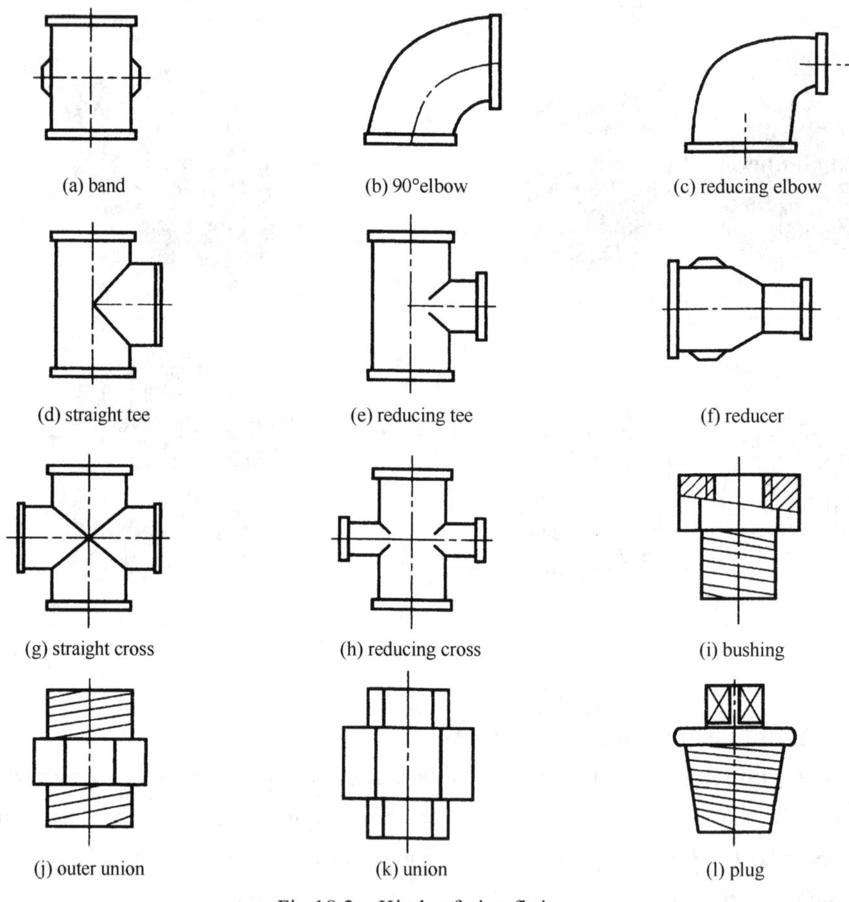

Fig.18.2 Kinds of pipe fittings

Knowledge extention

Advantages and disadvantages of pipe flange

Large-diameter pipe is put together by means of welding or pipe flanges. When flanges (Fig.18.3) are used they are usually welded to the ends of the pipe. Flanges have two **advantages**: first, welding can be done on the ground or in a shop; second, the pipe can be easily disassembled for inspection or replacement. The **disadvantages** are that the joints sometimes leak and the flanges are expensive. Welded pipe is harder to assemble because the welds must often be made while the pipe is high in the air or in awkward positions.

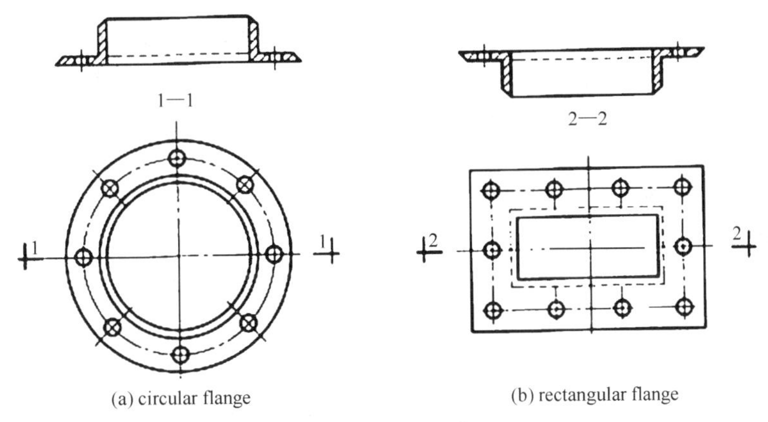

(a) circular flange (b) rectangular flange

Fig.18.3 Shapes of flanges

Exercises

1. Fill in the blanks with the proper words in the text.

(1) Flanges have two **advantages**: first, welding can be done on the ______ or in a ______; second, the pipe can be easily ______ for inspection or ______.

(2) The **disadvantages** are that the joints sometimes ______ and the flanges are ______. Welded pipe is harder to ______ because the welds must often be made while the pipe is ______ in the air or in ______ positions.

2. Match the English with the Chinese. Draw lines.

(1) flange a. 弯管
(2) teepipe b. 大小头
(3) elbow c. 四通
(4) reducer d. 接头
(5) cross e. 法兰
(6) union f. 三通
(7) bushing g. 丝堵
(8) plug h. 补心

3. Translate the words in the picture (Fig.18.4) of common section steel into English.

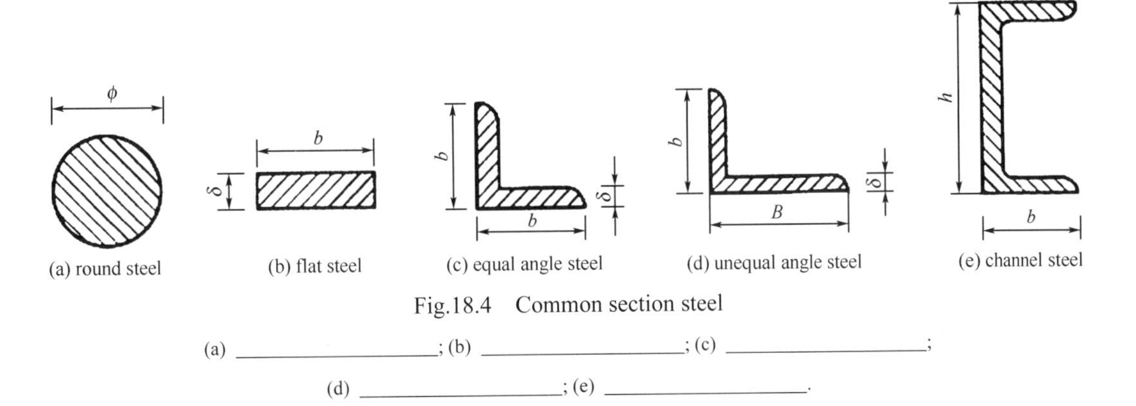

(a) round steel (b) flat steel (c) equal angle steel (d) unequal angle steel (e) channel steel

Fig.18.4 Common section steel

(a) ____________; (b) ____________; (c) ____________;

(d) ____________; (e) ____________.

4. Decide whether the following statements are true (T) or false (F) according to the text.

(1) Piping is an important part of any process plant because there is so little of it. (　　)

(2) Large-diameter pipe is not put together by means of welding or pipe flanges. (　　)

(3) The disadvantages are that the joints sometimes leak and the flanges are expensive. (　　)

(4) Welded pipe is easily to assemble because the welds must often be made while the pipe is high in the air or in awkward position. (　　)

(5) There are only one pipe fitting used with all piping systems. (　　)

Words and phrases

pipe [paɪp] 管道

teepipe 三通管

reducer [rɪ'dju:sə] 大小头

pipe in process plant 工厂管道

degree [dɪ'gri] 度数

advantage [əd'vɑ:ntɪdʒ] 优点

large-diameter [daɪ'æmɪtə(r)] 大直径

by means of 通过……方式

be welded to 被焊接到……

on the ground 在地面上

disassembled [ˌdisə'sembld] 拆装/卸

replacement [rɪ'pleɪsmənt] 替换，代替

leak [li:k] 泄漏

harder 较困难

weld 焊接，焊缝

high 高的

flange [flændʒ] 法兰

ell [el] 弯头

home pipe 家用管道

elbow ['elbəʊ] 弯管

cross [krɒs] 交叉，十字

disadvantage 缺点

be put together 连接在一起

welding [weldɪŋ] 焊接

end 端部

in a workshop 在车间里

inspection [ɪn'spekʃn] 检查

joint [dʒɔɪnt] 接头

expensive [ɪk'spensɪv] 昂贵

assemble [ə'sembl] 组装，拼装

while 当……时候

in the air 在高空

in awkward positions ['ɔ:kwəd] [pə'zɪʃənz] 在不方便的位置

Lesson 19 Pressure Vessels

Look and select

Look at the pictures and select the correct terms from the box.

scaffolding	site	safety	ASME

1. ____________________

2. ____________________

3. ____________________

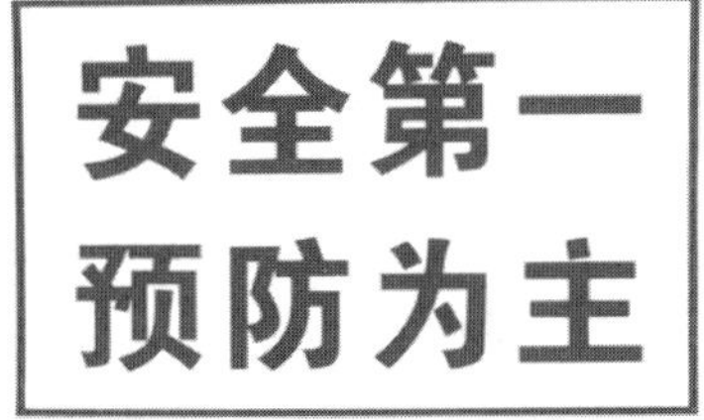

4. ____________________

Pipe

1. The conception and the sizes of pressure vessels

Pressure vessels are leakproof containers. They are made in all sizes and shapes (Fig.19.1). The smaller ones may be no larger than a fraction of an inch in diameter, whereas the larger vessels may be 150 ft or more in diameter. Some are buried in the ground or deep in the ocean; most are positioned on the ground or supported on platforms; and some actually are found in storage tanks and hydraulic units in aircraft.

2. Classification of pressure vessels

For the purpose of design and analysis, pressure vessels are sub-divided into two classes depending on the ratio of the wall thickness to vessel diameter: thin-walled vessels, with a thickness ratio of less than 1 / 10, and thick-walled above this ratio. Thick-walled vessels are used for high pressures.

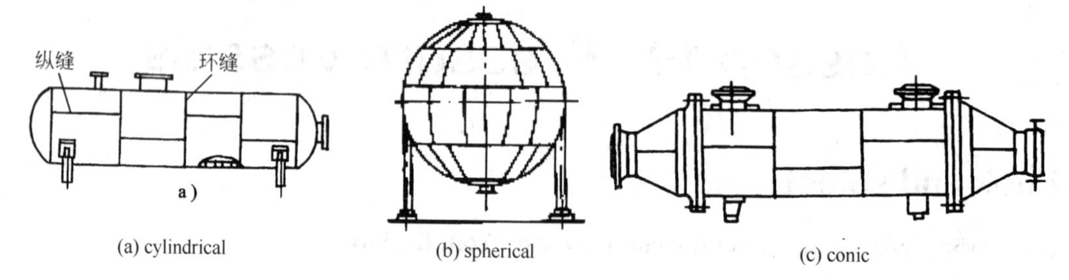

(a) cylindrical (b) spherical (c) conic

Fig.19.1 Typical styles of pressure vessels

Knowledge extention

Typical components of pressure vessels are as follows shown in Fig.19.2.

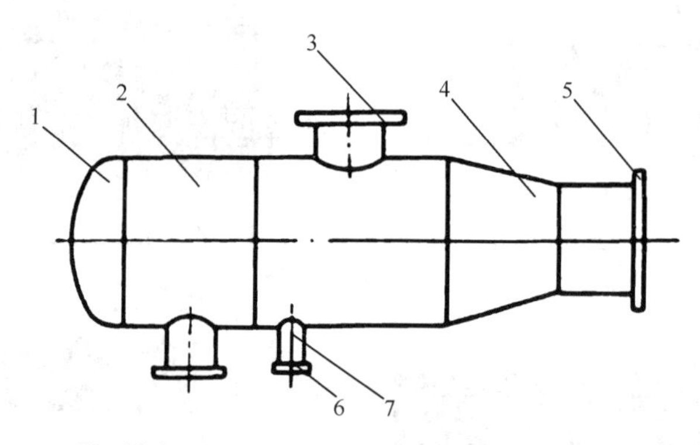

Fig.19.2 Combined structure of pressure vessels

1—gland; 2—barrel; 3—manhole flange; 4—conic barrel; 5—flange; 6—pipe flange; 7—nozzle

1. Cylindrical shell (barrel) is very frequently used for constructing pressure vessels in the petrochemical industry.

2. Formed heads (gland) A large variety of end closures and transition are available to the design engineer. Using one configuration versus another depends on many factors such as method of forming, material cost, and space restriction.

3. Blind flanges and **cover plates** Usually, the blind flange is bolted to a vessel flange with a gasket between two flanges.

4. Openings and nozzles All process vessels require openings to get the contents in and out. Foe some process vessels, the contents enter and exit through openings in the heads and shell to which nozzles and piping are attached.

5. Supports Most vertical vessels are supported by skirts (Fig.19.3). Skirts are economical because they generally transfer the loads from the vessel by shear action. They also transfer the loads to the foundation through anchor bolts and bearing plates. Horizontal vessels are normally supported by saddles. Stiffening rings may be required if the shell is too thin to transfer the loads to the saddles.

Fig.19.3 The kinds of pressure vessels

Exercises

1. Fill in the blanks with the proper words in the text.

(1) ASME is the short form of ______________________ (美国机械工程师协会).

(2) Pressure vessels are ___________ containers. They are made in all ___________ and shapes. The smaller ones may be no larger than a fraction of an ___________ in diameter, whereas the larger ___________ may be 150 ft or more in ___________.

(3) Some are buried in the ________ or deep in the ocean; most are positioned on the ground or supported on _________; and some actually are found in storage tanks and ________in aircraft.

2. Match the English with the Chinese. Draw lines.

(1) barrel	a. 压力容器
(2) pressure vessel	b. 筒体
(3) gland	c. 鞍座
(4) flange	d. 接管
(5) skirt	e. 法兰
(6) saddle	f. 封头
(7) nozzle	g. 裙座

3. Translate the names of each part of pressure (Fig.19.4) vessel into Chinese.

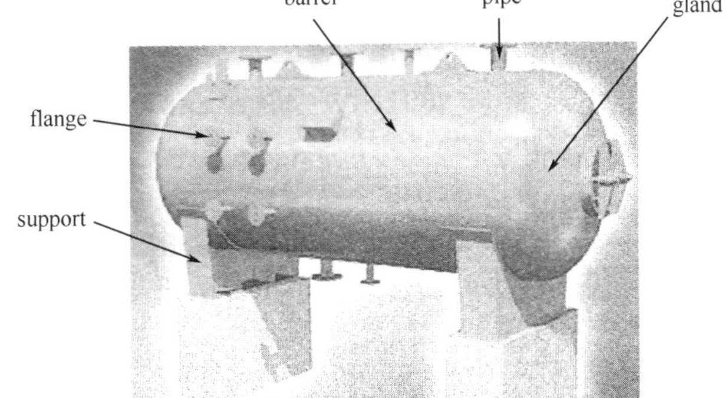

Fig.19.4 Structure of pressure vessel

support _________ flange _________ barrel _________ pipe _________ gland _________

4. Translate the following Chinese words into English.

安全第一　　施工现场　　脚手架　　圆柱形　　球形　　锥形

Words and phrases

ASME (American Society of Mechanical Engineers) 美国机械工程师协会

pressure ['preʃə(r)] 压力

vessel ['vesl] 容器

scaffolding ['skæfəldɪŋ] 脚手架

site [saɪt] 现场，工地

safety 安全

leakproof ['li:kpru:f] 防泄漏

container [kən'teɪnə(r)] 容器，集装箱

size 规格，尺寸

shape [ʃeɪp] 形状

no larger than 不比……大

inch 英寸

diameter [daɪ'æmɪtə(r)] 直径

bury ['berɪ] 埋，埋葬

in the ground 在地里/下

in the deep ocean 在深海里

most 多数

position [pə'zɪʃn] 位置，定位

on the ground 在地面上

support [sə'pɔ:t] 支架，支撑

on platform ['plætfɔ:m] 在平台上

in storage tank ['stɔ:rɪdʒ] [tæŋk] 在储罐里

hydraulic [haɪ'drɔ:lɪk] 液压的

unit 单元，单位，装置

in aircraft ['ɛəkrɑ:ft] 在航空器中

classification [ˌklæsɪfɪ'keɪʃn] 分类，类别

be sub-divided into 分类为……

thin-walled vessel 薄壁容器

thick-walled 厚壁

cylindrical shell [sə'lɪndrɪkl] [ʃel] 圆柱壳体

formed head 封头

opening and nozzle ['nɒzlz] 开口/孔及接管

flange [flændʒ] 法兰

skirt [skə:t] 裙子，裙座

saddle ['sædl] 马鞍，鞍座

Lesson 20 Heat Exchangers

Look and select

Look at the pictures and select the correct terms from the box.

safety guard	manhole	construction	health/ safety/environment

1. ________________

2. ________________

3. ________________

4. ________________

Text

1. The conception of heat exchanger

Heat exchangers are equipment primarily for transferring heat between hot and cold streams. They have separate passages for the two streams and operate continuously.

2. Classifications of heat exchangers

(1) Plate-and-Frame Exchangers (Fig.20.1) are assemblies of pressed corrugated plates on a frame.

(2) Spiral Heat Exchangers (Fig.20.2) In spiral heat exchangers, the hot fluid enters at the center of the spiral element and flows to the periphery; flow of the cold liquid is countercurrent, entering at the periphery and leaving at the center.

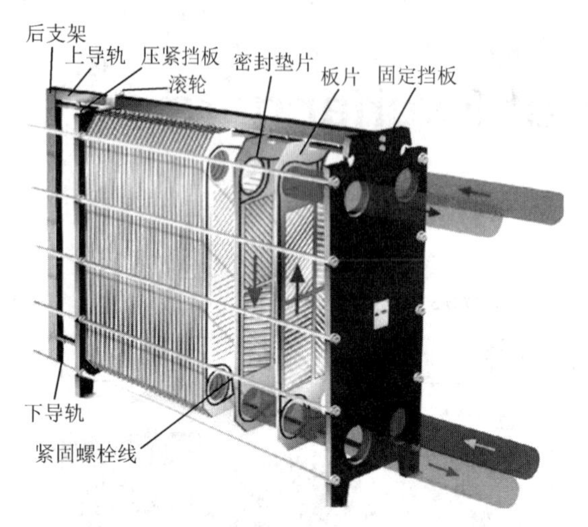

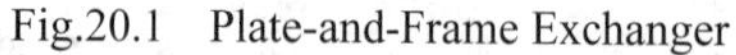

Fig.20.1 Plate-and-Frame Exchanger

Fig.20.2 Spiral Heat Exchanger

(3) Plate-Fin Exchangers (Fig.20.3) Compact exchangers are used primarily for gas service. Typically they have surfaces of the order of 1200m^2/m^3, corrugation height 3.8~11.8 mm, corrugation thickness 0.2 ~ 0.6 mm, and fin density 230~700 fins / m.

(4) Air coolers (Fig.20.4) In such equipment the process fluid flows through finned tubes and cooling air is blown across them with fans.

Fig.20.3 Plate-Fin Exchanger

Fig.20.4 Air cooler

(5) Double-pipe Exchangers (Fig.20.5) This kind of exchanger consists of a central pipe supported within a larger one by packing glands.

(6) Shell-and-tube Exchangers (Fig.20.6) are made up of a number of tubes in parallel and

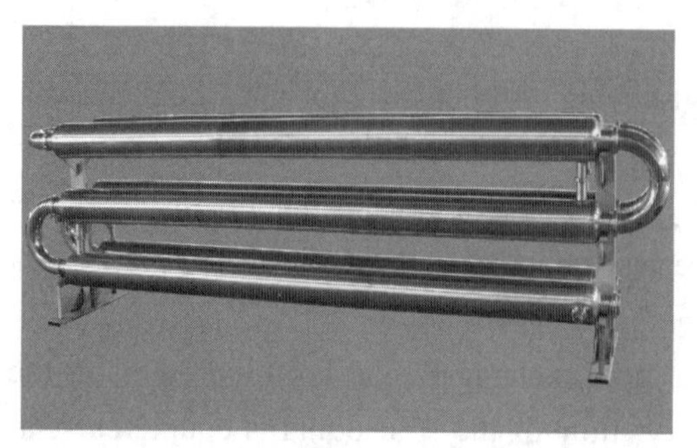

Fig.20.5 Double-pipe Exchanger

Fig.20.6 Shell-and-tube Exchanger

series through which one fluid travels and enclosed in a shell through which the other fluid is conducted. The shell side is provided with a number of baffles to promote high velocities and largely more efficient cross flow on the outside of tubes. They are most widely used.

Knowledge extention

Understand the structure of shell-and-tube exchanger (Fig.20.7).

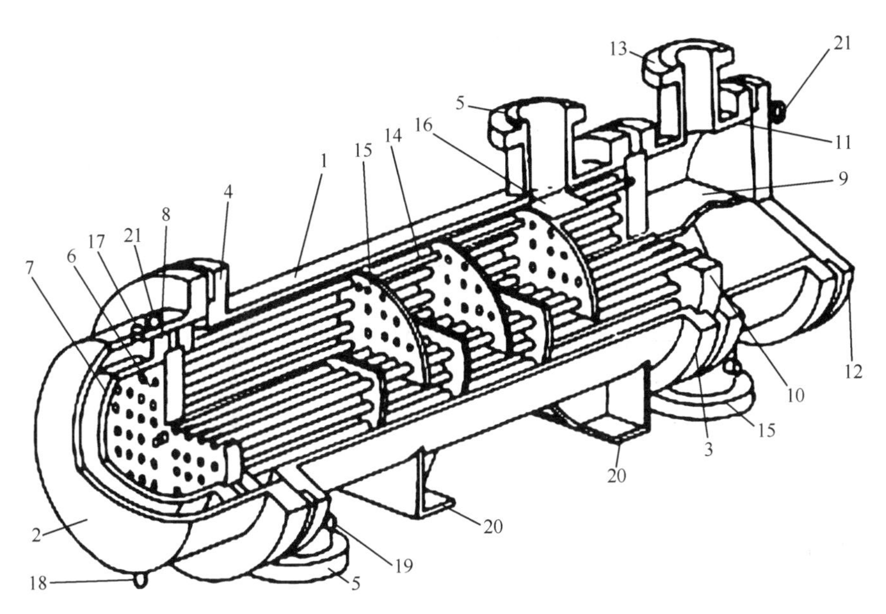

Fig.20.7 The structure of shell-and-tube exchanger

1—Shell; 2—Shell cover; 3—Shell channel; 4—shell cover end flange; 5—Shell nozzle; 6—Floating tubesheet; 7—Floating head; 8—Floating head flange; 9—Channel partition; 10—Stationary tubesheet; 11—Channel; 12—Channel cover; 13—Channel nozzle; 14—Tie rods and spacers; 15—Transverse baffles or supportplates; 16—Impingement baffle; 17—Vent connection; 18—Drain connection; 19—Test connection; 20—Support saddles; 21—Lifting ring

Exercises

1. Fill in the blanks with the proper words in the text.

(1) HSE is the short form of ____________ (健康，安全，环境).

(2) Heat exchangers are ______________ for transferring heat between _____________ and ________ streams. They have separate ________ for the two ________ and operate continuously.

2. Match the English with the Chinese. Draw lines.

(1) plate-and-frame exchanger	a. 盘管式换热器
(2) spiral heat exchanger	b. 板翅式换热器
(3) plate-fin exchanger	c. 空气冷却器
(4) air cooler	d. 管壳式换热器
(5) double-pipe exchanger	e. 板框式换热器
(6) shell-and-tube exchanger	f. 套管式换热器

3. Translate the words in the picture (Fig.20.8) into Chinese.

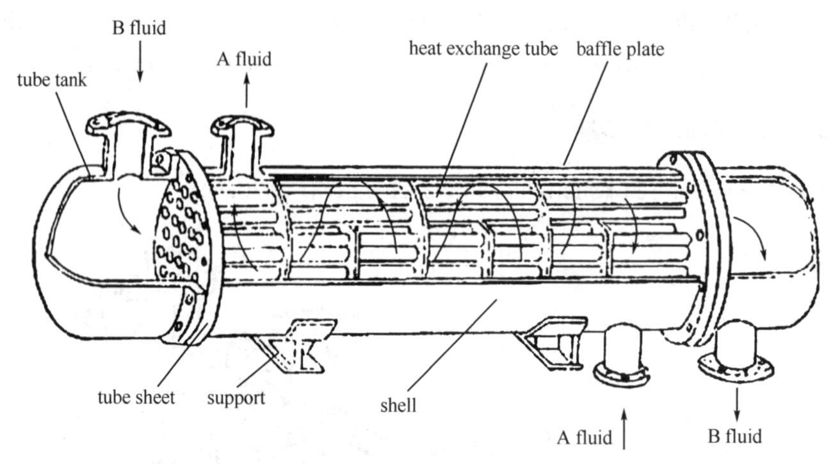

Fig.20.8 Fixed tube-plate exchanger

tube tank ____________________; heat exchange tube ____________________;
fluid ____________________; baffle plate ____________________;
tube sheet ____________________; support ____________________;
shell ____________________.

Words and phrases

health/safety/environment [helθ] [ɪn'vaɪrənmənt] 健康/安全/环境

heat [hi:t] 热，热量，加热

primarily for [praɪ'merəli] 主要用于

transfer [træns'fɜ:(r)] 传送

hot [hɒt] 热的

separate ['seprət] 各自的，分别的

continuously [kən'tɪnjʊəslɪ] 连续/不断地

plate [pleɪt] 板

spiral ['spaɪrəl] 螺旋式，盘管式

air [eə(r)] 空气

double ['dʌbl] 双的，两个

shell [ʃel] 壳，贝壳，壳牌

exchanger [ɪks'tʃeɪndʒə] 交换器

equipment [ɪ'kwɪpmənt] 设备

between 在……之间

cold stream [stri:m] 冷流体

passage ['pæsɪdʒ] 通道

operate ['ɒpəreɪt] 操作，运作

frame [freɪm] 框，框架

fin [fɪn] 翅，翅膀

cooler ['ku:lə(r)] 冷却器

pipe [paɪp] 管，管道

tube [tju:b] 管，小管

Lesson 21 Reactors

Look and select

Look at the pictures and select the correct terms from the box.

safety cap	crane	welding	workshop

1. ________________

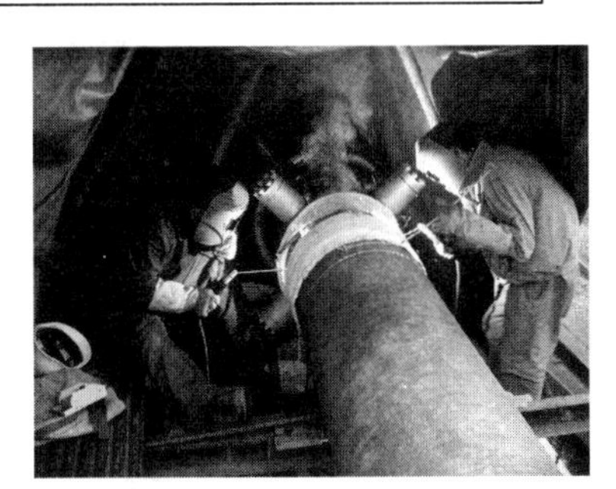

2. ________________

3. ________________

4. ________________

Text

1. General of reactor

Almost every kind of holding or contacting equipment has been used as a chemical reactor at some time, from mixing nozzles and centrifugal pumps to the most elaborate towers and tube assemblies.

2. Classification of reactors

(1) Stirred Tanks (Fig.21.1) are the most common type of batch reactor. Stirring is used to mix the ingredients initially, to maintain homogeneity during reaction, and to enhance heat transfer at a jacket wall or internal surfaces.

(2) Tubular Flow Reactors The ideal behavior of tubular flow reactors (TFR) is plug flow, in which all nonreacting molecules have equal residence times.

(3) Gas-liquid Reactors Except with highly volatile liquids, reactions between gases and liquids occur in the liquid phase, following a transfer of gaseous participants through gas and liquid films.

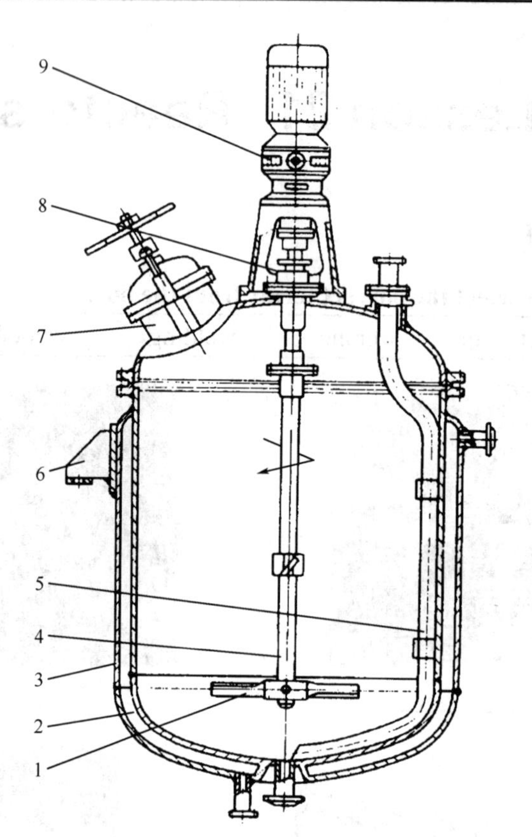

Fig.21.1 Structure of stirred reactor

1—agitator; 2—tank body; 3—jacket; 4—stirring shaft; 5—extruding pipe; 6—support; 7—manhole; 8—shaft seal; 9—transmission unit

(4) Fixed Bed Reactors (Fig.21.2) The fixed beds of concern here are made up of catalyst particles in the range of 2~5 mm dia.

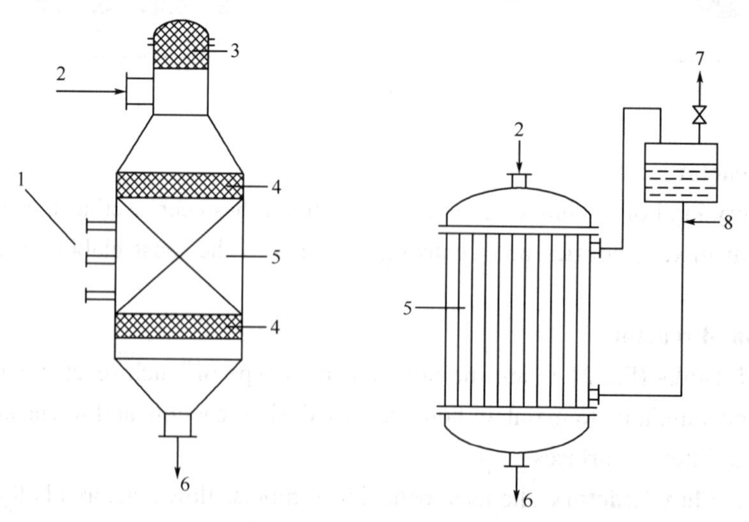

Fig.21.2 Fixed bed reactor

1—temperature monitoring holes; 2—feed gas; 3—slag wool; 4—porcelain ring; 5—catalyst; 6—product; 7—steam; 8—supplementary water

(5) Moving Bed Reactors In such vessels granular or lumpy material moves vertically downward as a mass. The solid may be a reactant or a catalyst or a heat carrier.

(6) Fluidized Bed Reactors This term is restricted here to equipment in which finely divided solids in suspension interact with gases.

(7) Kilns and Hearth Furnaces These units are primarily for high temperature services, the kilns up to 2500 °F and the furnaces up to 4000 °F. Usual construction is steel-lined with ceramics, sometimes up to several feet in thickness.

Knowledge extention

Understand the safety instructions (Fig.21.3).

Incorrect dressing and behaviors:

1. smoking
2. not wearing safety cap
3. drinking
4. not wearing safety clothes
5. not wearing safety shoes
6. not closing safety belt

Fig.21.3 Incorrect dressing and behaviors

Exercises

1. Fill in the blanks with the proper words in the text.

Almost every kind of holding or contacting ________ has been used as a chemical reactor at some time, from ________ and ________ to the most elaborate towers and tube assemblies.

2. Match the English with the Chinese. Draw lines.

(1) stirred tank	a. 气液反应器
(2) tubular flow reactor	b. 固定床反应器
(3) gas-liquid reactor	c. 管道式流动反应器
(4) fixed bed reactor	d. 流化床反应器
(5) moving bed reactor	e. 窑炉及床炉式反应器
(6) fluidized bed reactor	f. 湍流槽
(7) kilns and hearth furnace	g. 移动床反应器

3. Translate the phrases into English.

吸烟________________________; 未戴安全帽 ________________________;

饮酒 ________________________; 未穿安全/工作服 ________________________;

未穿安全/工作鞋 ________________________; 未系安全带 ________________________.

Words and phrases

reactor [ri'æktə(r)] 反应器
crane [kreɪn] 起重机
workshop ['wɜ:kʃɒp] 工作车间
most common type 最常见/普遍形式
ingredient 成分
molecule ['mɒlɪkju:l] 分子，微粒
enhance heat transfer 加强热传递
gas-liquid reactor 气液反应器
volatile liquid ['vɒlətaɪl] 挥发性液体
fixed bed reactor [fɪkst] 固定床反应器
fluidized bed reactor 流化床反应器
lumpy ['lʌmpɪ] 块料，粒状材
incorrect [ˌɪnkə'rekt] 不正确的
behavior ['dresɪŋ] 行动，行为
wearing ['weərɪŋ] 穿戴
close safety belt 系安全带
safety cap 安全帽
welding [weldɪŋ] 焊接
classification [ˌklæsɪfɪ'keɪʃn] 分类
stirred tank [stə:d] [tæŋk] 湍流槽
internal surface 内表面
residence ['rezɪdəns] 滞留
jacket wall ['dʒækɪt] 夹套/套筒壁
interact [ˌɪntər'ækt] 相互作用/影响
occur in the liquid phase 以液态出现
catalyst particle ['kætəlɪst] 触媒颗粒
granular ['grænjələ(r)] 颗粒状的
suspension [sə'spenʃn] 悬浮
dressing ['dresɪŋ] 着装
smoking ['sməʊkɪŋ] 抽烟
drinking ['drɪŋkɪŋ] 喝酒
safety shoes 安全靴/鞋
maintain homogeneity [ˌhɒmədʒə'ni:əti] 保持均匀，同质
tubular flow reactor ['tju:bjələ(r)] 管道式流动反应器
nonreacting molecules ['mɒlɪkju:l] 未发生反应的分子
have equal residence times ['rezɪdəns] 有相同的驻留时间
kilns and hearth furnaces [kɪln] [hɑ:θ] ['fɜ:nɪs] 窑炉及床炉式反应器

Lesson 22 Pumps

Look and select

Look at the pictures and select the correct terms from the box.

piston	motor	valve	blower

1. ________________ 2. ________________

3. ________________ 4. ________________

Text

1. General of pumps

Pumps, device used to raise, transfer, or compress liquids and gases. Four general classes of pumps for liquids are described below. In all of them, steps are taken to prevent cavitation (the formation of a vacuum), which would reduce the flow and damage the structure of the pump. Pumps used for gases and vapors are usually known as compressors.

2. Types

(1) Reciprocating pumps

Reciprocating pumps (Fig.22.1) consist of a piston moving back and forth in a cylinder that has valves to regulate the flow of liquid into and out of the cylinder. These pumps may be single or double acting.

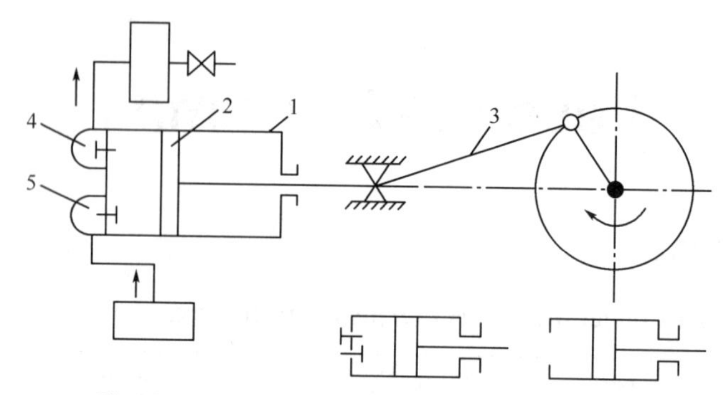

Fig.22.1 Working principle of reciprocating pump

1—cylinder; 2—piston; 3—crank link mechanism; 4—exhaust valve; 5—intake valve

(2) Centrifugal pumps

Also known as rotary pumps, centrifugal pumps (Fig.22.2) have a rotating impeller, also known as a blade, that is immersed in the liquid. Liquid enters the pump near the axis of the impeller, and the rotating impeller sweeps the liquid out toward the ends of the impeller blades at high pressure.

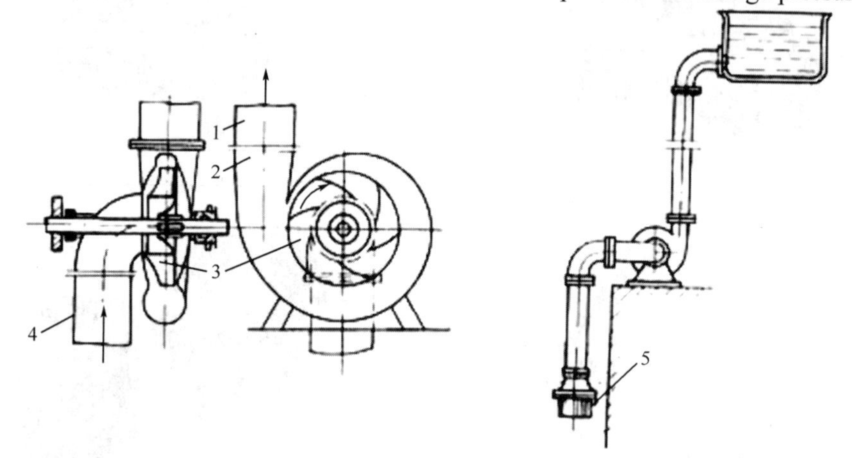

Fig.22.2 Working principle of centrifugal pump

1—exhaust pipe; 2—volute; 3—impeller; 4—suction pipe; 5—hydraulic valve

(3) Jet pumps

Jet pumps (Fig.22.3) use a relatively small stream of liquid or vapor, moving at high velocity, to move a larger flow of liquid.

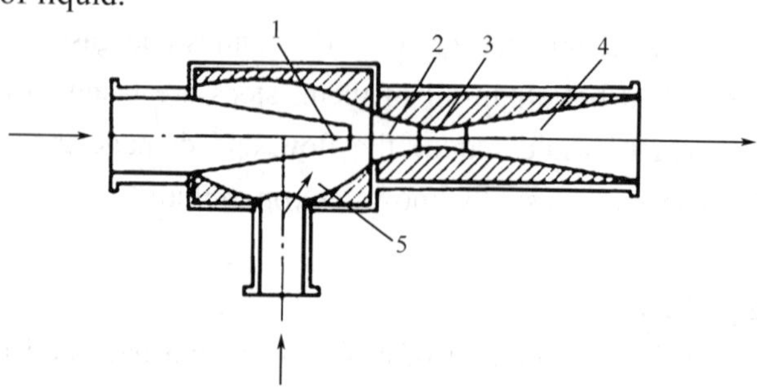

Fig.22.3 Structure of jet pump

1—nozzle; 2—mixing chamber; 3—throat; 4—diffuser chamber; 5—vacuum chamber

(4) Other types

Positive-displacement pump, gear pump (Fig.22.4), lift pump, screw pump etc.

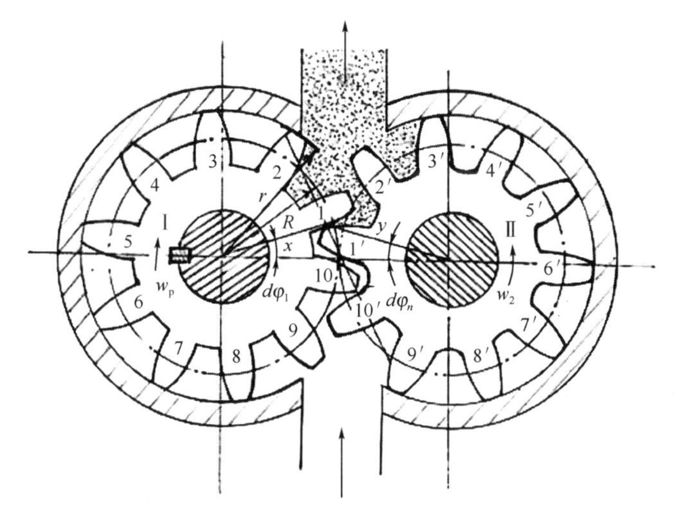

Fig.22.4 Working principle of gear pump

Knowledge extention

Two types of modern pumps used to move water are the positive-displacement pump and the centrifugal pump. Positive-displacement pumps use suction created by a vacuum to draw water into a closed space. Centrifugal pumps use motor-driven propellers that create a flow of water when they rotate. Early version of the centrifugal pump is the screw pump.

Exercises

1. Fill in the blanks with the proper words in the text.

(1) Pumps, device used to _______, _______, or _______ liquids and gases. Steps are taken to prevent _______ (the formation of a vacuum), which would reduce the _______ and damage the _______ of the pump. Pumps used for _______ and _______ are usually known as _______.

(2) Positive-displacement pumps use ________ created by a ________ to draw water into a closed space. Centrifugal pumps use ________ propellers that create a flow of ________ when they rotate.

2. Match the English with the Chinese. Draw lines.

A.

(1) reciprocating pump	a. 离心泵
(2) centrifugal pump	b. 往复式泵
(3) jet pump	c. 容积泵
(4) positive-displacement pump	d. 抽水机
(5) gear pump	e. 螺杆泵
(6) lift pump	f. 齿轮泵
(7) screw pump	g. 喷射泵

B.

(1) piston
(2) motor
(3) valve
(4) blower
(5) volute
(6) impeller
(7) chamber

a. 马达
b. 阀
c. 室
d. 叶轮
e. 蜗壳
f. 鼓风机
g. 活塞

Words and phrases

pump [pʌmp] 泵
motor ['məʊtə(r)] 电机,马达
blower ['bləʊə(r)] 鼓风机
raise [reɪz] 提升
compress [kəm'pres] 压缩
gas [gæs] 气体
are described below 描述如下
cavitation [ˌkævɪ'teɪʃən] 气穴
vacuum ['vækjʊəm] 真空
flow [fləʊ] 流动
structure ['strʌktʃə(r)] 结构
consist of 由……组成
cylinder ['sɪlɪndə(r)] 气瓶，缸体
single ['sɪŋgl] 单个，单程
rotary ['rəʊtərɪ] 旋转的
rotating impeller [ɪm'pelə] 旋转的叶轮
immerse [ɪ'mɜːs] 浸没
axis ['æksɪs] 轴
at high pressure ['preʃə(r)] 高压
relatively ['relətɪvli] 相对地
at high velocity [və'lɒsətɪ] 高速
lift pump [lɪft] 提升泵，抽扬泵，抽水机
reciprocating pump [rɪ'sɪprəkeɪtɪŋ] 往复式泵
positive-displacement pump ['pɒzətɪv] [dɪs'pleɪsmənt] 容积泵
small stream of liquid or vapor [striːm] ['lɪkwɪd] ['veɪpə] 液体或蒸汽的细流

piston ['pɪstən] 活塞
valve [vælv] 阀
device [dɪ'vaɪs] 仪器，装置
transfer [træns'fɚ]传送
liquid [□lɪkwɪd] 液体
general classe 一般类别
prevent [prɪ'vent] 阻止
formation [fɔː'meɪʃn] 形成
reduce [rɪ'djuːs] 减少
damage ['dæmɪdʒ] 损坏
compressor [kəm'presə(r)] 压缩机
moving back and forth 前后移动
regulate ['regjʊleɪt] 调节
double acting 双程运动
centrifugal pump [ˌsentrɪ'fjuːgl] 离心泵
blade [bleɪd] 叶片
enter ['entə(r)] 进入
sweep [swiːp] 吹扫，清扫
jet pump [dʒet] 喷射泵
screw pump 螺杆泵，螺旋泵
gear pump [gɪr] 齿轮泵

Lesson 23 Compressors

Look and select

Look at the pictures and select the correct terms from the box.

gas	turbine	petrochemical	manufacture

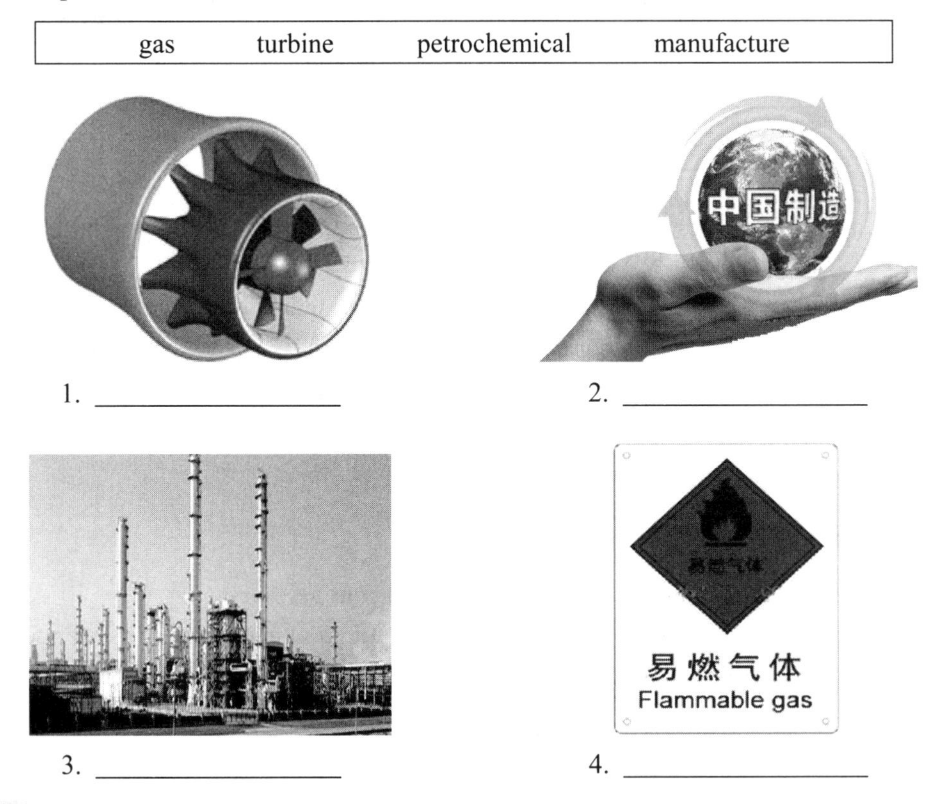

Text

1. General of compressors

The purpose of compressors is to move air and other gases from place to place. Gases, unlike liquids, are compressible and require compression devices. Compressors, blowers, and fans are such compression devices.

2. Types of compressors

(1) Reciprocating piston compressor

This type of compressor (Fig.23.1, Fig.23.2) which may consist of from one to twelve stages is the only one capable of developing very high pressures, such as for example over 350 MN/m^2 as required for polythene manufacture, as shown in Fig.23.3.

(2) Rotary blowers and compressors

These can be divided into two classes-those which develop a high compression ratio and those which have very low ratios.

Fig.23.1 Piston compressor

Fig.23.2 Centrifugal compressor

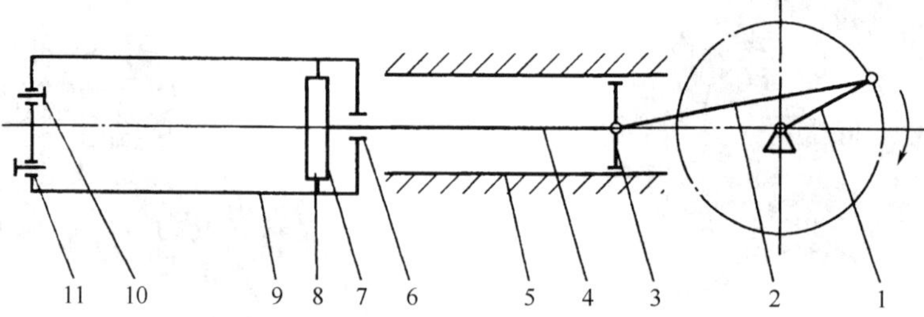

Fig.23.3 Working process of piston compressor

1—crank; 2—link rod; 3—crosshead; 4—piston rod; 5—slideway; 6—seal; 7—piston; 8—piston ring; 9—cylinder; 10—intake valve; 11—exhaust valve

Knowledge extention

Centrifugal blowers and compressors, including turbocom pressors

These depend on the conversion of kinetic energy into pressure energy. Fans are used for low pressures, and can be made to handle very large quantities of gases. For the higher pressure ratios

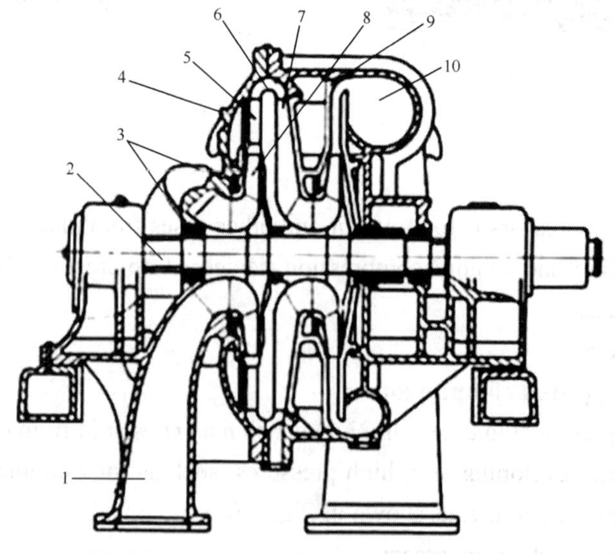

Fig.23.4 Structure of centrifugal compressor

1—intake chamber; 2—spindle; 3—sealing; 4—casing; 5—diffuser; 6—curve; 7—reflux condenser; 8—impeller; 9—partition plate; 10—volute chamber

now in demand, multistage centrifugal compressors (Fig.23.4) are mainly used, particularly for the requirements of high capacity chemical plants. Thus in catalytic reforming, petrochemical separation plants (ethylene manufacture), ammonia plants with a production rate of 12 kg /s. and for the very large capacity needed for natural gas fields, this type of compressor is now supreme (Fig.23.4).

Exercises

1. Fill in the blanks with the proper words in the text.

(1) The purpose of compressors is to move ________ and other ________ from place to place. Gases, unlike liquids, are ________ and require compression ________. Compressors, blowers and ________ are such compression devices.

(2) Reciprocating piston compressor which may consist of from ________ to ________ stages is the only one capable of developing very ________.

(3) Fans are used for ________, and can be made to handle very large quantities of gases. For the ________ ratios now in demand, ________ centrifugal compressors are mainly used.

2. Match the English with the Chinese. Draw lines.

A.

(1) gas	a. 压缩机
(2) turbine	b. 气体
(3) petrochemical	c. 制造
(4) manufacture	d. 石化
(5) compressor	e. 涡轮

B.

(1) intake chamber	a. 扩压器
(2) spindle	b. 进气室
(3) sealing	c. 回流器
(4) casing	d. 主轴
(5) diffuser	e. 密封
(6) reflux condenser	f. 叶轮
(7) impeller	g. 机壳
(8) partition plate	h. 涡室
(9) volute chamber	i. 隔板

Words and phrases

compressor [kəm'presə(r)] 压缩机
petrochemical [ˌpetrəʊ'kemɪkl] 石化
manufacture [ˌmænjʊ'fæktʃə(r)] 制造
purpose ['pɜːpəs] 目的
from place to place 从一个地方到另一个地方
gas 气体
turbine ['tɚbɪn] 涡轮
general ['dʒenrəl] 概述/论
move 移动，流动
unlike 与……不一样

compressible [kəm'presɪbl] 可压缩的
require [rɪ'kwaɪə(r)] 要求
type 类型，种类
stage 级，阶段
develop very high pressure 产生高压
be divided into 分为……
for example 例如
liquid ['lɪkwɪd] 液体
blower ['bləʊə(r)] 鼓风机
consist of 由……组成
be capable of 能够
such as 比如，诸如
two classes 两类
as required 按……所要求的
compression device [kəm'preʃn] [dɪ'vaɪs] 压缩装置
polythene manufacture ['pɒlɪθi:n] 聚乙烯制造/生产
compression ratio [kəm'preʃn] ['reɪʃɪəʊ] 压缩比
centrifugal blower [ˌsentrɪ'fju:gl] 离心式鼓风机
turbocompressor [tɜ:bəʊ'kɒmpresər] 涡轮压缩机
rotary blowers and compressors ['rəʊtərɪ] 旋转式鼓风机及压缩机
reciprocating piston compressor [rɪ'sɪprəkeɪtɪŋ] ['pɪstən] 往复式活塞压缩机

Lesson 24 Spherical Tanks

Look and select

Look at the pictures and select the correct terms from the box.

painting	air tightness	heat treatment	hydraulic

1. ____________________

2. ____________________

3. ____________________

4. ____________________

Text

1. Types of tanks (Fig.24.1~Fig.24.3)

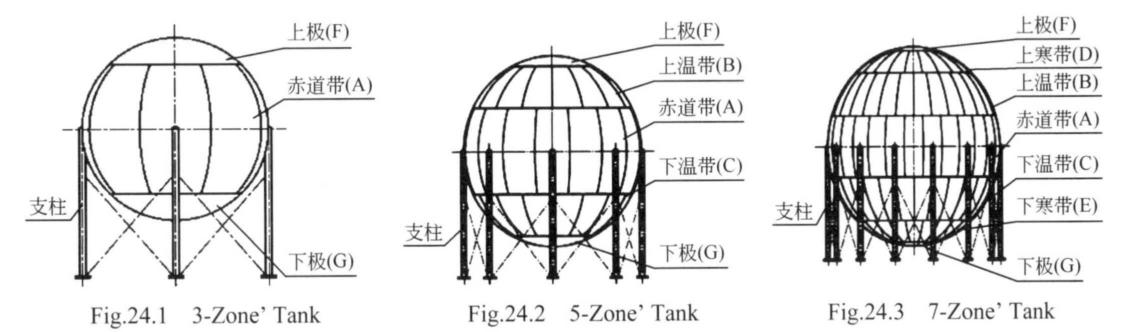

Fig.24.1 3-Zone' Tank Fig.24.2 5-Zone' Tank Fig.24.3 7-Zone' Tank

2. The structure of 5-zones tank

column → equator zone → lower temperate zone → lower crown → upper temperate zone → upper crown

3. Flow chart of spherical tank installation (Fig.24.4)

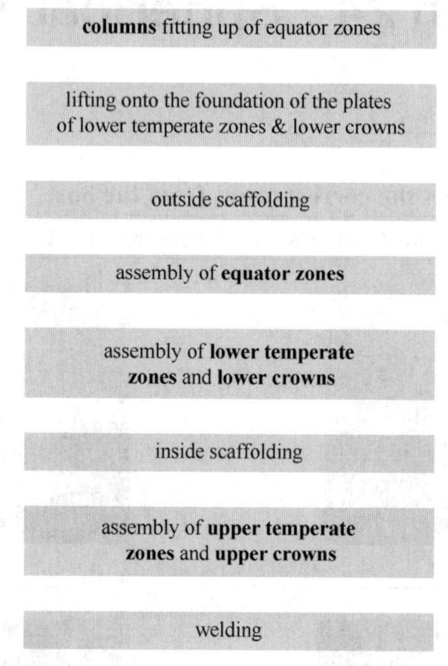

Fig.24.4 Main installation of spherical tank

4. QC Procedures (quality control)

a. NDT Inspections 无损探伤/检测

UT 超声波探伤 RT 射线探伤 MT 磁粉探伤 PT 渗透探伤

b. Tests 试验

Hydraulic test 水压试验 Air-tightness test 气密性试验

5. Hydraulic test (Fig.24.5)

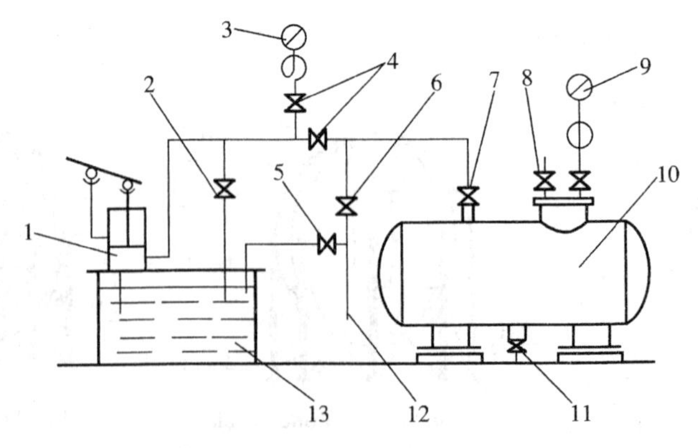

Fig.24.5 Facilities of hydraulic test

1—pump; 2,4,5,6—valves; 3,9—pressure meters; 7—water inlet valve; 8—exhaust valve; 10—equipment to be tested; 11—water outlet valve; 12—water inlet pipe; 13—water sink

Knowledge extention

Flow chart of spherical tank installation (Fig.24.6)

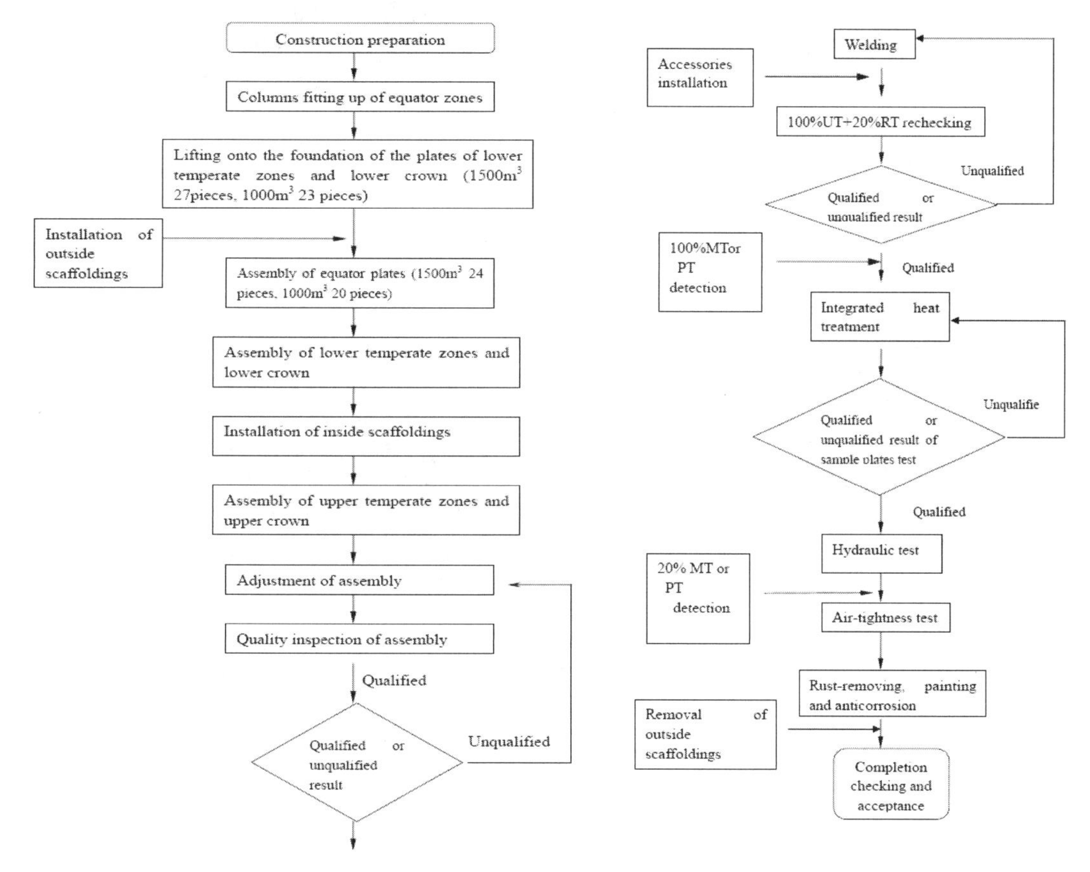

Fig.24.6 Flow chart of spherical tank installation

Exercises

1. Arrange the terms in the box according to the installation flow.

equator zones	upper temperate zones	upper crowns
lower temperate zones	columns	lower crowns

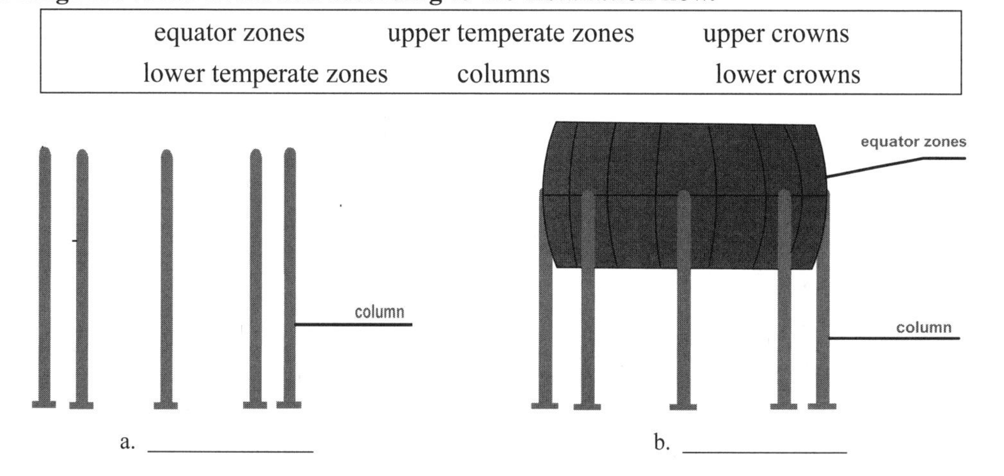

a. ______________ b. ______________

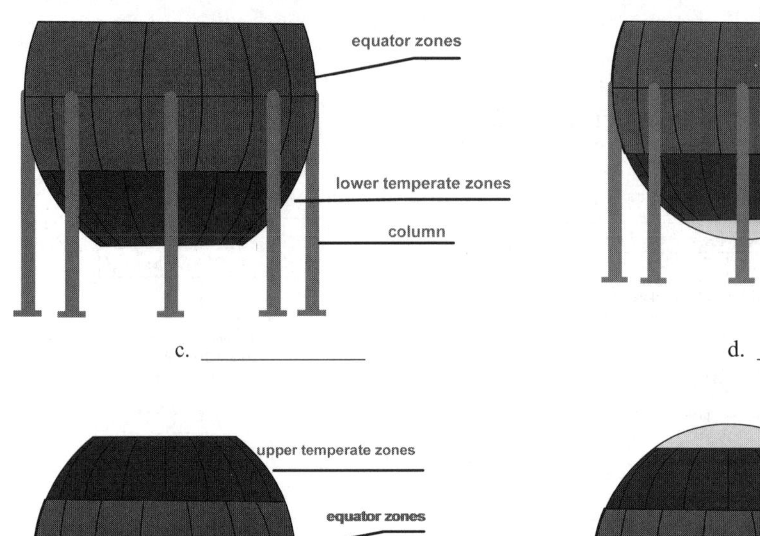

c. ______________ d. ______________

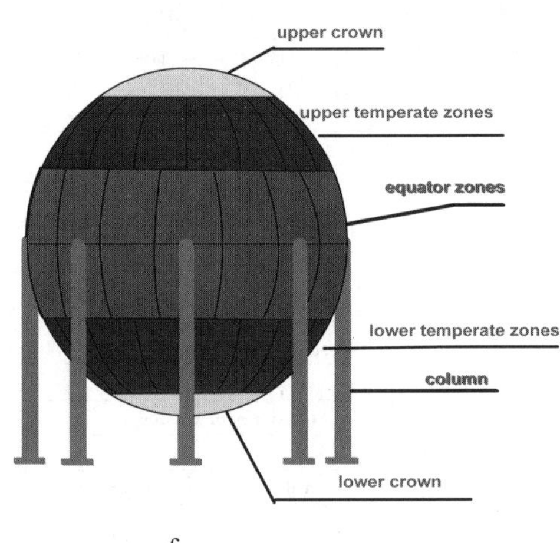

e. ______________ f. ______________

2. Match the English with the Chinese. Draw lines.

A.

(1) column	a. 下温带
(2) equator zones	b. 上温带
(3) lower temperate zone	c. 立柱
(4) lower crown	d. 赤道带
(5) upper temperate zone	e. 上极板
(6) upper crown	f. 下极板

B.

(1) NDT	a. 射线探伤
(2) UT	b. 磁粉探伤
(3) RT	c. 超声波探伤
(4) MT	d. 无损探伤
(5) PT	e. 水压试验
(6) Hydraulic test	f. 气密性试验
(7) Air-tightness test	g. 渗透探伤

3. Answer the following questions.

(1) How many main parts does the spherical tank include? What are they?

(2) After which procedure should welding be performed ? How about heat treatment?

(3) What are the last procedure during erection?

(4) What do the NDT inspections and quality tests include?

Words and phrases

spherical tank [tæŋk] 球罐
air tightness [eə(r)] [taɪtnəs] 气密性
hydraulic test [haɪ'drɔ:lɪk] [test] 水压试验
column ['kɔləm] 立柱，支柱
lower crown [kraʊn] 下极板
upper crown 上极板
lifting onto 吊装到……
foundation [faun'deiʃən] 基础
lower temperate zone 下温带
inside scaffolding 内脚手架
UT 超声波探伤
MT 磁粉探伤
hydraulic test 水压试验
painting ['peɪntɪŋ] 油漆
heat treatment [hi:t] ['tri:tmənt] 热处理
zone [zəʊn] 地带，地域
equator zone [ɪ'kweɪtə(r)] 赤道带
upper temperate zone ['ʌpə(r)] 上温带
fitting up 拼装
assembly 组装
plate [pleɪt] 板
lower crown 下极板
welding [weldɪŋ] 焊接
RT 射线探伤
PT 渗透探伤
air-tightness test 气密性试验
lower temperate zone ['ləʊə(r)] ['tɛmpərɪt, 'tɛmprɪt] 下温带
outside scaffolding [ˌaʊt'saɪd] ['skæfəldɪŋ] 外脚手架
NDT inspection [ɪn'spekʃn] 无损探伤/检测

Lesson 25 Installation Record of Vertical Equipment

Look and understand

1. Look at the picture (Fig.25.1) and understand each item concerning construction.

Fig.25.1 Safety instructions

2. Understand the organization chart (Fig.25.2).

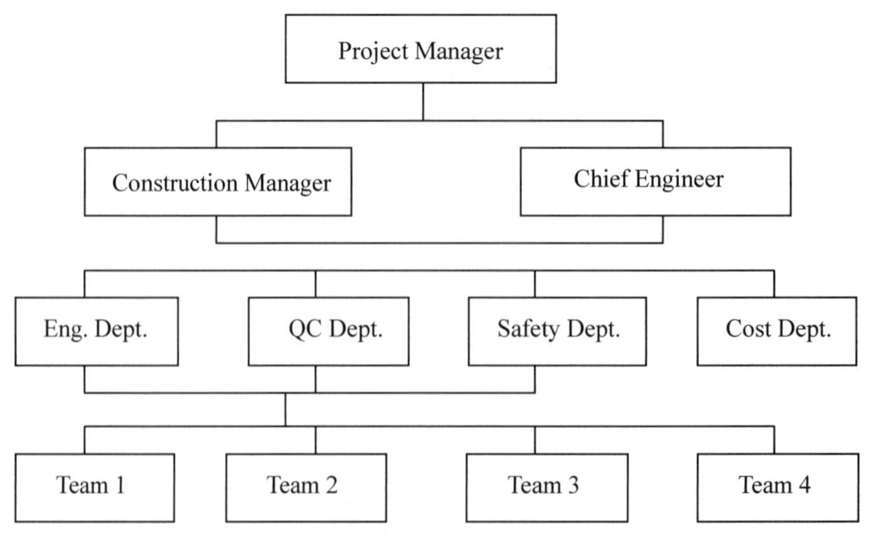

Fig.25.2 Organization chart

Text

Installation record of vertical equipment (Table 25.1)

Table 25.1 Installation record of vertical equipment

SH 3503-J302		立式设备安装记录 Installation Record of Vertical Equipment		工程名称： Project Description: 单元名称： Unit Description:	
设备名称 Name of Equipment		位号 Tag No.		型号及规格 Type and Specification	

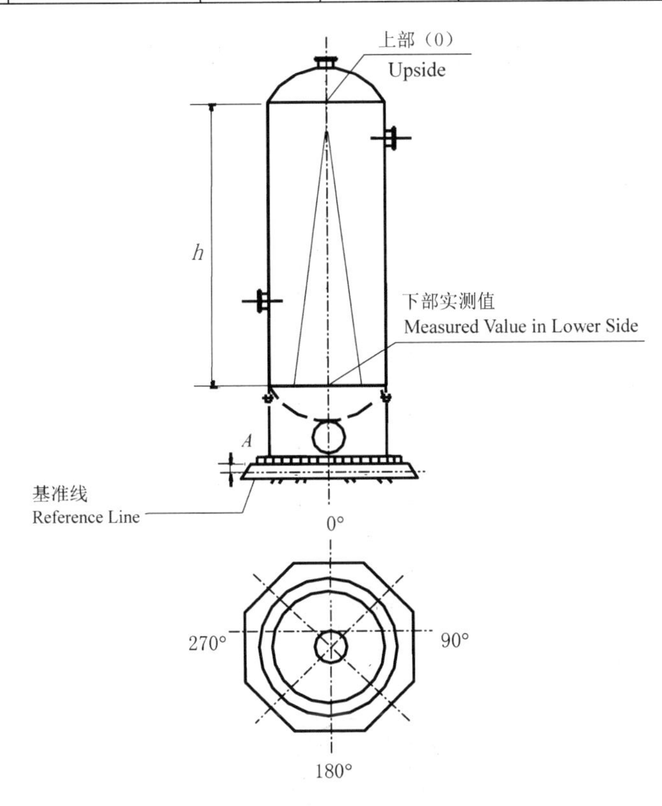

h 为两测点间的距离，*h* =　　　　mm。
h represents the distance between the two measured points, *h*=　　　　mm.

项目 Item	方位 Orientation	允许值 Allowable Value/mm	实测值 Measured Value/mm	项目 Item	测量部位 Measured Position	允许值 Allowable Value/mm	实测值 Measured Value/mm
标高偏差 Elevation Deviation (*A*)	0°			中心线位置偏差 Deviation of Central Line Position	0°		
	90°				90°		
	180°			铅垂度 Plumbness	0°		
	270°				90°		
结论 Conclusion							

续表

建设/监理单位 Owner / Supervision Unit	施工单位 Construction Unit
专业工程师： Discipline Engineer: 年 (Y) 月 (M) 日 (D)	技术负责人： Technical Responsible Person: 质量检查员： Quality Inspector: Team leader 队（组）长： 年 (Y) 月 (M) 日 (D)

Knowledge extention

Translate the English items in the record into Chinese (Table 25.2).

Table 25.2 Installation record of vertical equipment (English version)

SH 3503–J302	Installation Record of Vertical Equipment			Project Description: Unit Description:	
Name of Equipment		Tag No.		Type and Specification	

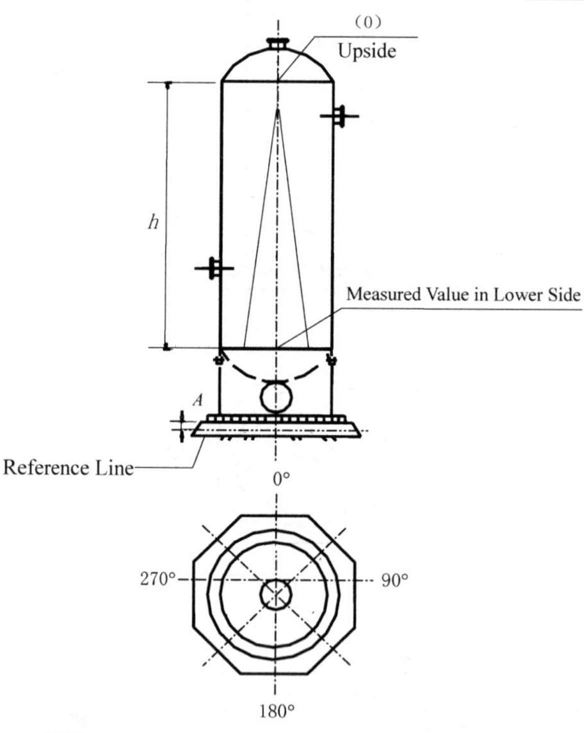

h 为两测点间的距离，h = mm。
h represents the distance between the two measured points, h = mm.

Item	Orientation	Allowable Value (mm)	Measured Value (mm)	Item	Measured Position	Allowable Value (mm)	Measured Value (mm)
Elevation Deviation (A)	0°			Deviation of Central Line Position	0°		
	90°				90°		
	180°			Plumbness	0°		
	270°				90°		
Conclusion							

续表

Owner / Supervision Unit	Construction Unit
Discipline Engineer: 年(Y) 月(M) 日(D)	Technical Responsible Person: Quality Inspector: Team Leader: (Y) (M) (D)

Words and phrases

record ['rekɔ:d] 记录
require [rɪ'kwaɪə(r)] 要求
ribbon ['rɪbən] 带，带状物
safety belt 安全带
fastener ['fa:stnəz] 紧固件，搭扣
Tag No. [tæg] ['nʌmbə] 位号，工号
type [taɪp] 类型，型号
upside ['ʌpsaɪd] 向上，上部
measured value ['meʒəd] ['vælju:] 实测值
reference line ['refrəns] [laɪn] 参考/基准线
allowable [ə'laʊəbl] 允许的
elevation [ˌelɪ'veɪʃn] 标高
central line 中心线
owner ['əʊnə(r)] 所有者，业主，建设方
supervision [ˌsju:pə'vɪʒn] 监理，监督
discipline ['dɪsəplɪn] 专业
responsible [rɪ'spɒnsəbl] 负责的
inspector [ɪn'spektə(r)] 检查员
vertical ['vɜ:tɪkl] 直立的，竖直的
fasten ['fɑ:sn] 系紧
uniform ['ju:nɪfɔ:m] 制服
safety shoes 安全皮鞋，工作皮鞋
project ['prɔdʒekt] 工程，项目
description [dɪ'skrɪpʃn] 明细，内容
specification [ˌspesɪfɪ'keɪʃn] 规范
lower side ['ləʊə(r)] [saɪd] 向下，下部
item ['aɪtəm] 项
orientation [ˌɔ:riən'teɪʃn] 方位，方向
position [pə'zɪʃn] 位置
deviation [ˌdi:vi'eɪʃn] 偏差
plumbness [p'lʌmnəs] 铅垂度
conclusion [kən'klu:ʒn] 结论
construction [kən'strʌkʃn] 施工
technical ['teknɪkl] 技术的
quality ['kwalɪti] 质量
team leader ['li:də(r)] 队/组长

Lesson 26 Safety

Look and select

Look at the pictures and select the correct terms from the box.

exploding parts	electric shock	risk of injury from burning
arc rays and noise	fumes and gases	risk of injury from hot coolant

1. ________________ 2. ________________ 3. ________________

4. ________________ 5. ________________ 6. ________________

Text

1. Describe what has happened to the man (Fig.26.1) in the picture.

Words and phrases for reference:

pound（砸中）

break one's leg （折断腿）

ambulance（救护车）

bind up a wound（包扎伤口）

stretcher（担架）

safety manual（安全手册）

Fig.26.1 Injured on construction site

2. Understand the safety signs (Fig.26.2):

当心电离辐射
Radiation（辐射）

禁止吸烟
No smoking（禁止吸烟）

Fig.26.2　Safety signs

Exercises

1. Match the English with the Chinese. Draw lines.

A.

(1) electric shock	a. 冷冻液
(2) arc rays and noise	b. 燃烧
(3) fumes and gases	c. 电击
(4) burning	d. 弧光及噪音
(5) coolant	e. 烟雾及气体
(6) injury	f. 救护车
(7) ambulance	g. 安全手册
(8) safety manual	h. 伤害

B.

(1) danger	a. 辐射
(2) caution	b. 坠落
(3) no smoking	c. 危险
(4) no touching	d. 爆炸的
(5) falling	e. 禁止吸烟
(6) flammable	f. 小心
(7) radiation	g. 禁止触摸
(8) explosive	h. 可燃的

2. Translate the following English words into Chinese.

no rolling ______________________; no welding ______________________;
crane operation ______________________; mechanical hurt ______________________;
safety helmet ______________________; protective goggles ______________________;
safety belt ______________________; toxic ______________________;
corrosive ______________________; break one's leg ______________________.

Words and phrases

risk of injury from burning [rɪsk] ['ɪndʒərɪ] ['bɜ:nɪŋ] 烧伤
risk of injury from hot coolant ['ku:lənt] 冷却液灼伤
exploding parts [iks'pləudɪŋ] [pa:ts] 爆炸物
fumes and gases 烟雾及气体
protective goggles [prə'tektɪv] ['gɔglz] 护目镜
arc rays and noise [nɔɪz] 弧光及噪音
electric shock [ɪ'lektrɪk] [ʃɒk] 触电，电击
operation [ˌɒpə'reɪʃn] 操作，运作
wear safety helmet ['seɪftɪ] ['helmɪt] 戴安全帽
toxic ['tɒksɪk] 有毒的
flammable ['flæməbl] 可燃的
object ['ɒbdʒɪkt] 物体
electric shock [ɪ'lektrɪk] [ʃɒk] 电击
danger ['deɪndʒə(r)] 危险
roll [rəʊl]旋转，滚动
touch [tʌtʃ] 触摸
caution ['kɔ:ʃn] 注意，小心
crane[kreɪn] 吊装，吊车
safety belt ['se:fti] [belt] 安全带
corrosive [kə'rəʊsɪv] 腐蚀的
fall [fɔ:l] 坠落
radiation [ˌreɪdi'eɪʃn] 辐射

Lesson 27 Resume

Look and select

Look at the pictures and select the correct terms from the box.

schedule	resume	awards	design

1. ____________

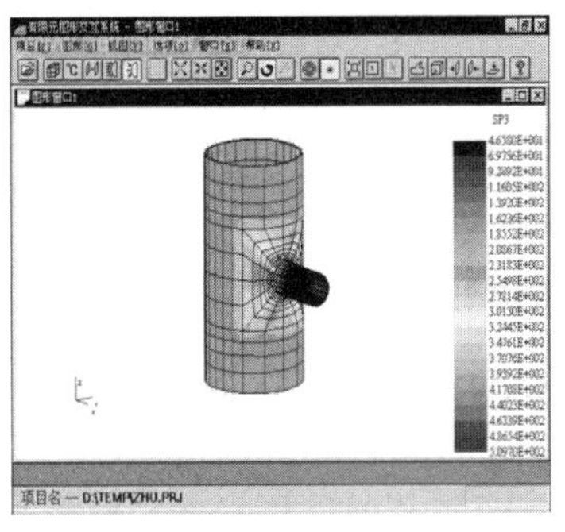

2. ____________

附件六

面试体检日程安排表

面试体检点	时 间	参 检 单 位
武 汉	7月27日	武汉市
	7月28日	咸宁市
	7月29日	黄石市、鄂州市
	7月30日	鄂州市、孝感市
	7月31日	孝感市、黄冈市
荆 州	7月27日—28日	荆州市
	7月29日	仙桃市、天门市
	7月30日	天门市、潜江市
	7月31日	潜江市、荆门市
宜 昌	7月27日—28日	宜昌市
	7月29日—30日	恩施州
襄 樊	7月27日—28日	襄樊市
	7月29日	随州市
	7月30日	十堰市 神农架林区
备 注	国防生面试体检时间均安排在各点日程安排最后一天。	

3. ____________

4. ____________

Text

Understand the resume of the student named Zhang Xiaoming (Table 27.1).

Table 27.1 Resume of Zhang Xiaoming

Name	Zhang Xiaoming	Gender	Male
Date of birth	Nov. 15th, 1995	Nationality	Han
Height	176cm	Degree	Vocational college

续表

Graduated from	Hunan Industrial Technician College		
E-mail		Telephone No.	
Address	Zhangshuxiang, Huabanqiao road, Yueyang City, Hunan PRC		
Job wanted	Piper	Full time/part time	Full time
Working place	Hunan	Salary	Discuss face to face
Self-recommendation	Strong ability of self-learning and hand-making; responsible and diligent for work; doing things with perseverance and innovation; amiable and amicable		
Education	2012.9~2015.7 Hunan Industrial Technician College, Installation Engineering Department		
Working experience	2013.8 working in Walmart Supermarket 2014.8 working in a steel factory		
Awards	2014.5 the first place of piping skill competition		

Knowledge extention

The questions you may be asked while meeting the interviewer

1. Please tell me about your work experience.
2. What is your greatest advantages?
3. Why do you feel you are qualified for this job?
4. If hired, when could you start working?
5. What kind of salary do you have in mind?
6. What kind of characters does the position involved?

Exercises

Try to complete the following blanks with your own information (Table 27.2).

Table 27.2 Resume

Name		Gender	
Date of birth		Nationality	
Height		Degree	
Graduated from			
E-mail		Telephone No.	
Address			
Job wanted		Full time/part time	
Working place		Salary	
Self-recommendation			
Education			
Working experience			
Awards			

Words and phrases

resume [rezju:mei] 简历
award [ə'wɔ:dz] 获奖
gender ['d□endə(r)] 性别
date of birth 出生日期
height [haɪt] 身高
technician college [vəʊ'keɪʃənl] 技师学院
industry ['ɪndəstrɪ] 工业
telephone 电话
want [wɔnt] 想要
salary ['sælərɪ] 薪酬
recommendation [ˌrekəmen'deɪʃn] 推荐
experience [ɪk'spɪərɪəns] 经历，经验
self-learning 自学
responsible [rɪ'spɒnsəbl] 负责的
perseverance [ˌpɜ:sə'vɪərəns] 恒心
amiable ['eɪmɪəbl] 和蔼可亲
engineering [ˌendʒɪ'nɪərɪŋ] 工程
Walmart Supermarket 沃尔玛超市
the first place 第一名
schedule ['ʃedju:l] 计划
design [dɪ'zaɪn] 设计
male [meɪl] 男，雄性
nationality [ˌnæʃə'næləti] 民族
degree [dɪ'gri:] 文化程度
graduated from ['grædʒueɪtɪd] 从……毕业
technology [tek'nɒlədʒɪ] 技术
address [ə'dres] 地址
piper ['paɪpə(r)] 管工
discuss [dɪ'skʌs] 讨论
education [ˌedʒu'keɪʃn] 教育
ability [ə'bɪlətɪ] 能力
hand-making 手工制作，动手
diligent ['dɪlɪdʒənt] 勤奋的
innovation [ˌɪnə'veɪʃn] 发明创造，创新
amicable ['æmɪkəbl] 和善，友好
department [dɪ'pɑ:tmənt] 系部，部门
steel factory 钢铁厂
skill competition 技能竞赛

Welding Part

Part I
Basic Knowledge of Welding

Lesson 28 What Is Welding

Look and select

Look at the pictures and select the correct terms from the box.

electrode holder	safety shoes	electrodes	gloves	helmet

1. ________________ 2. ________________ 3. ________________

4. ______________ 5. ______________

Text

Formerly welding was "**the joining of metals by fusion**," that is , by melting, but this definition will no longer do. Even though fusion methods are still the most common, they are not always used.

Welding was next defined as the "**joining of metals by heat**," but this is no longer a proper definition either. Not only metals can be welded, so can many of the plastics.

Furthermore, several welding methods do not require heat, such as cold pressure welding. Besides, we can weld with sound and even with light from the famous laser. Faced with a diversity of welding methods that increase year by year, we must adopt the following definition of welding: **"welding is the joining of metals and plastics by methods that do not employ fastening devices.**"

Knowledge extention

1. The basic methods of joining of two or more metal parts (see Fig.28.1 and Fig.28.2).

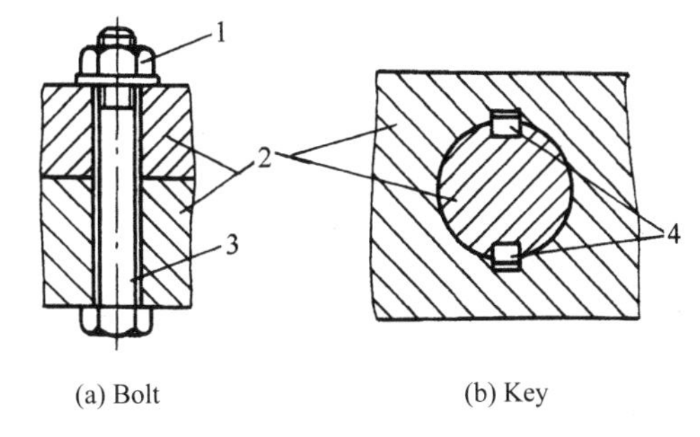

(a) Bolt (b) Key

Fig.28.1 Able-dismantled joining

1—nut; 2—components; 3—screw; 4—keys

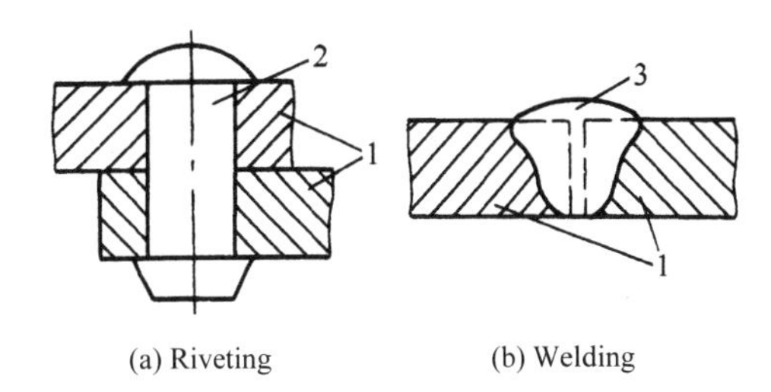

(a) Riveting (b) Welding

Fig.28.2 Unable dismantled joining

1—components; 2—rivet; 3—weld

2. Self - introduction

My name is Liu Xiaobing. I'm 18 years old. I am a student in Hunan Industrial Technician College (Central South Industrial School). My major is welding. I like my major very much. I want to become a good welder in the future.

Exercises

1. Answer the following questions according to the hints given above.

a. What is welding?

b. What's your name?

c. What's your major?

d. What's your job?

e. Try to make a simple self-introduction by filling in the blanks:

My name is_名字_. I come from_家乡_. I'm_年龄_years old. I am a student in_学校_ (Central South Industrial School). My hobby is_爱好_. My major is_专业_. I like my major very much.

I want to become 职业 in the future.

2. Match the English with the Chinese. Draw lines.

A.

(1) electrode	a. 连接
(2) helmet	b. 金属
(3) join	c. 焊条
(4) metal	d. 加热
(5) fusion	e. 拆卸
(6) heat	f. 熔化
(7) dismantle	g. 面罩

B.

(1) low-alloy steel	a. 碳钢
(2) carbon steel	b. 焊条药皮
(3) tensile strength	c. 焊接位置
(4) deposited metal	d. 美国焊接协会
(5) welding position	e. 抗拉强度
(6) electrode covering	f. 低合金钢
(7) American Welding Society	g. 熔敷金属

Words and phrases

welding [weldɪŋ] 焊接
safety ['seɪftɪ] 安全
helmet ['helmit] 面罩
metal ['metl] 金属
heat [hi:t] 热量，加热
employ [ɪm'plɔɪ] 使用，雇用
major ['meɪdʒə(r)] 专业
technician [tɛk'nɪʃən] 技师
welder ['weldə(r)] 焊工，焊机
safety shoes 安全鞋/靴
electrode [ɪ'lektrəʊd] 焊条
gloves [g'lʌvz] 手套
join [dʒɔɪn] 连接
fusion ['fju:ʒn] 熔化
plastic ['plæstɪk] 塑料
method ['meθəd] 方法
industrial [ɪn'dʌstriəl] 工业的
college ['kɒlɪdʒ] 学院
electrode holder 焊钳
fastening device 紧固装置

Lesson 29 Classification of Welding Processes

Look and select

Look at the pictures and select the correct terms from the box.

flux	power source	welding machine	welding rod

1. ________________

2. ________________

3. ________________

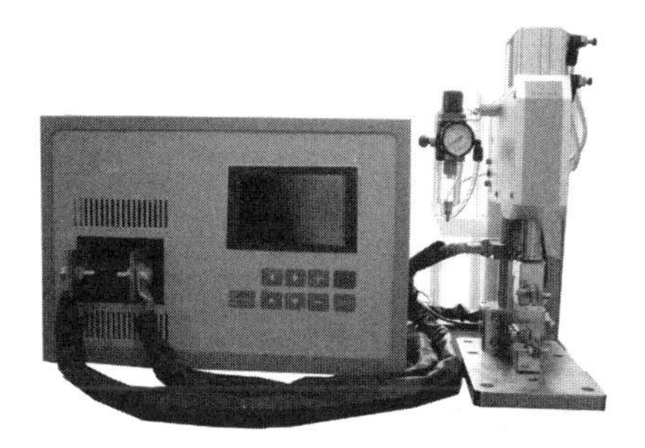

4. ________________

Text

The classification of welding processes is listed below:

1. Fusion welding 熔焊

① oxyacetylene welding 氧乙炔焊

② arc welding 电弧焊

③ electroslag welding 电渣焊

④ thermit welding 热剂焊

⑤ laser beam welding 激光束焊

⑥ electron beam welding 电子束焊

2. Pressure welding 压焊（resistance 电阻焊）

① spot welding 点焊
② seam welding 缝焊
③ butt welding 对焊
④ cold pressure welding 冷压焊
⑤ friction welding 摩擦焊
⑥ ultrasonic welding 超声波焊
⑦ explosion welding in vacuum 真空爆炸焊
⑧ diffusion welding 扩散焊
⑨ high frequency welding 高频焊

3. Brazing welding 钎焊

Knowledge extention

1. Basic characteristics of welding materials

The basic characteristics that make the metals so very useful are their weldability, hardness, stiffness, and ductility (that is the property to be shaped easily). These characteristics of metals are of great importance to a welder.

2. Understand the weld seam shown in Fig.29.1.

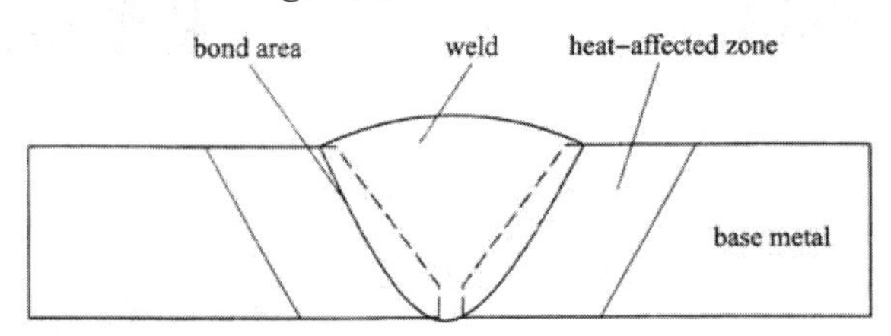

Fig.29.1 Heat-affected zone

base metal— ________________ heat-affected zone— ________________
bond area— ________________ weld— ________________

Exercises

1. Answer the following questions according to the text.

(1) What are the main three kinds of welding?
(2) What kinds of welding processes are included in the fusion welding?
(3) Are butt welding and spot welding included in the pressure welding?

2. Match the English with the Chinese. Draw lines.

A.

(1) pressure welding	a. 熔焊
(2) arc welding	b. 氧乙炔焊
(3) laser beam welding	c. 电弧焊
(4) spot welding	d. 钎焊
(5) oxyacetylene welding	e. 激光束焊

(6) resistance welding f. 压焊
(7) butt welding g. 电阻焊
(8) cold pressure welding h. 点焊
(9) fusion welding i. 对焊
(10) friction welding j. 二氧化碳焊
(11) brazing welding k. 摩擦焊
(12) CO_2 welding l. 冷压焊

B.

(1) characteristic a. 硬度
(2) weldability b. 刚性
(3) hardness c. 特性
(4) stiffness d. 可焊性
(5) ductility e. 展延性

3. Fill in the blanks with the proper words in the Fig.29.1.

In the picture of weld seam, four elements are involved: __________, weld, bond area and __________.

Words and phrases

process ['prəʊses] 工艺，方法，过程
classification [ˌklæsɪfɪ'keɪʃn] 分类，类别
power source ['paʊə(r)] [sɔ:s] 电源
oxyacetylene [ˌɒksiə'setəli:n] 氧乙炔
electroslag [ɪ'lektrəʊslæg] 电渣
laser beam ['leɪzə(r)] [bi:m] 激光束
electron beam [ɪ'lektrɒn] 电子束
seam [si:m] 缝
cold [kəʊld] 冷
ultrasonic [ˌʌltrə'sɒnɪk] 超声波
diffusion welding [dɪ'fju:ʒn] 热剂焊
frequency ['fri:kwənsi] 频率
characteristic [ˌkærəktə'rɪstɪk] 特性
material [mə'tɪəriəl] 材料
weldability [weldə'bɪlɪtɪ] 可焊性
hardness [hɑ:dnəs] 硬度
ductility [dʌk'tɪlɪtɪ] 展延性
be of great importance to 对……很重要
flux [flʌks] 焊剂
arc [ɑ:k] 电弧
pressure ['preʃə(r)] 压力
fusion ['fju:ʒn] 熔化
thermit ['θɜ:mɪt] 热剂
spot welding [spɒt] 点焊
resistance [rɪ'zɪstəns] 电阻
butt welding [bʌt] 对焊
friction ['frɪkʃn] 摩擦
explosion [ɪk'spləʊʒn] 爆炸
vacuum ['vækjuəm] 真空
brazing [breɪzɪŋ] 钎焊
basic ['beɪsɪk] 基本的
metal ['metl] 金属
useful ['ju:sfl] 有用的
stiffness [stɪfnəs] 强度
property ['prɒpəti] 性能

Lesson 30 Welding Positions and Symbols

Look and select

Look at the pictures and select the correct terms from the box.

overhead position	flat position	vertical down position welding
horizontal position	vertical position	vertical up position welding

1. ______________ 2. ______________ 3. ______________

progress of welding

progress of welding

4. ______________ 5. ______________ 6. ______________

Text

There are four basic welding positions. They are flat position (1G), horizontal position(2G), vertical position(3G) and overhead position(4G). For fillet welding (Table 30.1), there is fillet welding in the flat position(1F), horizontal vertical position welding(2F), fillet welding in the vertical position(3F), and horizontal overhead position welding(4F). For pipe butt welding (Table 30.1), the welding position are 1G, 2G and 5G. The welding position must be mentioned in welding procedure specification (WPS).

Table 30.1 Welding positions and symbols

Figure	Welding positions	Symbol of welding position
	horizontal vertical position	2F
	fillet welding in the vertical position	3F

续表

Figure	Welding positions	Symbol of welding position
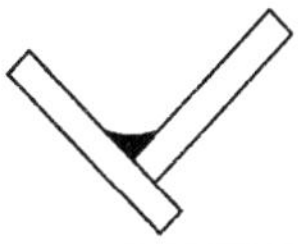	fillet welding in the flat position	1F
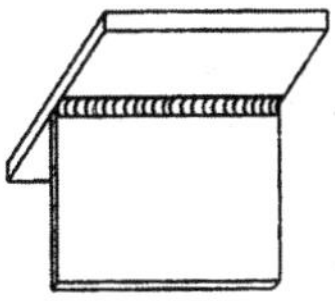	horizontal overhead position	4F
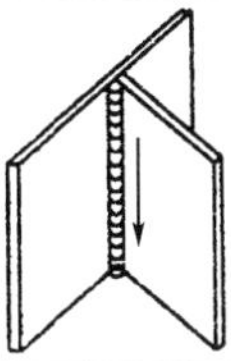 progress of welding	vertical downward position	PG
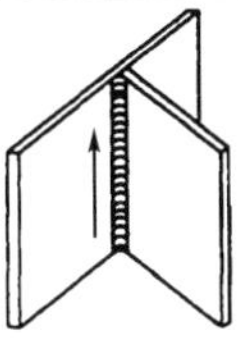 progress of welding	vertical upward position	PF, 3F-up
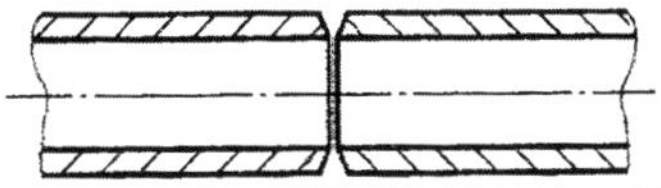	fixed pipe, horizontal axis	5G
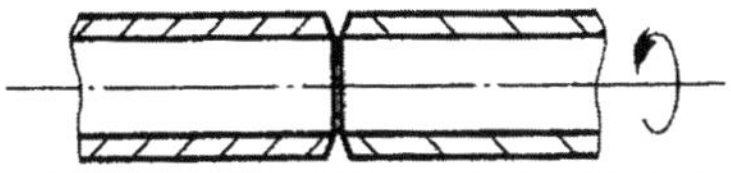	rotating pipe, horizontal axis	1G
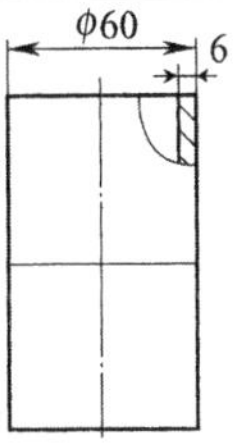 $\phi 60$ 6	fixed pipe, vertical axis	2G

Knowledge extention

Learn the welding methods (Table 30.2).

Table 30.2 Welding methods

English	Chinese
Forehand welding	左焊法
Backhand welding	右焊法
Inclined position welding	倾斜焊
Upward welding in the inclined position	上坡焊
Downward welding in the indined position	下坡焊

The backhand welding and forehand welding are shown in Fig.30.1.

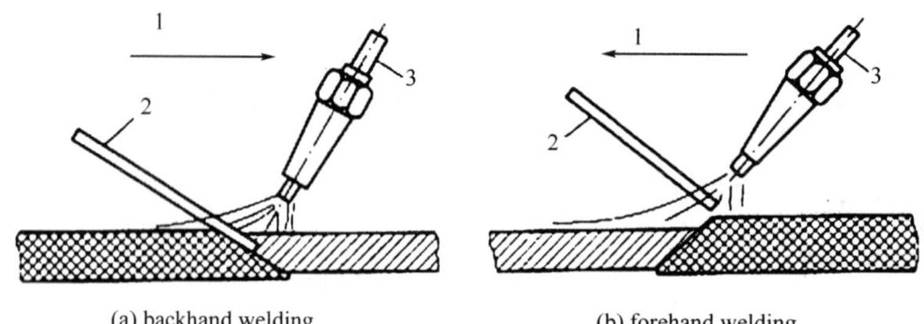

(a) backhand welding (b) forehand welding

Fig.30.1 Backhand welding and forehand welding

1—welding direction; 2—welding wire; 3—welding torch

Exercises

1. Point out the welding positions of the following welds (Fig.30.2).

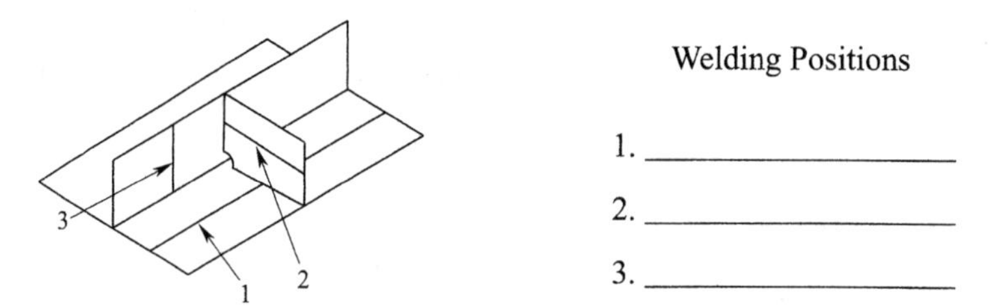

Welding Positions

1. ____________________
2. ____________________
3. ____________________

Fig.30.2 Three kinds of welding positions

2. Answer the following questions according to the text.

(1) What are the four basic welding positions?

(2) What are the four kinds of fillet welding?

(3) What are the three kinds of pipe butt welding?

3. Match the English with the Chinese. Draw lines.

A.

(1) flat position	a. 立焊位
(2) horizontal position	b. 平焊位
(3) vertical position	c. 横焊位
(4) overhead position	d. 仰焊位

B.

(1) forehand	a. 反手
(2) backhand	b. 正手
(3) upward	c. 向下
(4) downward	d. 向上

Words and phrases

position [pə'zɪʃn] 位置

overhead ['əuvəhed] 上面的，头顶上的

horizontal [ˌhɔri'zɔntəl] 水平的

vertical ['vɜːtɪkl] 竖立的

downward ['daʊnwəd] 向下的

fixed [fɪkst] 固定的

rotating [rəʊ'teɪtɪŋ] 旋转的

backhand ['bækhænd] 反手

direction [də'rekʃn] 方向

symbol ['sɪmbl] 符号

flat [flæt] 平的，平面

figure ['fɪgə(r)] 图

fillet welding ['fɪlɪt] 角焊

upward ['ʌpwəd] 向上的

axis ['æksɪs] 轴

forehand ['fɔːhænd] 正手

inclined [ɪn'klaɪnd] 倾斜的

pipe [paɪp] 管道

welding procedure specification (WPS) 焊接工艺规程

Lesson 31 Welding Joints and Welds

Look and translate

Translate the terms in Fig.31.1 into Chiese.

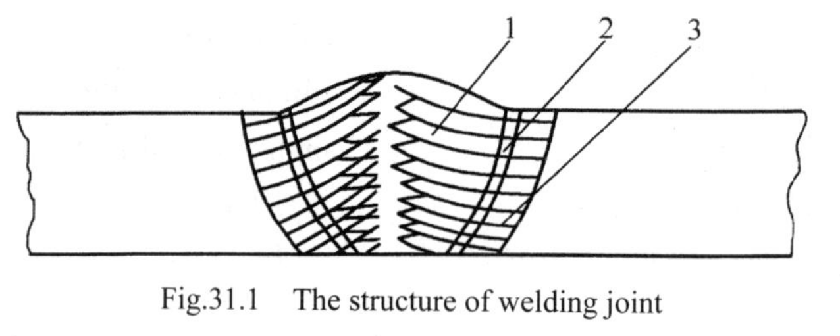

Fig.31.1 The structure of welding joint

1—weld __________; 2—fusion zone __________; 3—heat affected zone __________

Text

1. There are five basic types of joints (Fig.31.2) for bring two members together.

Butt joint: 对接接头 two parts in approximately the same plate.

Corner joint: 角接接头 two parts located at right angles to each other.

T joint: T 形接头 parts at right angles, in the form of T.

Lap joint: 搭接接头 between overlapping parts in parallel planes.

Edge joint: 端接接头 between the edges of two or more parallel parts.

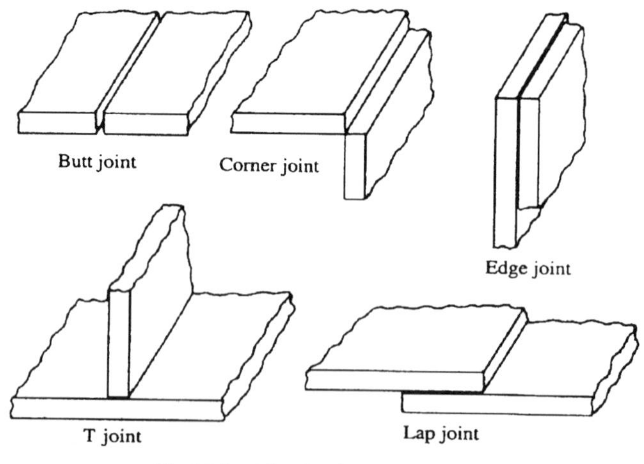

Fig.31.2 Five basic types of joints

2. The kinds of welds.

The most common welds are fillet weld and butt weld. There are other types of welds: plug weld, slot weld, spot weld, seam weld, surfacing weld, back bead, backing bead, and so on (Table 31.1).

Table 31.1 Kinds of welds

Weld form	Figure
butt weld	
fillet weld	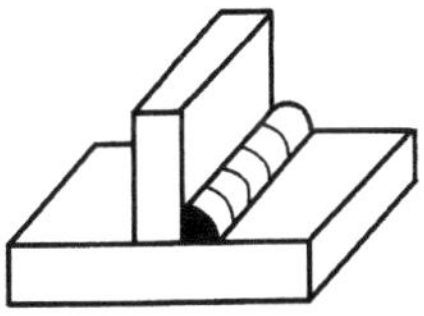
plug weld	
spot weld	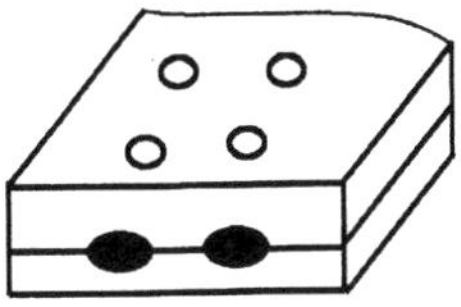
seam weld	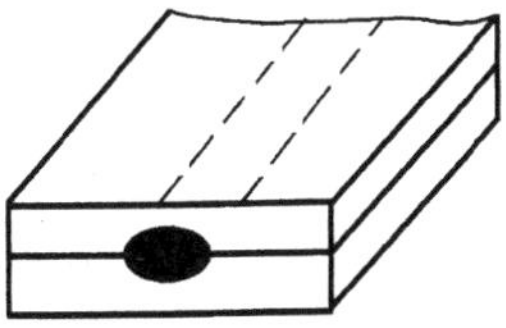
back bead	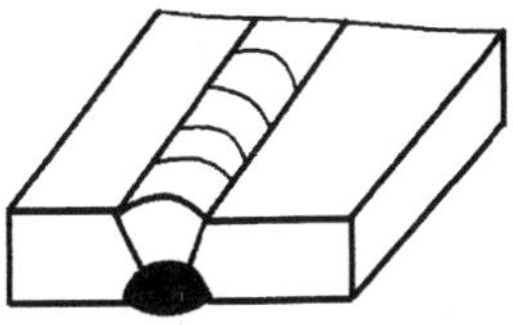
surfacing weld	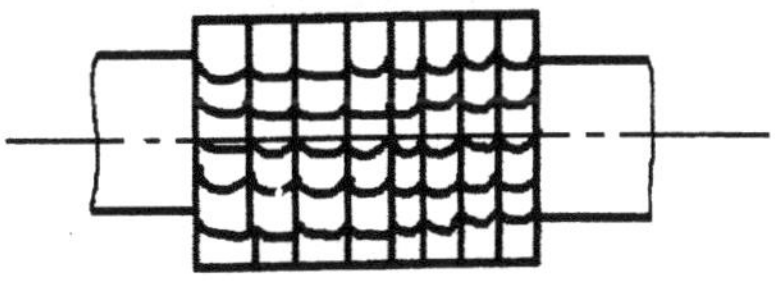

Knowledge extention

Try to recite the following phrases (Table 31.2).

Table 31.2 Other kinds of welds

English	Chinese	English	Chinese
tack weld	定位焊缝	transverse weld	横向焊缝
edge weld	端接焊缝	longitudinal weld	纵向焊缝
flat fillet weld	平行角焊缝	concave fillet weld	凹形角焊缝
seal weld	密封焊缝	convex fillet weld	凸形角焊缝

Exercises

1. Match the English with the Chinese. Draw lines.

A.

(1) butt joint	a. 角接头
(2) corner joint	b. 搭接接头
(3) edge joint	c. T 形接头
(4) tee joint	d. 对接接头
(5) lap joint	e. 端接接头

B.

(1) butt weld	a. 塞焊焊缝
(2) fillet weld	b. 封底焊缝
(3) plug weld	c. 对焊焊缝
(4) spot weld	d. 堆焊焊缝
(5) seam weld	e. 点焊焊缝
(6) back bead	f. 角焊焊缝
(7) surfacing weld	g. 缝焊焊缝

2. Look at the pictures and select the correct terms from the box.

welding by one side	back bead	multi-layer and multi-pass welding
multi-layer welding	backing bead	welding by both sides

1. ____________________ 2. ____________________

3. ____________________ 4. ____________________

5. ____________________ 6. ____________________

3. Look at the pictures (Fig.31.3) and translate the terms into Chinese.

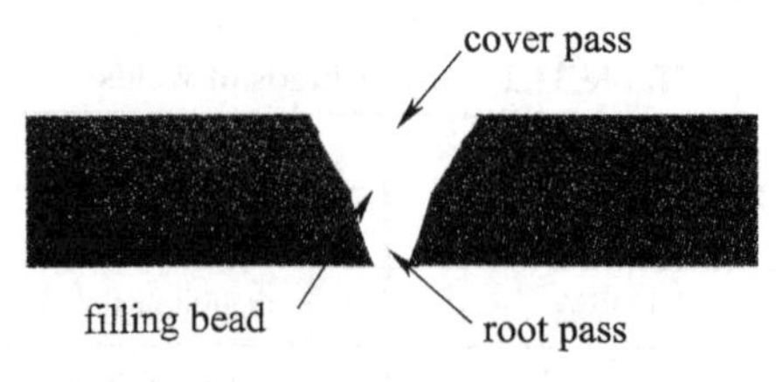

Fig.31.3 Weld

root pass ________; filling bead ________; cover poss ________

Words and phrases

joint [dʒɔɪnt] 接头
butt [bʌt] 对接
lap [læp] 搭接，部分重叠
kind[kaɪnd]种类
other ['ʌðə(r)] 其他的
plug [plʌg] 塞子
seam [si:m] 缝
back bead [bæk] 封底焊缝
one side 一边，一侧
filling bead ['fɪlɪŋ] 填充焊道
multi-layer / pass ['mʌlti] ['leiə] 多层/多道
weld [weld] 焊缝
corner ['kɔ:nə(r)] 角落
edge [edʒ] 边缘， 端部
common ['kɒmən] 普遍
type [taɪp] 类型
slot [slɒt] 槽
surfacing welding ['sɜ:fɪsɪŋ] 堆焊
backing bead ['bækɪŋ] 打底焊缝
both [bəuθ] 两者
root [ru:t] 根部
cover ['kʌvə(r)] 覆盖

Lesson 32 Welding Symbols

Look and translate

Look at the pictures (Fig.32.1) and translate the terms into Chinese.

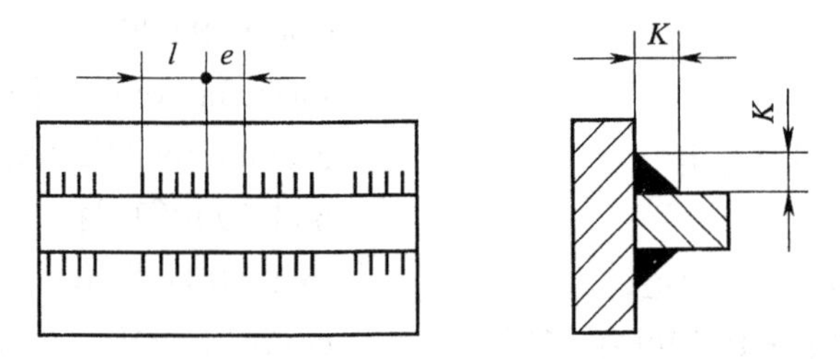

Fig.32.1 Welding terms

l—weld length; *e*—weld spacing; *K*—fillet weld size

l—__________; *e*—__________; *K*—__________

Text

Understand the weld forms and symbols (Table 32.1).

Table 32.1 Weld forms and symbols

Weld form	Figure	Symbol
fillet weld		
plug weld		
spot weld		
seam weld		
back bead		

续表

Weld form	Figure	Symbol
surfacing weld	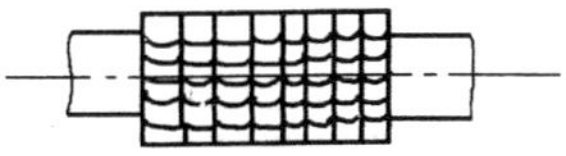	
I (square) groove	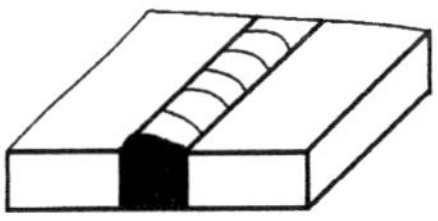	
V groove	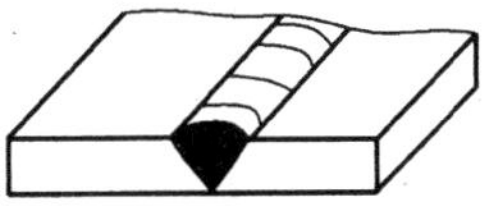	
V groove with root face	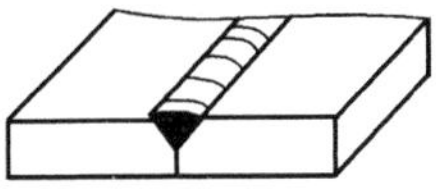	
single V (bevel) groove	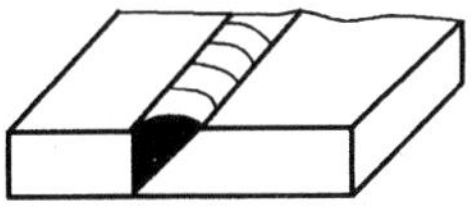	
single V groove (bevel) with root face	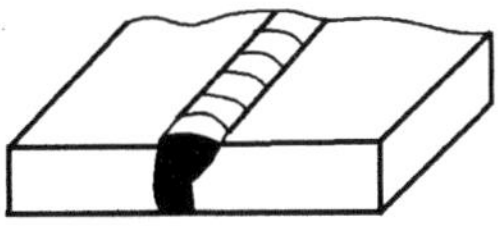	
U groove	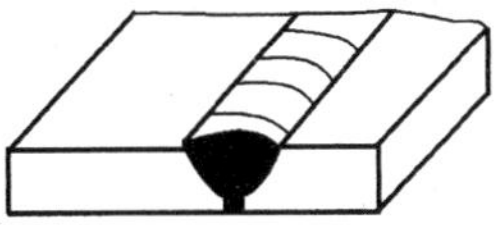	
single U groove	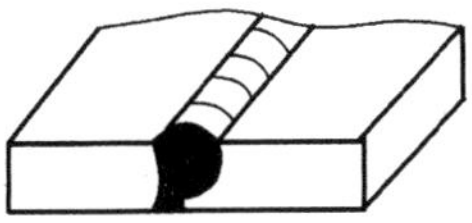	
flanged edge weld	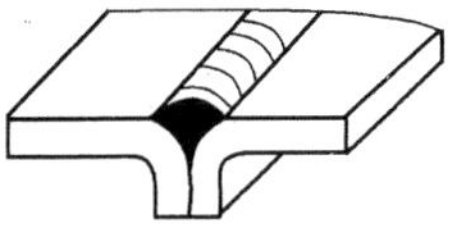	
single flanged edge weld	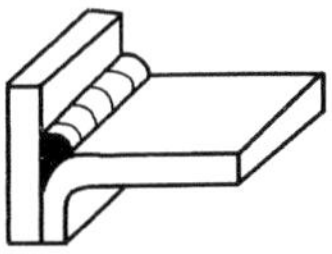	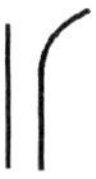

Knowledge extention

Learn the following welding methods and codes (Table 32.2).

Table 32.2 Welding methods and codes

焊接方法	代号	焊接方法	代号
电弧焊 Arc welding	1	电阻焊 Resistance welding	2
气焊 Oxyfuel gas welding	3	压焊 Pressure welding	4
焊条电弧焊 Shielded metal arc welding	111	点焊 Spot welding	21
埋弧焊 Submerged arc welding	12	缝焊 Seam welding	22
熔化极惰性气体保护焊 MIG welding	131	钨极惰性气体保护焊 TIG welding	141

Exercises

1. Translate the welding terms in Fig.32.2 into English.

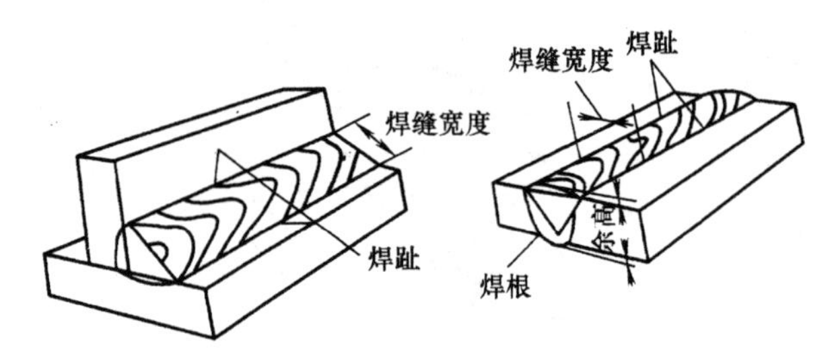

Fig.32.2 Welding terms

焊缝宽度________________；焊趾________________；焊根________________；余高________________

2. Explain the welding symbols in Fig.32.3.

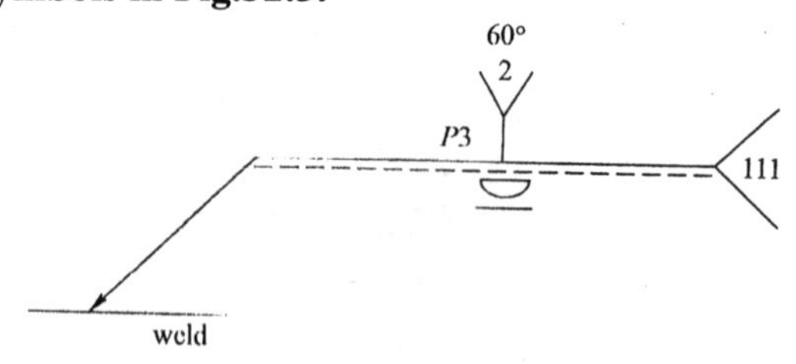

Fig.32.3 Welding symbols

*P*3 indicates ____.

"60° " indicates ____.

Digit 2 indicates ____.

Symbol Y indicates ____.

The number 111 indicates ____.

Symbol ⌓ indicates ____.

3. Match the symbols with the relative welding seams.

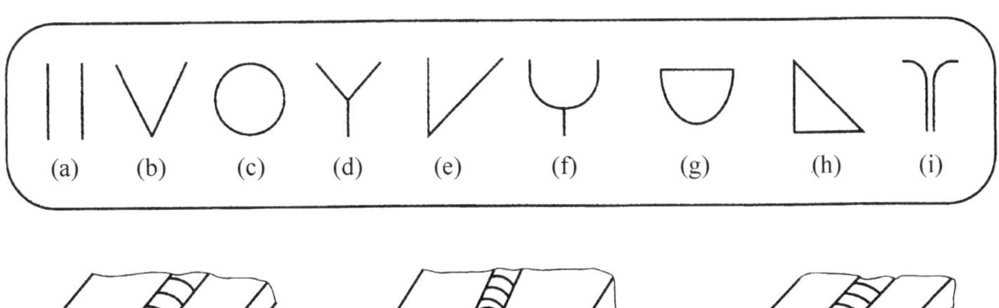

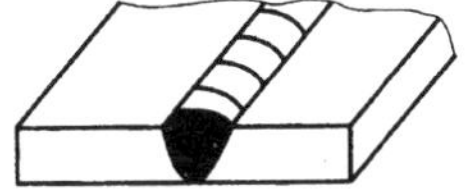

single V groove

1. ________________

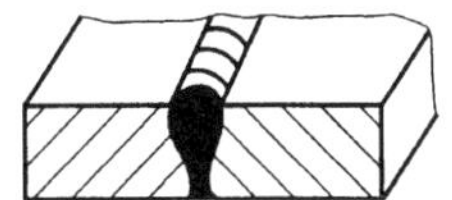

single U groove with root face

2. ________________

square groove

3. ________________

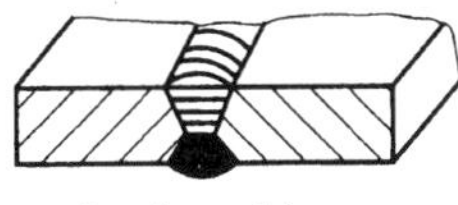

back weld

4. ________________

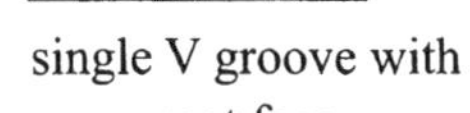

single V groove with root face

5. ________________

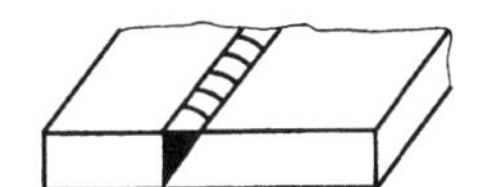

single bevel groove

6. ________________

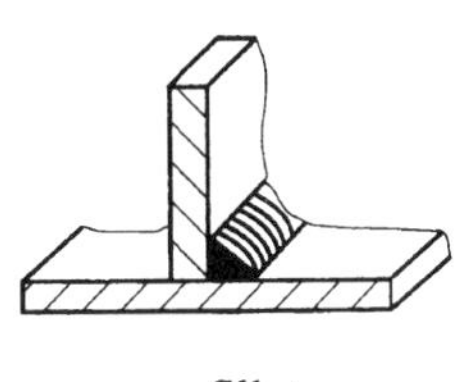

fillet

7. ________________

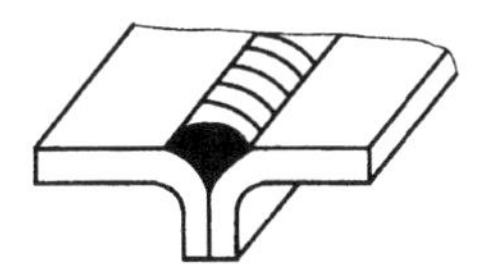

flanged edge weld

8. ________________

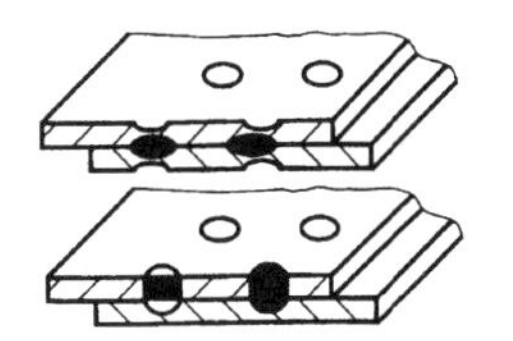

spot weld

9. ________________

Words and phrases

symbol ['sɪmbl] 符号

weld spacing ['speɪsɪŋ]焊缝间距

single ['sɪŋgl] 单个的

root face [ru:t] [feɪs] 坡口钝边

single bevel groove ['bevl] 单边 V 形坡口

flanged edge weld [flændʒd] [edʒ] 卷边焊

spot weld [spɒt] 点焊

digit ['dɪdʒɪt] 数字

explain [ɪk'spleɪn] 解释

weld length [lɛŋkθ] 焊缝长度

fillet weld size ['fɪlɪt] 焊脚

groove [gru:v] 坡口

square groove [skweə(r)] 平头坡口

back weld [bæk] 封底焊道

fillet 填角焊缝

indicate ['ɪndɪkeɪt] 表示，指示

figure ['figə] 图表

Lesson 33 Welding Currents and Polarities

Look and translate

Look at the picture (Fig.33.1) and translate the words into Chinese.

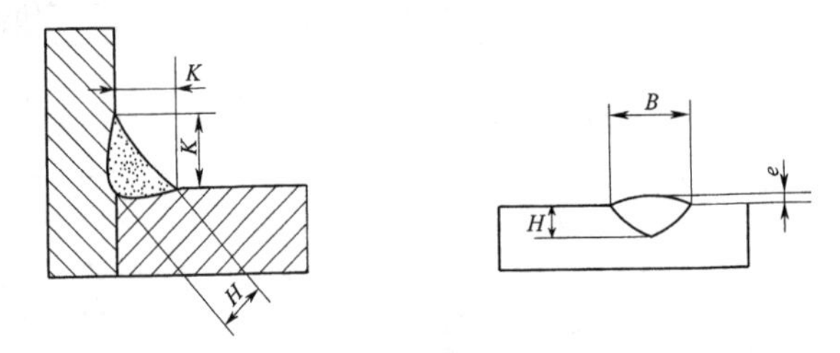

Fig.33.1 Welding dimensions

K—fillet weld size ____________; *H*—theoretical throat ____________;

B—weld width ____________; *e*—weld reinforcement ____________

Text

Look at the pictures (Fig.33.2) and select the correct polarities from the box.

electrode positive	reversed polarity	electrode negative	straight polarity

Generator
\+ −
Electrode cable
Electrode
Arc
Work
Work cable

1. ______________________________

Generator
\+ −
Electrode cable
Electrode
Arc
Work
Work cable

2. ______________________________

Fig.33.2 Welding polarities

Polarity indicates the direction of the current in the circuit. The polarity of the electric current, or the direction of flow, is important when direct current welding generators are used. When the electrode cable is fastened to the negative pole of the generator and the work to the positive pole, the polarity is commonly referred to as straight polarity. If the electrode cable is attached to the positive terminal of the generator and the work to the negative pole, the circuit is called reversed polarity.

Knowledge extention

Search ASME specification from web and fill in the blanks (Table 33.1).

Table 33.1 Welding current and polarity

Metals or alloys to be welded with TIG welding	The current and polarity used		
	DCEN	DCEP	AC
Aluminum and its alloys thickness≤2.5mm	acceptable	1.	2. ________
Aluminum and its alloys thickness>2.5mm	3. ________	not recommended	4. ________
carbon steels and low alloys steels	best	5. ________	6. ________
stainless steels	7. ________	8. ________	9. ________
copper	10. ________	11. ________	12. ________

Exercises

1. Match the English with the Chinese. Draw lines.

A.

(1) and so on　　a. 不锈钢
(2) cleaning action of the the cathode　　b. 碱性焊条
(3) basic electrode　　c. 熔深
(4) spatter　　d. 熔敷速度
(5) penetration　　e. 飞溅
(6) stainless steel　　f. 等等
(7) deposition rate　　g. 碳钢
(8) carbon steel　　h. 阴极清理作用

B.

(1) generator　　a. 工件电缆
(2) electrode cable　　b. 发电机
(3) work cable　　c. 焊条电缆
(4) positive　　d. 负极
(5) negative　　e. 正极
(6) straight polarity　　f. 反接
(7) reversed polarity　　g. 正接

2. Translate the words in the picture (Fig.33.3) into Chinese.

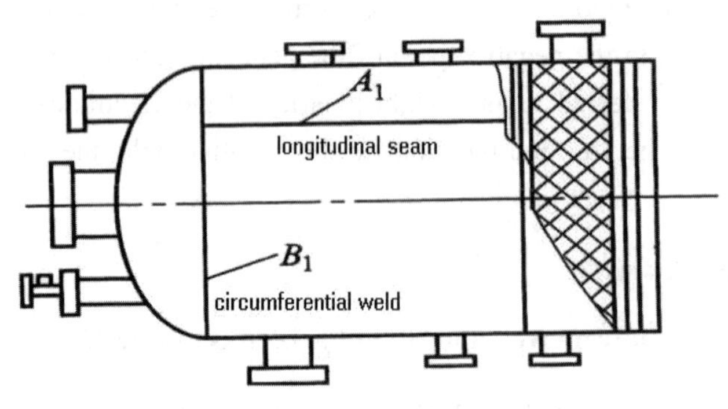

Fig.33.3 Structure of HP heater

A_1—longitudinal seam ________; B_1—circumferential weld ________

3. Answer the following questions.

(1) What does polarity indicate?

(2) What is positive polarity?

(3) What is reversed polarity?

Words and phrases

theoretical throat [ˌθɪə'retɪkl] [θrəʊt] 焊缝计算厚度

reinforcement [ˌri:ɪn'fɔ:smənt] 加强，增援，余高

current ['kʌrənt] 电流

polarity [pə'lærəti] 极性

size [saɪz] 尺寸，规格

width [wɪdθ] 宽度

positive ['pɒzətɪv] 肯定的，阳性/极

negative ['negətɪv] 否定的，阴性/极

reversed [rɪ'vɜ:st] 反，颠倒，反接

straight [streɪt] 直接的，正接

generator ['dʒenəreɪtə(r)] 发电机，焊机

electrode [ɪ'lektrəʊd] 焊条

cable ['keɪbl] 电缆

arc [ɑ:k] 电弧

indicate ['ɪndɪkeɪt] 表示，显示

direction [də'rekʃn] 方向

flow [fləʊ] 流动

important [ɪm'pɔ:tnt] 重要的

clean[kli:n]清洁，弄干净

action ['ækʃn] 行动，动作

cathode ['kæθəʊd] 阴极

basic ['beɪsɪk] 基本的

spatter ['spætə(r)] 飞溅

penetration [ˌpenɪ'treɪʃn] 渗透，熔透

stainless ['steɪnlɪs] 不锈的

steel [sti:l] 钢材

deposition [ˌdepə'zɪʃn] 熔敷

rate [reɪt] 率

Lesson 34 Welding Procedure Specification (WPS)

Look and translate

Look at the picture (Fig.34.1) and translate the terms into Chinese.

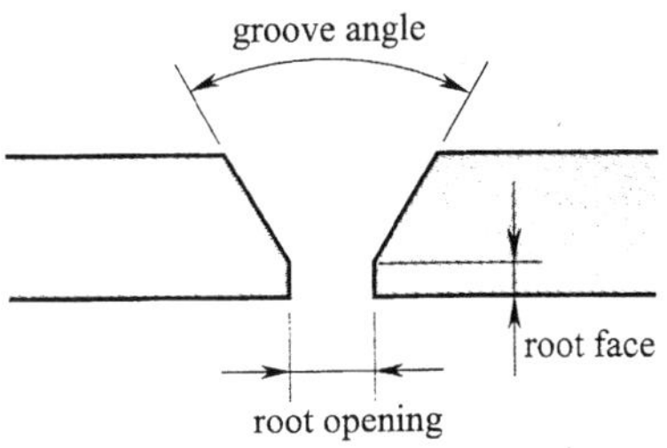

Fig.34.1 Groove

groove angle ___________

root opening ___________

root face ___________

Text

Fill in the WPS according to the conditions given below (Table 34.1).

Table 34.1 Welding procedure specification (WPS)

WPS No.__× × ×__ Date:__× × ×__

Company Name __× × ×__
Supporting PQR No. (s) __× × ×__ By:__× × ×__

Welding place: __welding of pipe system__
Welding process(es):__________
Welding position and direction:__________ Welding current and polarity:__________
Base metal: __________ Thickness range: __2~7 mm__
Diameter range: __$\phi \geqslant 17$ mm__
Filler metal spec.:__________ Backing: __none__
Shielding gas:__Ar__ Flow rate: __5~15 L/min__
Root treatment:__penetrate bead__
Single or multiple pass:__________
Weaving (yes/no):__________ Max.:__________
Preheat temperature:__none__ Interpass temperature: __<100℃__
PWHT:__none ℃__ Max.:__— ℃__ Holding time:__— min/mm__

	Pass No.		Filler metal dia. /mm	Welding parameter		
				Welding Cur. /A	Arc Voltage /V	Speed /(mm/min)
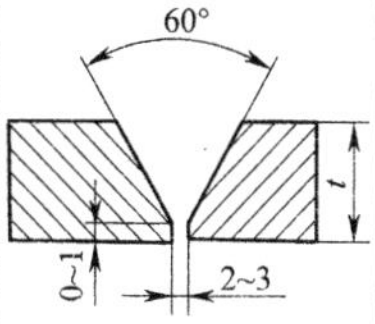	TIG	Root pass		80 ~ 110	9 ~ 13	50 ~ 80
		Filling bead		100 ~ 130	9 ~ 13	50 ~ 80
		Cover pass		100 ~ 130	9 ~ 13	50 ~ 80

采用 TIG 焊焊接 1Cr18Ni9Ti 的不锈钢管，管的尺寸为ϕ60mm×6mm，水平固定，采用直流正接，背面不加衬垫，填充焊丝为 ER308L，直径为 1.6mm，保护气体的流量为 5～15L/min，多道焊，摆动的最大幅度为 7mm，不需预热和后热，应保证道间温度不高于 100℃（注：表中的“PQR”是“procedure qualification record”的缩写，指工艺评定报告；焊接参数如焊接电流、电弧电压、焊接速度等已在表中给出）。

Knowledge extention

Translate the words in Table 34.2 into Chinese.

Table 34.2 Procedure qualification record (PQR)

<table>
<tr><td>No.</td><td colspan="4"></td><td colspan="2">Date</td><td colspan="2"></td></tr>
<tr><td colspan="7">No. of the relative welding process guidance</td><td colspan="2"></td></tr>
<tr><td>Method</td><td colspan="4"></td><td colspan="2">Type of joint</td><td colspan="2"></td></tr>
<tr><td rowspan="3">Parent materials of specimens</td><td rowspan="3">Plate</td><td colspan="2">material</td><td></td><td rowspan="3">Pipe</td><td>material</td><td colspan="2"></td></tr>
<tr><td colspan="2">class No.</td><td></td><td>class No.</td><td colspan="2"></td></tr>
<tr><td colspan="2">specfication</td><td></td><td>specification</td><td colspan="2"></td></tr>
<tr><td>Quality certification</td><td colspan="4"></td><td colspan="2">No. of recheck report</td><td colspan="2"></td></tr>
<tr><td>Electrode mode</td><td colspan="4"></td><td colspan="2">Electrode specification</td><td colspan="2"></td></tr>
<tr><td>Position</td><td colspan="4"></td><td colspan="2">Electrode drying Tepm.</td><td colspan="2"></td></tr>
<tr><td rowspan="2">Welding parameters</td><td colspan="2">Arc voltage/V</td><td colspan="2">Current/A</td><td colspan="2">Speed（cm/min）</td><td>Welder Name</td><td></td></tr>
<tr><td colspan="2"></td><td colspan="2"></td><td colspan="2"></td><td>Stamp No.</td><td></td></tr>
<tr><td rowspan="3">Result</td><td rowspan="2">Visual inspection</td><td rowspan="2">RT</td><td colspan="2">Tensile test</td><td colspan="2">Bending test</td><td rowspan="2">Macro metallographic test</td><td rowspan="2">Impacs toughness test</td></tr>
<tr><td></td><td></td><td>face</td><td>back</td></tr>
<tr><td></td><td></td><td></td><td></td><td colspan="2"></td><td></td><td></td></tr>
<tr><td>Report No.</td><td colspan="8"></td></tr>
<tr><td>Result of PQR</td><td colspan="8"></td></tr>
<tr><td>Approved by</td><td colspan="4"></td><td colspan="2">Reported by</td><td colspan="2"></td></tr>
</table>

Exercises

1. Match the English with the Chinese. Draw lines.

A.

(1) company　　a. 规范
(2) pipe system　　b. 位置和方向
(3) position and direction　　c. 管道系统
(4) base metal　　d. 公司
(5) diameter range　　e. 流量
(6) specification　　f. 厚度范围
(7) thickness range　　g. 母材
(8) flow rate　　h. 直径范围

B.

(1) groove angle　　a. 单个的
(2) root face　　b. 双的
(3) single　　c. 斜边/面
(4) double　　d. 钝边
(5) bevel　　e. 根部开口
(6) root opening　　f. 位置
(7) place　　g. 试件

(8) specimen h. 坡口角度

2. Fill in the blanks with the proper words in the text.

(1) WPS is the short form of ______________________________.

(2) PQR is the short form of ______________________________.

(3) PWHT is the short form of ______________________________.

3. Look at the pictures and select the correct terms from the box.

single V groove with root face	double bevel groove	single J groove
double V groove with root face	square groove	single bevel groove

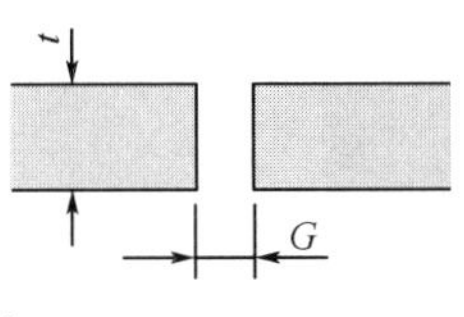

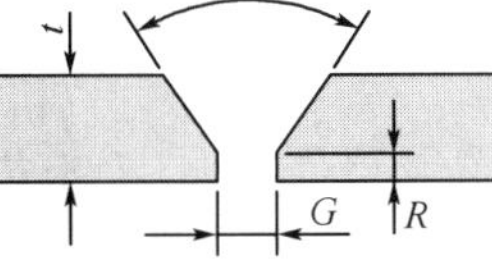

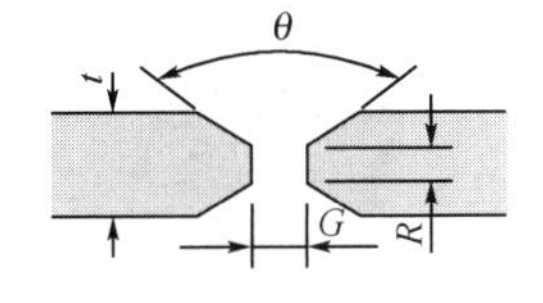

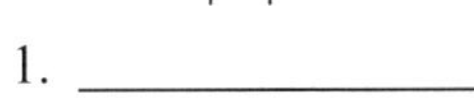

1. ________________

2. ________________

3. ________________

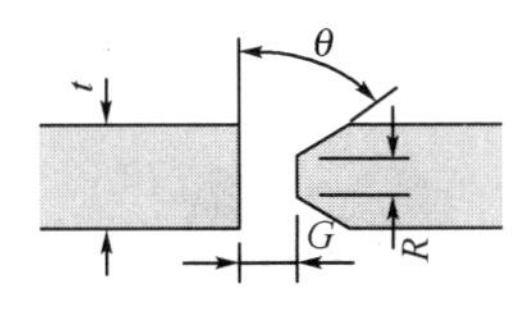

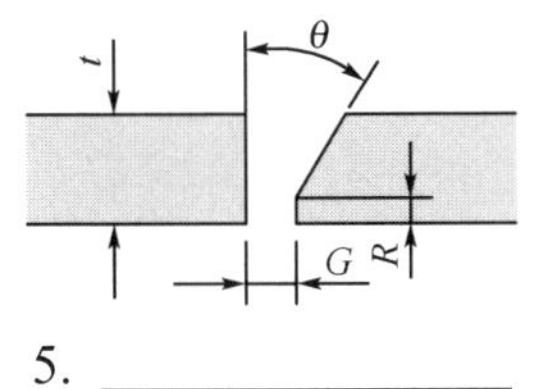

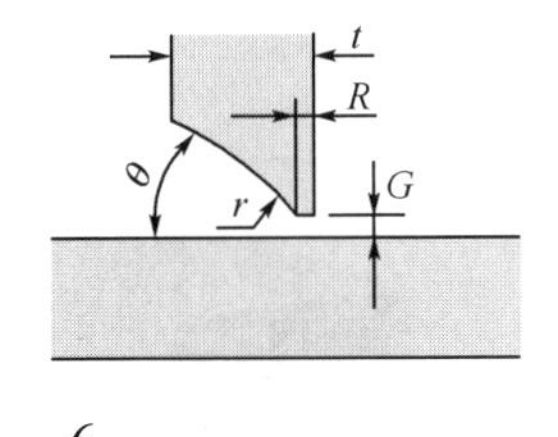

4. ________________

5. ________________

6. ________________

Words and phrases

specification [ˌspesɪfɪ'keɪʃn] 规范，规格

groove [gru:v] 坡口

opening ['əʊpnɪŋ] 开口

company ['kʌmpənɪ] 公司

system ['sɪstəm] 系统

thickness range ['θɪknəs] [reɪndʒ] 厚度范围

filler ['fɪlə(r)] 填充物

gas [gæs] 气体

penetrate bead ['penətreɪt] 底部熔透

multiple ['mʌltɪpl] 多个的

temperature ['temprətʃə(r)] 温度

inter pass [ɪntɜ:'pɑ:s] 道间

min/mm 分/毫米

holding ['həʊldɪŋ] 维持，保持

speed [spi:d] 速度

square [skweə(r)] 方形，直角

procedure [prə'si:dʒə(r)] 程序,工艺

angle ['æŋgl] 角度，角

face [feɪs] 面，脸

support [sə'pɔ:t] 支持

process [prə'ses] 过程，工艺

diameter [daɪ'æmɪtə(r)] 直径

shield [ʃi:ld] 屏蔽，保护

treatment ['tri:tmənt] 处理，对待

single ['sɪŋgl] 单个的

weaving ['wi:vɪŋ] 摆动，编织

preheat [ˌpri:'hi:t] 预热

Max [mæks] 最大

PWHT 焊后热处理

arc voltage ['vəʊltɪdʒ] 电弧电压

mm/min 毫米/分

bevel ['bevl] 斜边/斜面

PQR (procedure qualification record) [ˌkwɒlɪfɪ'keɪʃn] ['rekɔ:d] 工艺评定报告

Lesson 35 Welding Machines

Look and select

Look at the pictures and select the correct terms from the box.

TIG welder	SMAW welder	SAW welder	CO_2 welder

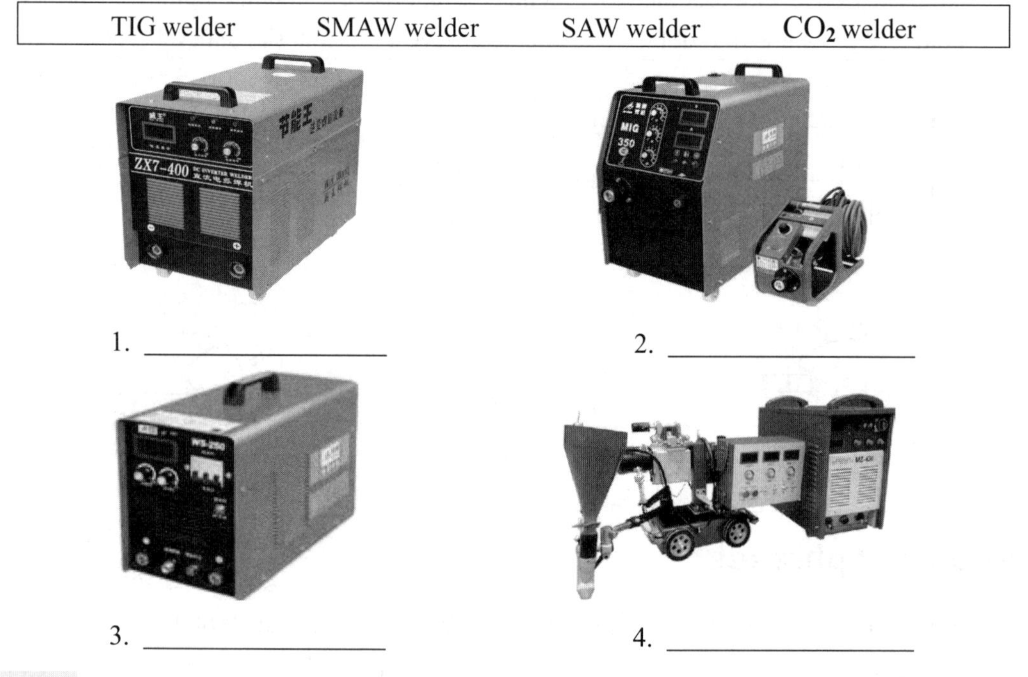

1. ____________ 2. ____________

3. ____________ 4. ____________

Text

1. Look at the picture (Fig.35.1) and translate the terms into Chinese.

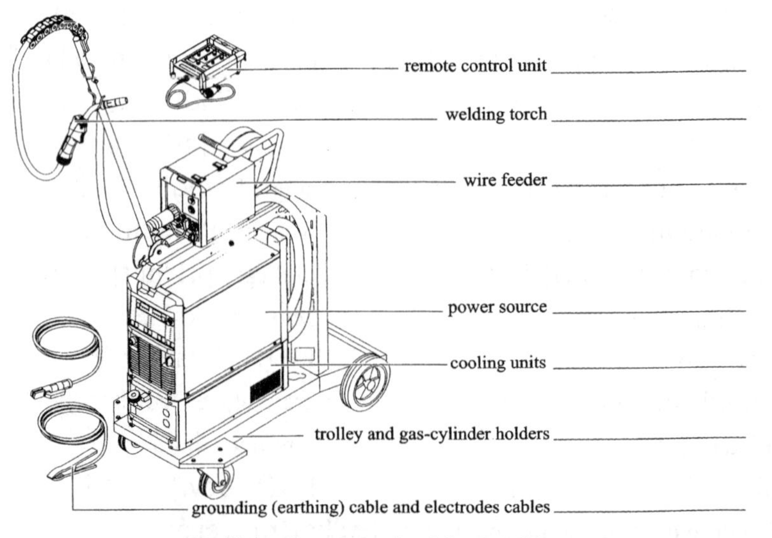

Fig.35.1 Structure of welding machine

2. Look at the pictures (Fig.35.2) and select the correct terms from the box.

Keep the wire feeder away from the spatter	Check dust or rust on the feed roll
Confirm tight connection at the output terminals	Don't drag the wire feeder

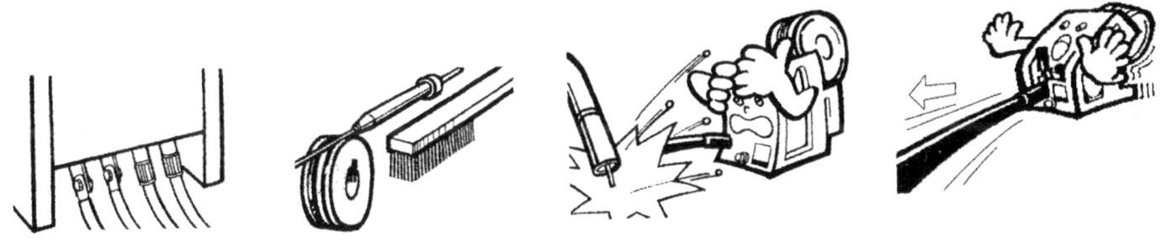

Fig.35.2 Instructions during welding

1. ________; 2. ________; 3. ________; 4. ________.

Knowledge extention

Understand the technical parameters of welding machine in Table 35.1.

Table 35.1 Technical parameters of welding machine

English	Chinese
No-load voltage	空载电压
Conventional load voltage	约定负载电压
Rated welding current	额定焊接电流
Current regulated range	电流调节范围
Rated duty cycle	额定负载持续率
Diameter of electrode	适用焊丝直径
Steps of voltage adjustment	电压调节级数
Input voltage	输入电压
frequency	频率
Rated input energy	额定输入容量
Class of insulation	绝缘等级
Cooling mode	冷却方式
Exterior dimension	外形尺寸

Exercises

1. Match the English with the Chinese. Draw lines.

(1) remote control	a. 焊枪
(2) welding torch	b. 电源
(3) wire feeder	c. 冷却
(4) power source	d. 遥控
(5) cooling	e. 接地
(6) trolley	f. 送丝机
(7) grouding	g. 焊接小车

2. Look at the pictures and select the correct terms from the box.

diameter/index button	purge button	inch forward button
parameter selection button	adjusting dial	store button

INCH FORWARD

1. ______________

PURGE

2. ______________

STORE

3. ______________

DIAMETER INDEX DEL

4. ______________

5. ______________

6. ______________

3. Look at the picture (Fig.35.3) and translate the terms into Chinese.

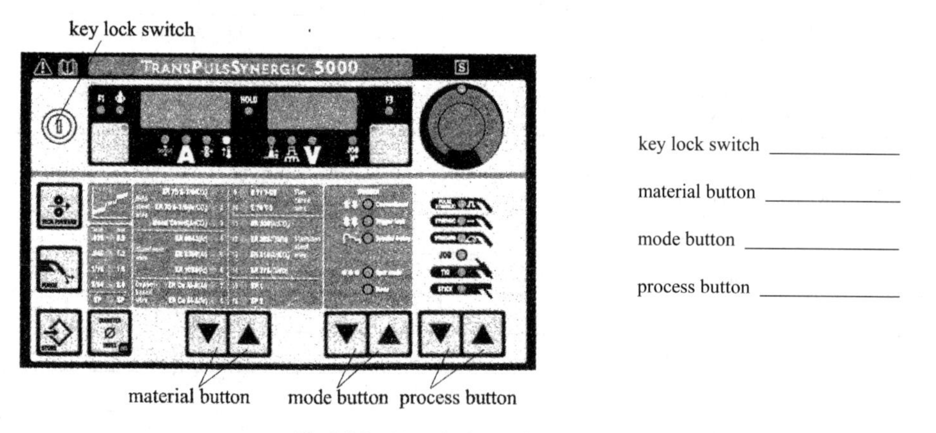

Fig.35.3 Panel of welding machine

Words and phrases

machine [mə'ʃi:n] 机器

wire feeder ['waɪə(r)] ['fi:də(r)] 送丝机构

cooling unit ['ku:lɪŋ] ['ju:nɪt] 冷却装置

gas-cylinder ['sɪlɪndə(r)] 气瓶

keep ... away from 使……远离

dust [dʌst] 尘

feed roll [fi:d] [rəʊl] 送丝辊

check [tʃek] 检查

power source ['paʊə(r)] [sɔ:s] 电源

trolley ['trɒlɪ] 移动小车

holder ['həʊldə(r)] 架子

spatter ['spætə(r)] 飞溅

rust [rʌst] 锈

confirm [kən'fɜ:m] 确定，明确

tight connection [taɪt] [kə'nekʃn] 紧密连接
at the output terminal ['tɜ:mɪnlz] 输出端
process button [prə'ses] 焊接方式选择键
mode button [məʊd] 焊枪操作模式键
material button [mə'tɪəriəl] 焊接材料选择键
diameter button [daɪ'æmɪtə(r)] 直径选择键
adjusting dial [ə'dʒʌstɪŋ] ['daɪəl] 调谐钮
remote control unit [rɪ'məʊt] [kən'trəʊl] ['ju:nɪt] 遥控器
grounding / earthing ['graʊndɪŋ] ['ɜ:θɪŋ] 地线，接地
store button [stɔ:(r)] ['bʌtn] 存储键
gas test 气体试验/检测
exit ['eksɪt] 出口
key lock switch [swɪtʃ] 开关锁
purge button [pɜ:dʒ] 气体检测键
inch forward button [ɪntʃ] 点动送丝键

Lesson 36 Welding Robots

Look and learn

Learn the parts of robot (Fig.36.1) according to the names of axes:

Name of axes	RT 轴 Rotating Trunk	UA 轴 Upper Arm	FA 轴 Forearm	RW 轴 Rotating Wrist	BW 轴 Bending Wrist	TW 轴 Twisting Wrist
Meaning	躯体转体	上臂	前臂	手腕转动	手腕弯曲	手腕扭转

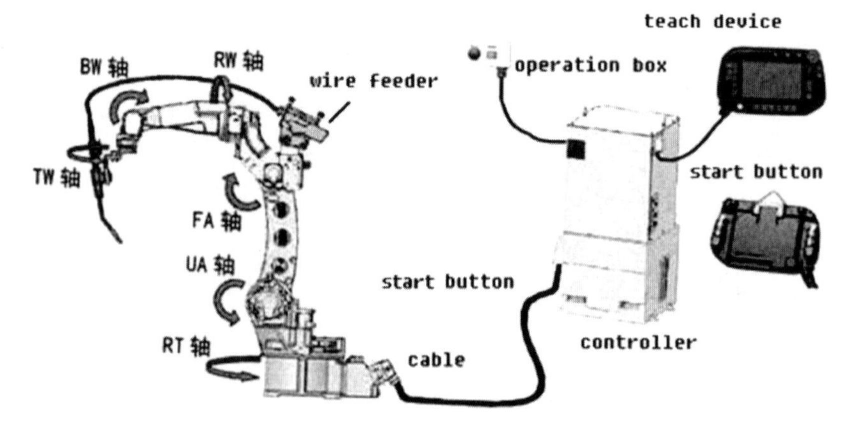

Fig.36.1 Structure of robot body

Text

Welding robots

Welding robot is an industrial robot who's engaged in welding works including cutting, assembly and spraying. The most common applications of welding robots are spot welding and arc welding.

Welding robot mainly consists of two parts: the robot and welding equipment. The robot is made up a robot body and a control cabinet (hardware and software). The welding equipment (spot welding or arc welding) includes the power source (including its control system), wire feeding machine and welding gun (torch) etc.

Current welding robot basically belongs to the joint robot. It has 6 axes: RT (rotating trunk), UA (upper arm), FA (forearm), RW(rotating wrist), BW(bending wrist), TW(twisting wrist).

Welding robots are widely used in various industries because of the following advantages:

(1) reduce the skills required for the technical workers;

(2) ease labor intensity, promote welding quality and efficiency;

(3) be able to work in the hazardous working places with poor conditions, heavy smog and heat radiation etc.

(4) do the repetitive and heavy work.

Knowledge extention

1. Spot welding robot

Spot welding robot, shown in Fig.36.2, consists of robot body, computer control system, teach-box and spot welding system. It has 6 freedom degrees: waist turn, big arm turn, forearm, wrist turn, wrist weaving and wrist twist. The driving modes are hydraulic drive and electrical drive. Spot welding system includes spot welding machine and welding torch. Operator can demonstrate the robot moving positions and action process through the teach-box and computer panel keys, and set the moving speed and welding parameters.

2. Arc welding robot

Arc welding robot (Fig.36.3) consists of robot body, control panel, teach-box, wire feeder, welding torch and welding power, illustrated in Fig.36.4. It can realize the continuous path control and spot position control under the control of computer and also perform the space welding by using the functions of linear interpolation and circular interpolation. Its operation types include consumable electrode welding and non-consumable electrode welding.

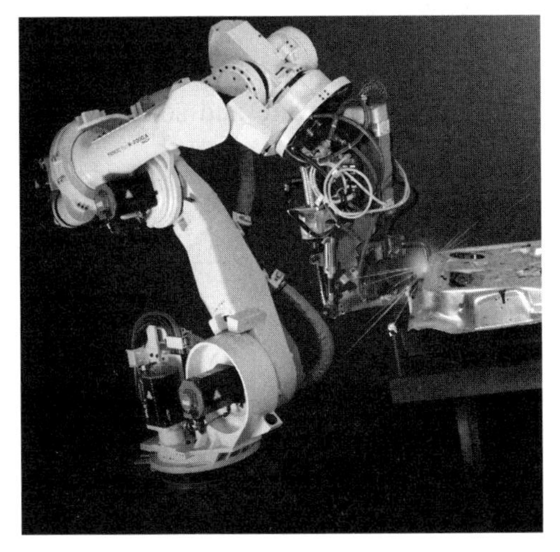

Fig.36.2 Spot welding robot

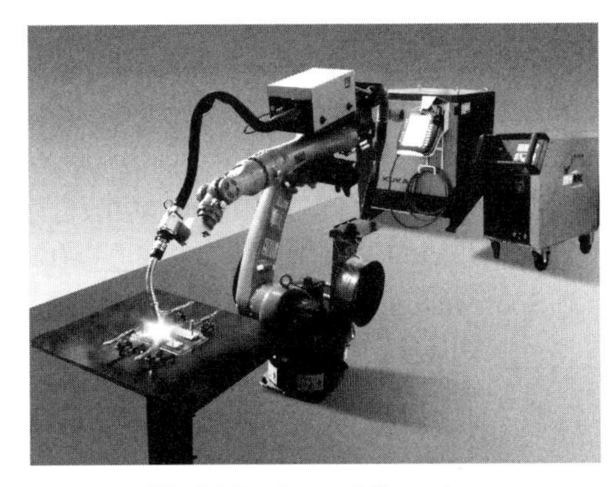

Fig.36.3 Arc welding robot

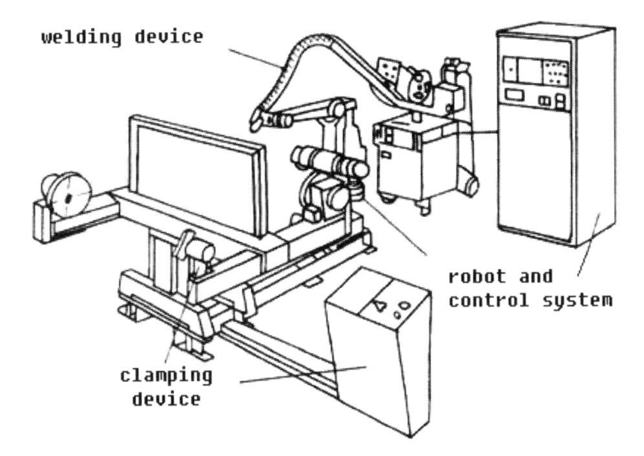

Fig.36.4 Structure of arc welding robot

Exercises

1. Match the English with the Chinese. Draw lines.

A.

(1) robot body　　a. 操作盒
(2) operation box　　b. 示教器
(3) teaching device　　c. 机器人本体
(4) start button　　d. 控制器
(5) controller　　e. 启动开关

B.

(1) Rotatig Trunk　　a. 上臂
(2) Upper Arm　　b. 手腕转动
(3) Forearm　　c. 躯体转体
(4) Rotating Wrist　　d. 手腕扭转
(5) Bending Wrist　　e. 前臂
(6) Twisting Wrist　　f. 手腕弯曲

2. Fill in the blanks with the proper words in the text.

(1) A robot is made up a robot body and a __________ (hardware and software) and the welding equipment contain the ____________ (including its control system), ____________ and __________ (torch) etc.

(2) Current welding robot basically belongs to the joint robot. It has __________ axes: RT (rotating trunk), __________ (upper arm), __________ (forearm), __________ (rotating wrist), __________ (bending wrist), __________ (twisting wrist).

(3) Spot welding robot consists of ____________, computer ____________, teach-box and __________ system.

(4) Arc welding robot can realize the ______________ and spot position control under the ______________.

3. Judge the following true or false.

(1) The most common applications of welding robots are butt welding robot and arc welding robot. (　　)

(2) Welding robots mainly consist of two parts: the robots and wire feeder. (　　)

(3) Current welding robots basically belong to the joint robots. It has 5 axes. (　　)

(4) The driving modes of spot welding robots are hydraulic drive and electrical drive. (　　)

(5) The arc welding robots can perform the space welding by using the functions of linear interpolation and circular interpolation. (　　)

4. Answer the following questions.

(1) What are the advantages of robots?

(2) What can the robots do in welding industry?

Words and phrases

welding robot ['rəʊbɒt] 焊接机器人
forearm ['fɔ:rɑ:m] 前臂
bend [bend] 弯曲
wire feeder ['waɪə(r)] ['fi:də(r)] 送丝机
teach device [dɪ'vaɪs] 示教仪
start button [swɪtʃ] 启动开关
robot body 机器人本体
assembly [ə'sembli] 组装
application [ˌæplɪ'keɪʃn] 应用，适用
arc welding 弧焊
equipment [ɪ'kwɪpmənt] 设备
hardware 硬件
intelligent [ɪn'telɪdʒənt] 智能的
current ['kʌrənt] 当前的
advantage [əd'vɑ:ntɪdʒ] 优势
skill [skɪl] 技能
labor intensity ['leɪbə(r)] [ɪn'tensəti] 劳动强度
quality ['kwɒləti] 质量
poor condition [pɔ:(r)] [kən'dɪʃn] 恶劣条件
heat radiation [hi:t] [ˌreɪdi'eɪʃn] 热辐射
repetitive [rɪ'petətɪv] 重复的
trunk [trʌŋk] 躯干
wrist [rɪst] 手腕
twist [twɪst] 扭曲
operation box [ˌɒpə'reɪʃn] 操作盒
controller [kən'trəʊlə(r)] 控制器
structure ['strʌktʃə(r)] 结构
cut [kʌt] 切割
spray [spreɪ] 喷射
spot welding [spɒt] 点焊
consist of [kən'sɪst] 由……组成
cabinet ['kæbɪnət] 柜
software 软件
sensing system 感应系统
joint [dʒɔɪnt] 关节，结合处
reduce [rɪ'dju:s] 降低
ease [i:z] 减轻
promote [prə'məʊt] 提高
efficiency [ɪ'fɪʃnsi] 效率
hazardous ['hæzədəs] 有害的
smog [smɒg] 烟雾
heavy work 繁重工作

Lesson 37 Cutting

Look and select

Look at the pictures and select the correct terms from the box.

T-steel	channel steel	angle bar	round steel

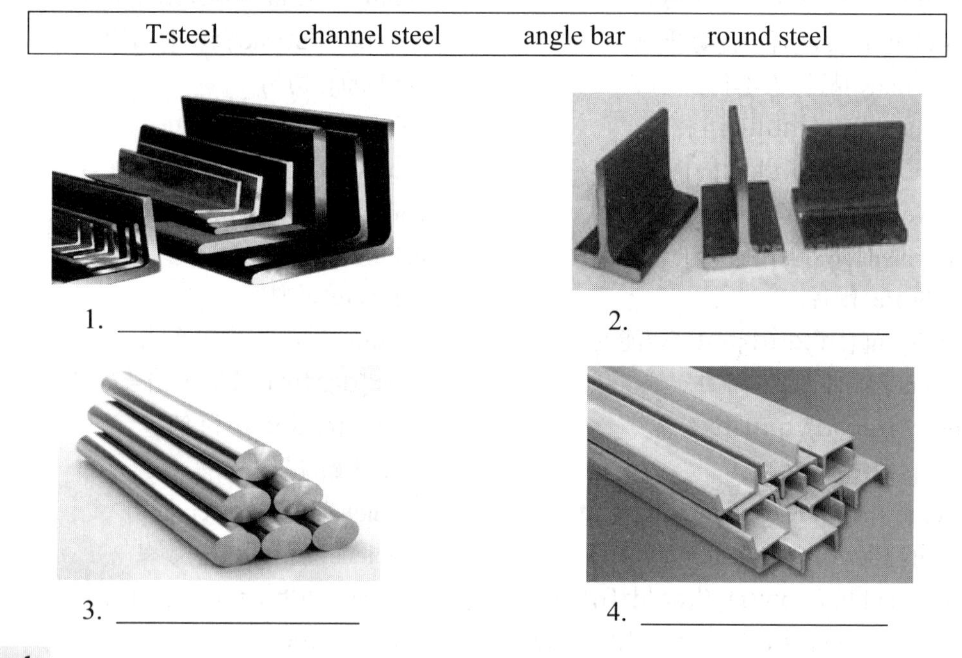

1. ________ 2. ________

3. ________ 4. ________

Text

Flame cutting

Flame cutting (Fig.37.1) is done by preheating a spot on ferrous metal to its ignition temperature and then burning it with a stream of oxygen. Machine flame cutting provides greater speed, accuracy, and economy.

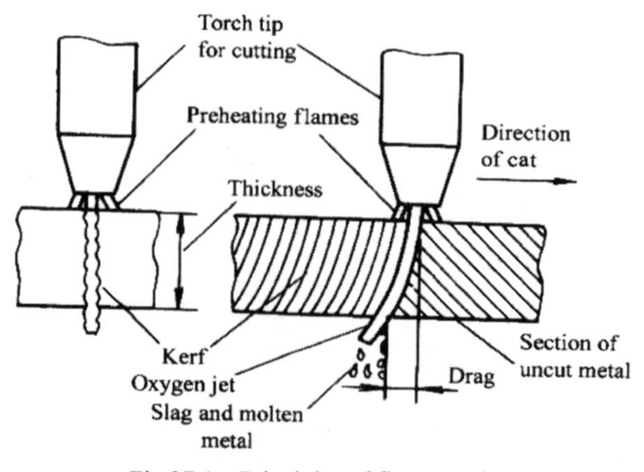

Fig.37.1 Principles of flame cutting

The cylinders of oxygen and acetylene are as shown in Fig.37.2 and Fig.37.3.

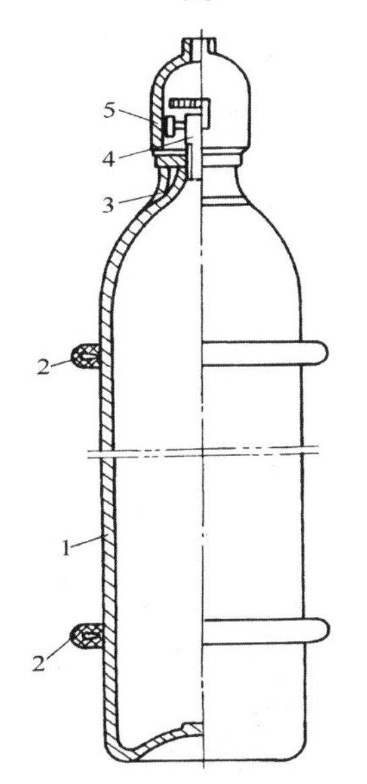

Fig.37.2 Structure of oxygen cylinder

1—body; 2—shockproof rubber ring;

3—hoop; 4—valve; 5—cap

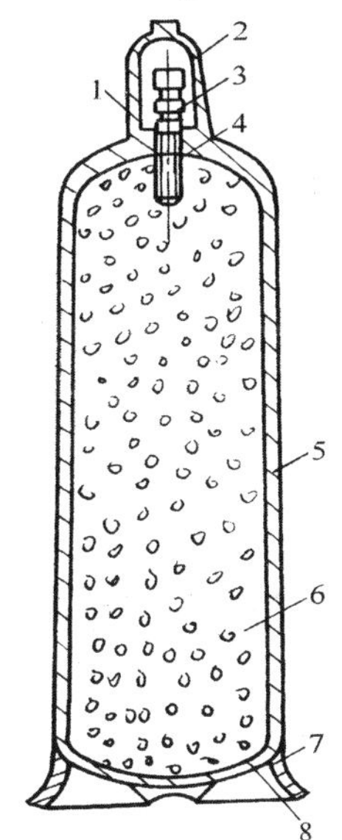

Fig.37.3 Structure of acetylene cylinder

1—neck; 2—cap; 3—valve; 4—asbestos;

5—body; 6—porous; 7—base; 8—bottom

Knowledge extention

1. Plasma arc cutting

The arc formed between the electrode and the workpiece is constricted by a fine bore, copper nozzle. This increases the temperature and velocity of the plasma emanating from the nozzle. The temperature of the plasma is in excess of 20,000℃ and the velocity can approach the speed of sound. When used for cutting, the plasma gas flow is increased so that the deeply penetrating plasma jet cuts through the material and molten material is removed. And the plasma arc is formed, shown in Fig.37.4.

2. Carbon arc air gouging

Carbon arc air gouging (Fig.37.5) can be used for cleaning the root of weld, repairing the weld defects, gouging the weld grooves, cleaning the casting trimmings.

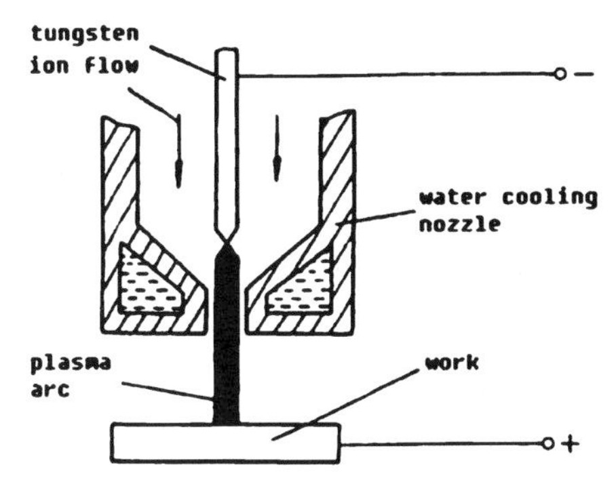

Fig.37.4 Principle of plasma arc

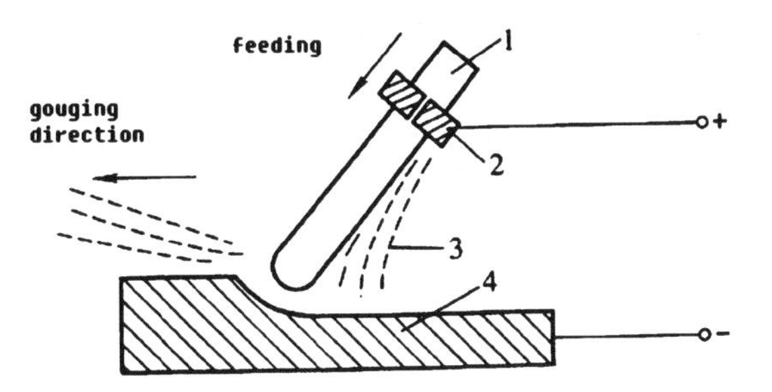

Fig.37.5 Carbon arc air gouging

1—carbon electrode; 2—holder; 3—compressive flow; 4—work slag

3. Structure and shapes of oxyacetylene flame (Fig.37.6)

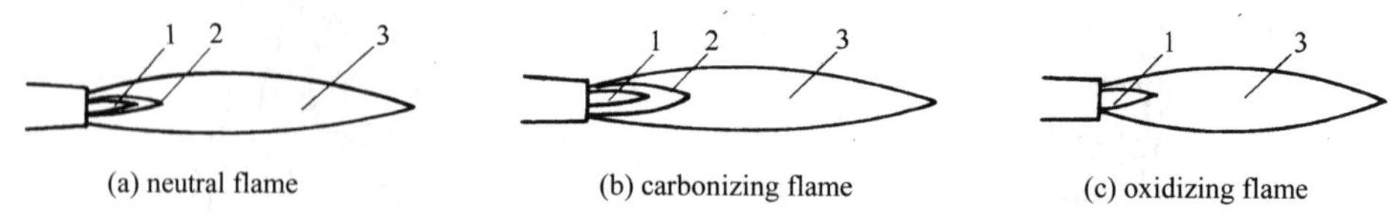

Fig.37.6 Structure and shapes of oxyacetylene flame

1—flame core; 2—inner flame; 3—outer flame

Exercises

1. Match the English with the Chinese. Draw lines.

(1) flame cutting	a. 预热火焰
(2) torch tip	b. 割炬喷嘴
(3) preheating flame	c. 氧气流
(4) kerf	d. 火焰切割
(5) oxygen jet	e. 割口
(6) slag and molten metal	f. 后拖量
(7) drag	g. 未切割的金属
(8) section of uncut metal	h. 焊渣及熔化金属

2. Fill in the blanks with the proper words in the text.

(1) The structure of oxygen cylinder includes body, ________________, hoop, valve and ____________.

(2) The structure of acetylene cylinder includes neck, ________, valve, ________, body, porous, __________ and __________.

(3) The structure of oxyacetylene flame contarns _________, inner flame and _________.

3. Answer the following questions.

(1) What is flame cutting?

(2) Describe the principle of plasma arc cutting.

(3) What are the roles of Carbon arc air gouging?

(4) Brief the structure and shapes of oxyacetylene flame.

Words and phrases

cutting ['kʌtɪŋ] 切割

half round steel [hɑ:f] [raʊnd] 半圆钢

angle bar ['æŋgl] [bɑ:(r)] 角钢

flame cutting [fleɪm] 火焰切割

ferrous metal ['ferəs] ['metl] 铁类（黑色）金属

machine flame cutting 机械火焰切割

greater speed [greɪtə] [spi:d] 速度更快

T-steel [sti:l] T 形钢

channel steel ['tʃænl] 槽钢

double ['dʌbl] 双的，两个

preheat [ˌpri:'hi:t] 预热

spot 点

burn [bɜ:n] 燃烧

provide [prə'vaɪd] 提供，供应

accuracy ['ækjərəsi] 精确
torch tip for cutting [tɔ:tʃ] [tɪp] 割炬喷嘴
preheating flame ['pri:hi:tɪŋ] 预热火焰
direction of cut 切割方向
section of uncut metal ['sekʃn] 未切割的金属
plasma arc cutting ['plæzmə] 等离子切割
carbon arc air gouging ['kɑ:bən] [ɑ:k] [eə(r)] ['gaʊdʒɪŋ] 碳弧气刨
ignition temperature [ɪg'nɪʃn] ['temprətʃə(r)] 点燃/燃烧温度
with a stream of oxygen [stri:m] ['ɒksɪdʒən] 用氧气流
slag and molten metal [slæg] ['məʊltən] 渣和熔化金属
economy [ɪ'kɒnəmɪ] 经济
thickness ['θɪknəs] 板厚
kerf [kɜ:f] 割口
oxygen jet [dʒet] 氧气流
drag [dræg] 后拖量

Part Ⅱ
Welding Methods

Lesson 38 Shielded Metal Arc Welding

Look and translate

Look at the picture (Fig. 38.1) and translate the words into Chinese:

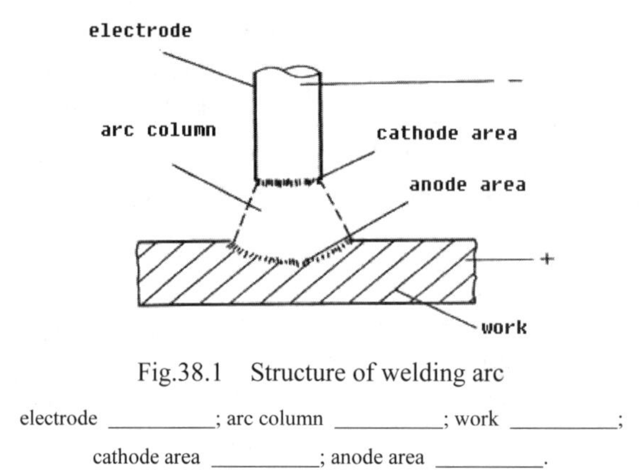

Fig.38.1 Structure of welding arc

electrode ________; arc column ________; work ________;

cathode area ________; anode area ________.

Text

1. The definition of shielded metal-arc welding

Shielded metal-arc welding (Fig.38.2) is an arc welding process wherein coalescence is produced by heating with an electric arc between a covered or "coated" metal rod called the electrode and the work. Shielding is obtained from decomposition of the electrode covering, and

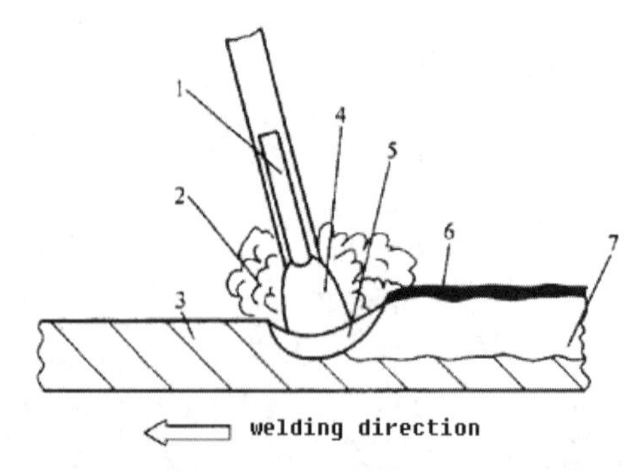

Fig.38.2 Process of shielded metal-arc welding

1—electrode; 2—shielded gas; 3—base metal; 4—arc; 5—molten pool; 6—slag; 7—weld filler metal is obtained from the electrode's metal core and metallic particles in the covering. The source of the shielding and filler metal varies with electrode design.

2. Elements of a typical welding circuit for shielded metal-arc welding

This circuit (Fig.38.3) includes a source of power, welding cables, an electrode holder, a ground clamp and the consumable welding electrode.

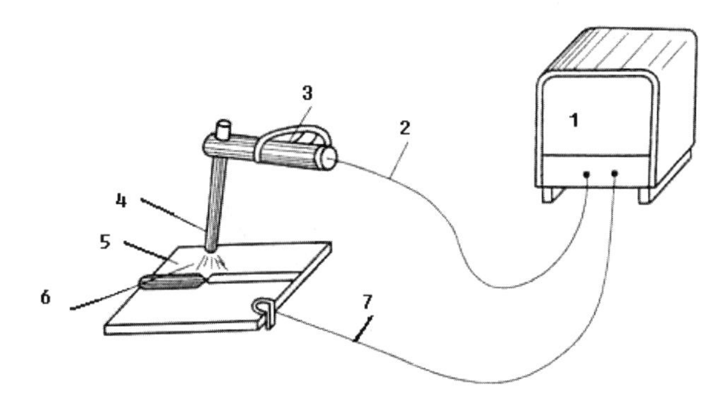

Fig.38.3 Circuit for shielded metal-arc welding

1—power supply; 2—electrode cable; 3—welding holder; 4—electrode; 5—work; 6—arc; 7—work cable

Knowledge extention

1. Electrode composition (Fig.38.4)

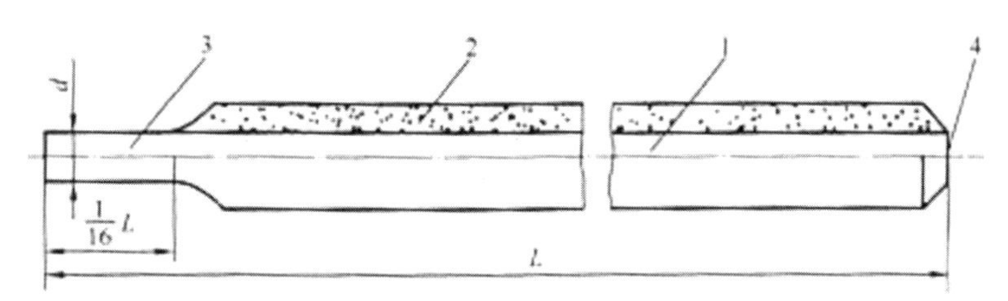

Fig.38.4 Diagram of electrode composition

1—wire core; 2—covering; 3—holding end; 4—arc ignition end

2. Methods of arc ignition (Fig.38.5)

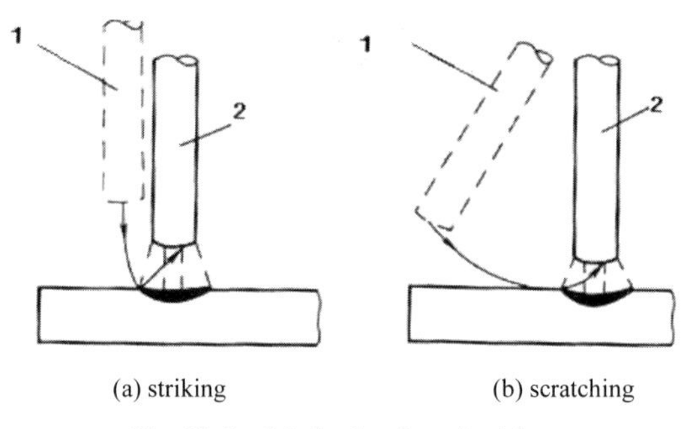

(a) striking (b) scratching

Fig.38.5 Methods of arc ignition

1—Before ignition; 2—After ignition

3. Arc ending (Fig.38.6)

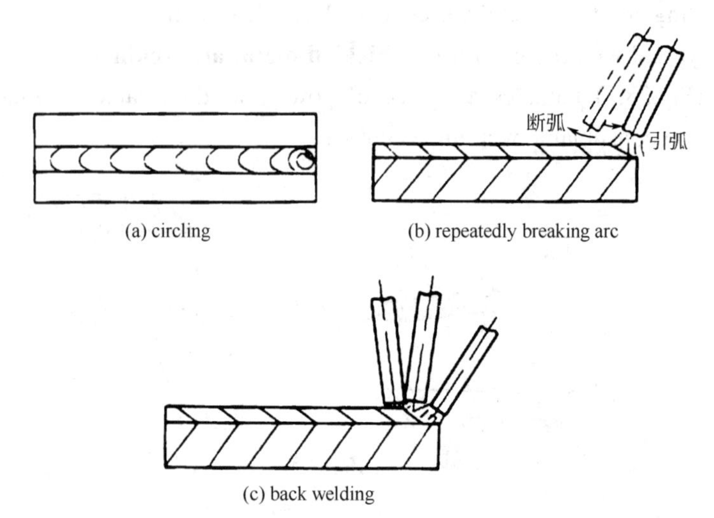

(a) circling (b) repeatedly breaking arc

(c) back welding

Fig.38.6 Methods of arc ending

4. Three basic moving directions of electrode (Fig.38.7)

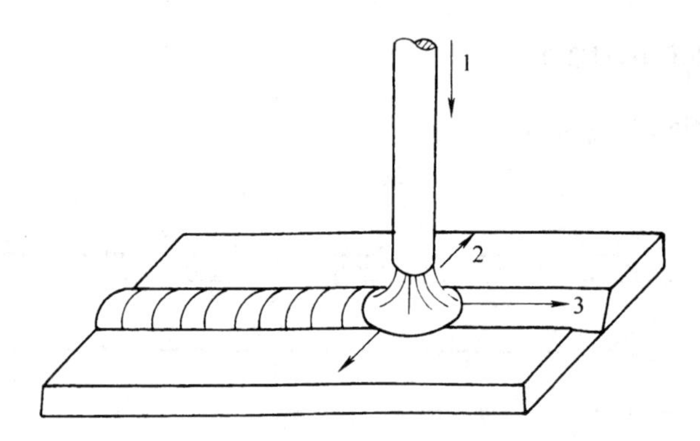

Fig.38.7 Three basic moving directions of electrode

1—electrode feeding; 2—electrode swinging; 3—moving along weld

Exercises

1. Match the English with the Chinese. Draw lines.

(1) welding machine	a. 焊枪/把
(2) electrode torch	b. 焊机
(3) electrode	c. 电弧
(4) arc	d. 接地电缆
(5) ground cable	e. 焊条
(6) electrode cable	f. 工件
(7) work	g. 焊条电缆

2. Fill in the blanks about electrodes' manipulation (Fig.38.8) with the words from the box.

circular	sawtooth	triangle
crescent	oblique sawtooth	figure eight shape

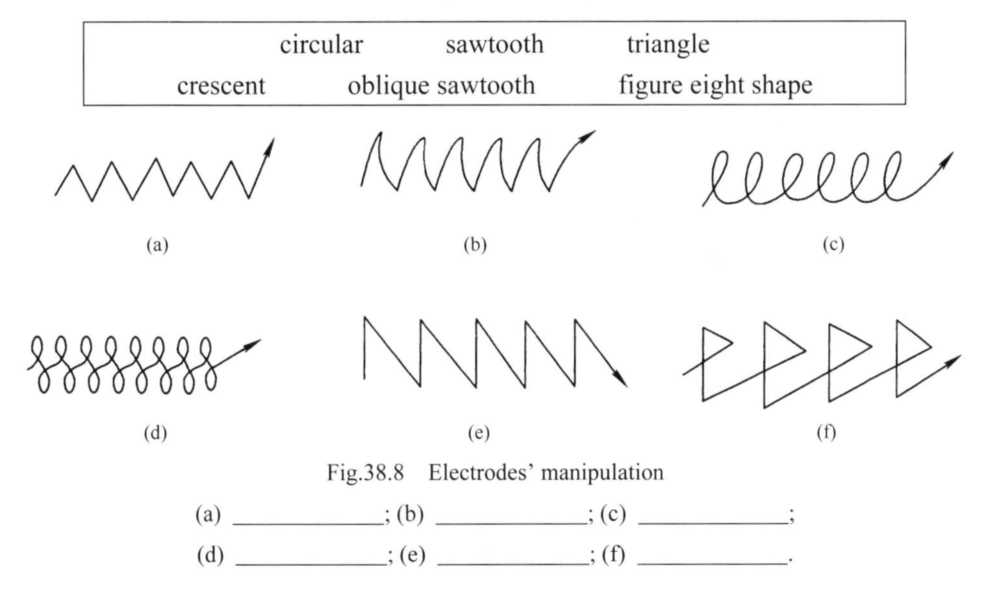

Fig.38.8 Electrodes' manipulation

(a) ____________; (b) ____________; (c) ____________;
(d) ____________; (e) ____________; (f) ____________.

3. Fill in the blanks with the proper words in the text.

(1) Process of shielded metal-arc welding in Fig.38.2 involves ________________, shielded gas, __________, arc, __________ pool, __________ and weld.

(2) The circuit for shielded metal-arc welding in Fig.38.3 includes ___________, electrode cable, ________, electrode, work, arc and ________.

(3) An electrode consists of four parts: core wire, ________, holding end and ________.

4. Answer the following questions.

(1) What is the Shielded metal-arc welding?

(2) Describe the methods of arc ignition and arc ending.

(3) What are the three basic moving directions of electrode?

Words and phrases

shielded ['ʃi:ldɪd] 屏蔽的，保护的
metal ['metl] 金属
anode ['ænəʊd] 阳极
cathode ['kæθəʊd] 阴极
coalescence [ˌkəʊə'lesns] 连接，结合
process [prə'ses] 工艺，方法
be produced by [prə'dju:st] 通过……产生
heat [hi:t] 加热，热量
metal rod ['metl] [rɒd] 金属棒
welding machine 焊机
ground cable [graʊnd] ['keɪbl] 接地电缆
work [wɜ:k] 工件
electrode cable [ɪ'lektrəʊd] 焊条电缆
electric arc [ɪ'lektrɪk] [ɑ:k] 电弧
wire core ['waɪə(r)] [kɔ:(r)] 焊芯
covering ['kʌvərɪŋ] 覆盖物，药皮
holding end ['həʊldɪŋ] [end] 夹持端
arc ignition end [ɪg'nɪʃn] 引弧端
strike [straɪk] 直击
scratch [skrætʃ] 刮擦
arc ending 收弧
breaking arc 断弧

feed [fiːd] 进给
move 移动
circular ['sɜːkjələ(r)] 环形
triangle ['traɪæŋgl] 三角形
manipulation [məˌnɪpjʊ'leɪʃn] 运作，操作
covered / coated ['kʌvəd] ['kəʊtɪd] 覆盖的，有涂层的
electrode torch [ɪ'lektrəʊd] [tɔːtʃ] 焊钳，焊枪
swing [swɪŋ] 摆动
figure eight shape 8 字形
sawtooth ['sɔːtuːθ] 锯齿形
crescent ['kresnt] 月牙形
oblique [ə'bliːk] 斜的

Lesson 39 Submerged Arc Welding

Look and select

Look at the following pictures and select the correct terms from the box.

pressure vessel	bridge	bird nest	ship

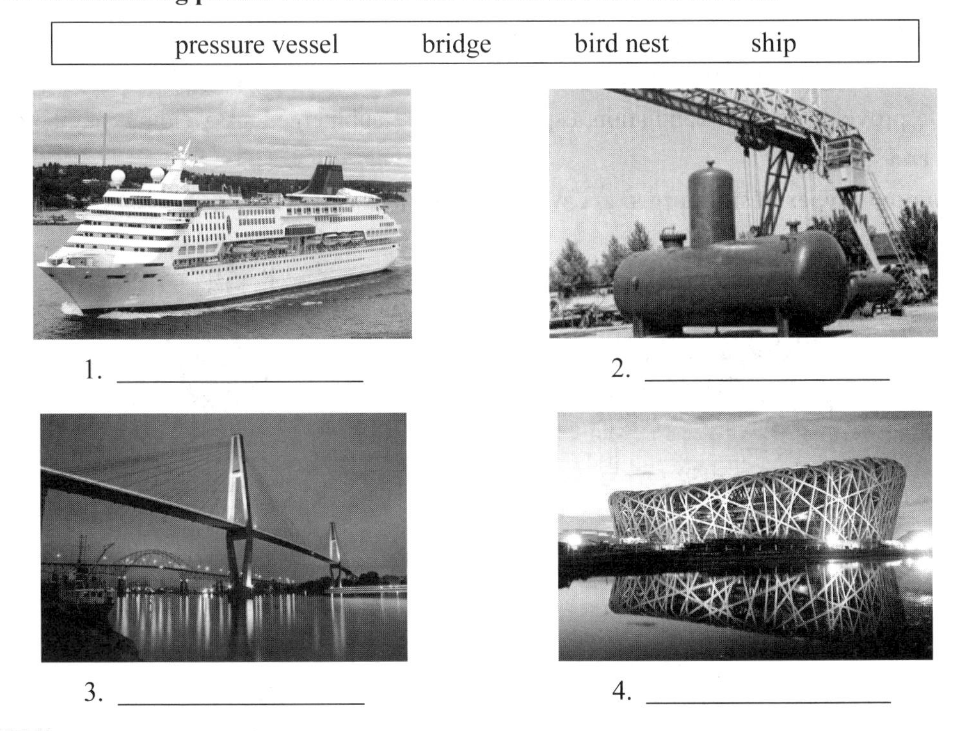

1. ________________ 2. ________________

3. ________________ 4. ________________

Text

The definition of submerged arc welding

Submerged arc welding (Fig.39.1) is defined by the American Welding Society as an arc welding process wherein coalescence is produced by heating with an arc or arcs between a bare metal electrode, or electrodes, and the work.

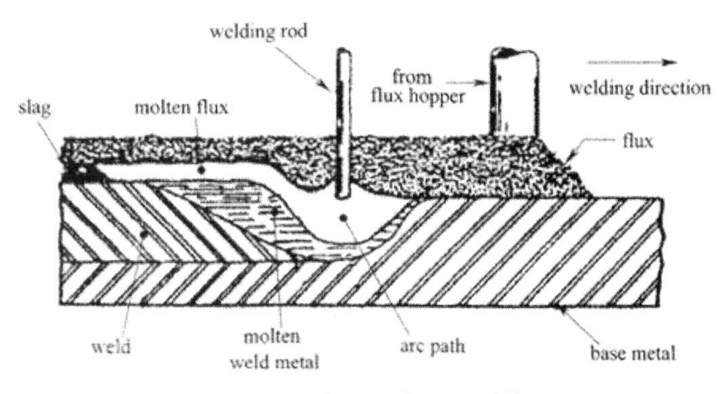

Fig.39.1 Submerged Arc Welding

Knowledge extention

1. Why don't the sparks or flash appear during the submerged arc welding operation?

Since the end of the electrode and the welding zone are completely covered at all times during the actual welding operation, the weld is made without the sparks, spatter, smoke or flash commonly observed in other arc welding processes. No protective shields or helmets are necessary.

2. During the submerged arc welding operation, what will influence the welder's health?

Since welding in general may produce fumes and gases hazardous to health, it is common practice to provide adequate ventilation, especially where submerged arc welding may be done in confined areas.

3. The common types of submerged arc welding machines (Fig.39.2).

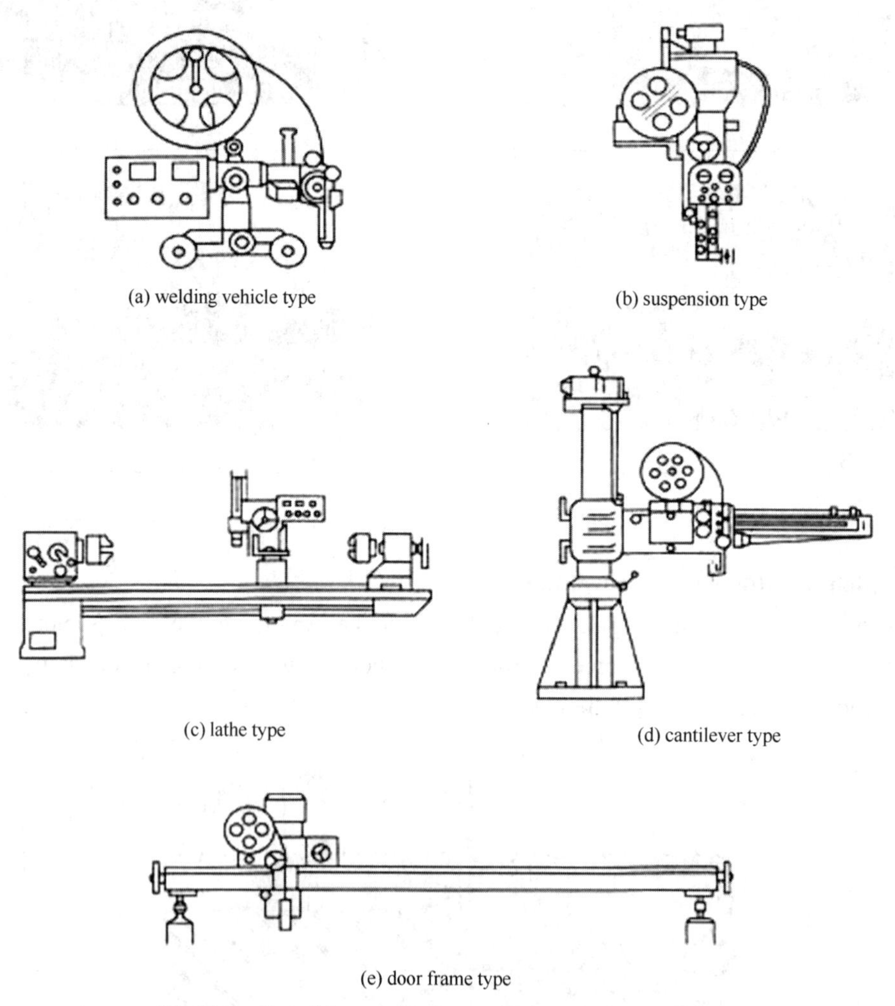

(a) welding vehicle type (b) suspension type

(c) lathe type (d) cantilever type

(e) door frame type

Fig.39.2 Travelling mechanisms of submerged arc welding

Exercises

1. Translate the words in the picture (Fig.39.3) into Chinese.

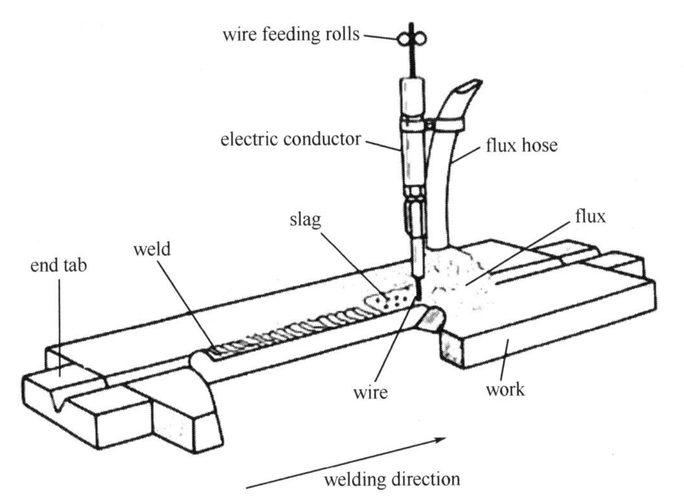

Fig.39.3 The process of submerged arc welding

end tab ________________; weld ________________________;

slag ____________________; electric conductor ________________;

flux hose ________________; wire feeding rolls ________________;

work ____________________; welding direction ________________.

2. Fill in the blanks with the proper words in the text.

(1) Submerged arc welding is defined by the ________ as an arc welding process wherein __________ is produced by heating with an __________ or arcs between a __________, or electrodes, and ________.

(2) The travelling mechanisms of submerged arc welding consist of five types: welding vehicle type, ____________ type, ___________ type, cantilever type and ___________ type.

3. Translate the sentences into Chinese.

(1) Since the end of the electrode and the welding zone are completely covered at all times during the actual welding operation, the weld is made without the sparks, spatter, smoke or flash commonly observed in other arc welding processes.

(2) Since welding in general may produce fumes and gases hazardous to health, it is common practice to provide adequate ventilation, especially where submerged arc welding may be done in confined areas.

Words and phrases

submerged [səb'mɜːdʒd] 埋伏的，潜伏的

pressure vessel ['preʃə(r)] ['vesl] 压力容器

be defined by ... as [dɪ'faɪnd] 被……定义为

bird nest [bɜːd] 鸟巢

bridge [brɪdʒ] 桥

ship [ʃɪp] 船

bare metal electrode [beə(r)] ['metl] 裸焊条，焊丝
molten flux ['məʊltən] [flʌks] 熔渣
flux hopper ['hɒpə(r)] 焊剂漏斗
welding direction [də'rekʃn] 焊接方向
arc path [pɑ:θ] 电弧空间
spark [spɑ:k] 火花，火星
operation [ˌɒpə'reɪʃn] 操作
fume [fju:m] 烟雾
hazardous ['hæzədəs] 有害的
confined area [kən'faɪnd] ['ɛriə] 有限的区域
American Welding Society [ə'merɪkən] [sə'saɪətɪ] 美国焊接协会
arc welding process 弧焊工艺
welding rod [rɒd] 焊丝
weld 焊缝
molten weld metal 熔化金属
base metal [beɪs] 母材
flash [flæʃ] 闪光
spatter ['spætə(r)] 飞溅
gas 气体
ventilation [ˌventɪ'leɪʃn] 通风
slag [slæg] 焊渣

Lesson 40 CO_2 Welding

Look and select

Look at the pictures and select the correct expressions from the box.

certificate of PETS	certificate of welding
certificate of computer	certificate of inspection

1. ____________________

2. ____________________

3. ____________________

4. ____________________

Text

The outline of CO_2 welding

CO_2 Shielded consumable electrode arc welding is a new welding technique which came into being in 1950's. As compared with other welding processes, CO_2 welding is high in efficiency, low in cost and ready for automatization. It can ease the intensity of labour and improve the quality of welding. CO_2 welding is mainly applied to low-carbon steels, low-alloy steels, stainless steels, high-temperature steels, etc. With all its advantages CO_2 welding will in many cases displace other types of welding processes. It promises a great prospect in welding 0.5~2.5 mm sheets instead of gas welding and argon shielded arc welding.

Knowledge extention

1. What are the problems with CO_2 welding that must be solved?

Three basic problems had to be solved before CO_2 could be used for gas shielded metal arc welding of steel and "CO_2 welding" became an everyday welding process: porosity in the welded metal; the problem of spatter; positional welding.

2. CO_2 welding equipment (Fig.40.1).

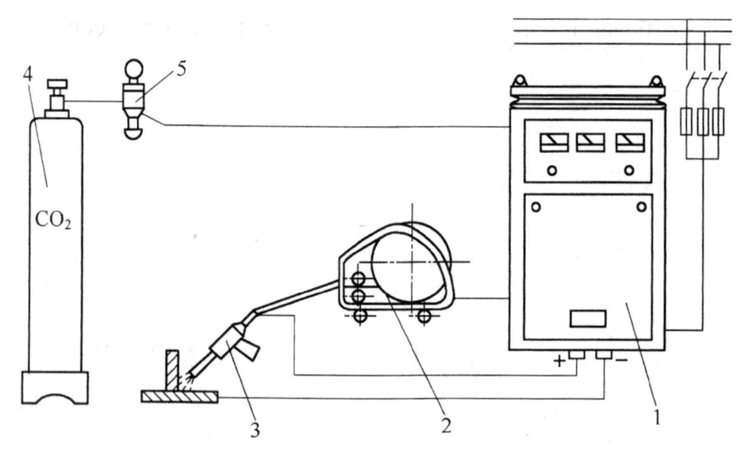

Fig.40.1 Diagram of semi-auto CO_2 welding equipment

1—power supply; 2—wire feeder; 3—welding gun; 4—CO_2 cylinder; 5—flow regulator

Exercises

1. Fill in the blanks (Table 40.1) with the words in the text.

Table 40.1 Outline of CO_2 welding

特性	As a shielded ______
产生时间	Came into being in ______
优点	High in ______, low in ______, ready for automatization, it ease the ______
应用	Is mainly applied to ______, low-alloy steel, stainless steel, ______ steel
前景	In welding ______ mm sheets instead of ______ and ______ welding
问题	Porosity in the welded metal, the problem of ______, positional ______.

2. Translate the words in the picture (Fig.40.2) into English.

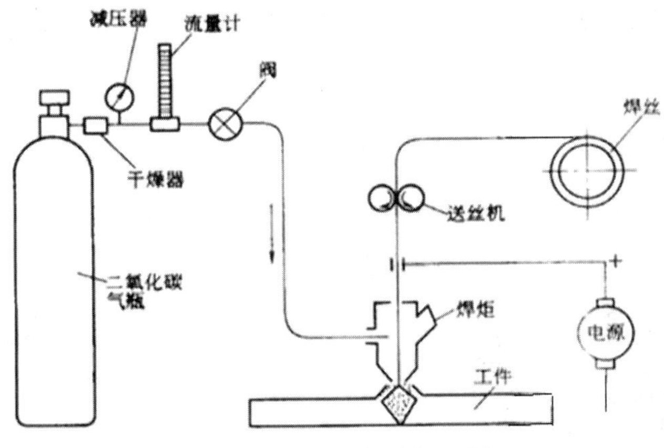

Fig.40.2 Process of CO_2 welding

二氧化碳气瓶 ______________； 送丝机 ______________；

工件 ______________； 焊炬 ______________；

焊丝 ______________； 阀 ______________；

电源 ______________； 流量计 ______________。

Words and phrases

CO_2 (carbon dioxide) ['kɑ:bən] [daɪ'ɒksaɪd] 二氧化碳

PETS(Public English Test System) ['pʌblɪk] ['sɪstəm] 全国英语等级考试

certificate of welding inspection 焊接检验资格证

certificate [sə'tɪfɪkət] 资格证

shielded medium ['ʃi:ldɪd] ['mi:dɪəm] 屏蔽介质

high 高

come into being 产生,问世

ease [i:z] 使……轻松，减轻

efficiency [ɪ'fɪʃnsɪ] 效率

cost [kɒst] 费用

mainly ['meɪnli] 主要地

be applied to [ə'plaɪd] 应用于

low-carbon steel 低碳钢

low-alloy steel 低合金钢

high-temperature steel ['temprətʃə(r)] 高温钢

stainless steel 不锈钢

sheet [ʃi:t]（纸）张，薄板

instead of [ɪn'sted] 替代

argon shielded arc welding ['ɑ:gɒn] 氩弧焊

porosity [pɔ:'rɒsətɪ] 气孔

flow meter [fləʊ] ['mi:tə(r)] 流量计

advantage [əd'vɑ:ntɪdʒ] 优点

positional welding [pə'zɪʃənəl] 全位置焊接

cylinder ['sɪlɪndə(r)] 气瓶

wire feeder ['waɪə(r)] 送丝机

flux [flʌks] 焊剂

work (workpiece) ['wɜ:kˌpi:s] 工件

valve [vælv] 阀

power source ['paʊə(r)] [sɔ:s] 电源

filler wire ['fɪlə(r)] 焊丝

intensity of labour [□n'tensətɪ] ['leɪbə(r)] 劳动强度

low 低

ready for automatization ['redɪ] ['ɔ:təʊ] [mætɪ'teɪʃən] 易于自动化

problem of spatter ['prɒbləm] ['spætə(r)] 飞溅问题

Lesson 41 Gas Metal Arc Welding

Look and select

1. Look at the pictures and select the correct terms from the box.

flow and gas regulator	welding gun	wire feeder	gas cylinder

1. ________________ 2. ________________

3. ________________ 4. ________________

2. Look at the picture (Fig.41.1) and translate the terms into Chinese.

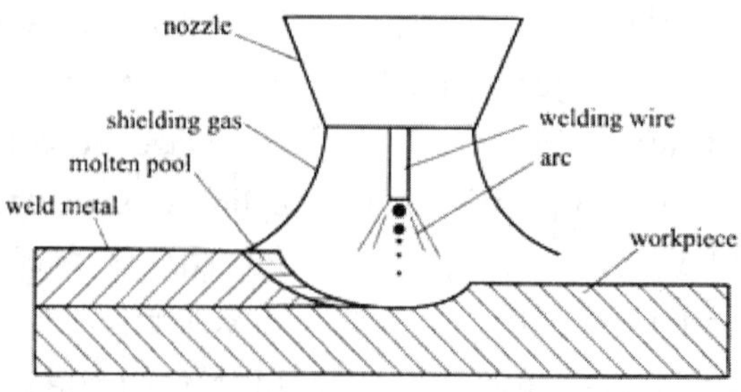

Fig.41.1 Gas metal arc welding

nozzle ________________; workpiece ________________;

shielding gas ________________; arc ________________;

molten pool ________________; welding wire ________________;

weld metal ________________.

Text

The outline of gas metal arc welding

Two welding materials used for gas metal arc welding (GMAW) (Fig.41.2) are welding wire and shielded gas. The welding wires include solid wire and flux cored electrode. The shielding gas may be an inert gas, active gas, or their mixture.

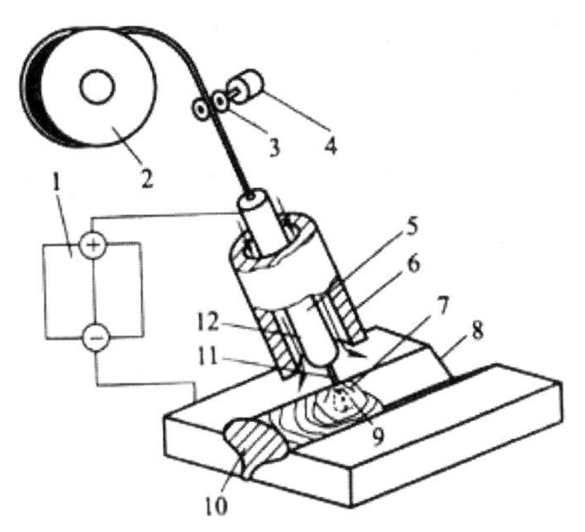

Fig.41.2 Process of gas metal arc welding

1—power supply; 2—wire reel; 3—wire feed rolls; 4—wire feed motor; 5—current contact; 6—nozzle; 7—arc; 8—base metal; 9—molten pool; 10—weld metal; 11—welding wire; 12—shielding gas

Knowledge extention

1. What is the first consideration in any welding operation?

The primary consideration in any welding operation is to produce a weld that has the same properties as the base metal. Such a weld can only be made if the molten puddle is completely protected from the atmosphere during the welding process.

2. What are the automatic welding and semi-automatic welding?

When completely automatic, the wire feed, power setting, gas flow and travel over the workpiece are pre-set and function automatically. When semi-automatic, the wire feed, power setting and gas flow are pre-set, but the torch is manually operated. The welder directs the torch over the weld seam, holding the correct arc-to-work distance and speed.

Exercises

1. Translate the words in the picture (Fig.41.3) into Chinese.

2. Match the English with the Chinese. Draw lines.

(1) welding gun	a. 气瓶
(2) gas cylinder	b. 焊枪
(3) nozzle	c. 保护气体
(4) shielding gas	d. 喷嘴
(5) welding wire	e. 气源

(6) gas supply　　f. 控制装置

(7) control unit　　g. 焊丝

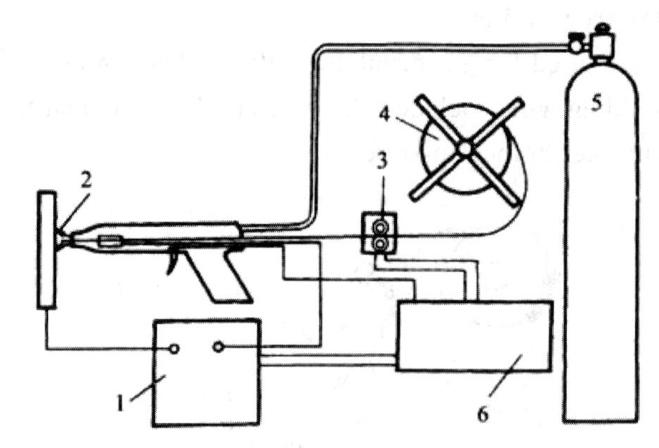

Fig.41.3 Equipments of Gas Metal Arc Welding

1—power supply __________; 2—shielded gas ______________;

3—wire feed rolls _________; 4—wire feed mechanism _________;

5—gas supply ___________; 6—control unit _______________.

3. Answer the following questions.

(1) What are the two welding materials used for gas metal arc welding (GMAW)?

(2) What is the shielding gas in the gas metal arc welding?

(3) Describe the classification of welding wires in the gas metal arc welding.

4. Translate the sentences into Chinese.

(1) The primary consideration in any welding operation is to produce a weld that has the same properties as the base metal.

(2) When completely automatic, the wire feed, power setting, gas flow and travel over the workpiece are pre-set and function automatically.

(3) The welder directs the torch over the weld seam, holding the correct arc-to-work distance and speed.

Words and phrases

GMAW (gas metal arc welding) 熔化极气体保护电弧焊

nozzle ['nɔzəl] 喷嘴

shielding gas [ʃi:ldɪŋ] [gæs] 保护气体

molten pool ['məʊltən] [pu:l] 熔池

weld metal ['metl] 焊缝金属

welding material [mə'tɪərɪəlz] 焊接材料

flux cored electrode [flʌks] [kɔ:d] 药芯焊丝

solid wire ['sɒlɪd] ['waɪə(r)] 实心焊丝

first consideration 首要考虑的

welding operation [ˌɒpə'reɪʃn] 焊接操作

workpiece ['wɜ:kˌpi:s] 工件

arc [ɑ:k] 电弧

welding wire ['waɪə(r)] 焊丝

gas supply [sə'plaɪ] 气源

power supply ['paʊə(r)] 电源

inert gas [ɪn'ɜ:t] 惰性气体

mixture ['mɪkstʃə(r)] 混合

property ['prɒpətɪ] 性能

active gas ['æktɪv] 活性的

molten puddle ['məʊltən] ['pʌdl] 熔池
atmosphere ['ætməsfɪə(r)] 大气
automatic welding [ˌɔ:tə'mætɪk] 自动焊
semi-automatic ['semi] 半自动
power setting ['paʊə(r)] ['setɪŋ] 电源设置
wire feed ['waɪə(r)] [fi:d] 送丝
gas flow [fləʊ] 气流
travel ['trævl] 移动，运动
pre-set [p'ri:s'et] 预设置
function ['fʌŋkʃn] 功能
automatically [ˌɔ:tə'mætɪklɪ] 自动地
torch [tɔ:tʃ] 焊炬，焊枪
be manually operated ['mænjʊəlɪ] ['ɒpəreɪtɪd] 手动操作

The welder directs the torch over the weld seam, holding the correct arc-to-work distance and speed. 焊工持焊炬/枪在焊缝上操作，保持着电弧与工件适当的距离与焊接速度。

Lesson 42 Gas Tungsten Arc Welding

Look and select

Look at the pictures and select the correct the terms from the box:

filler wire for TIG welding	electrode holder
tungsten electrode	contact tube

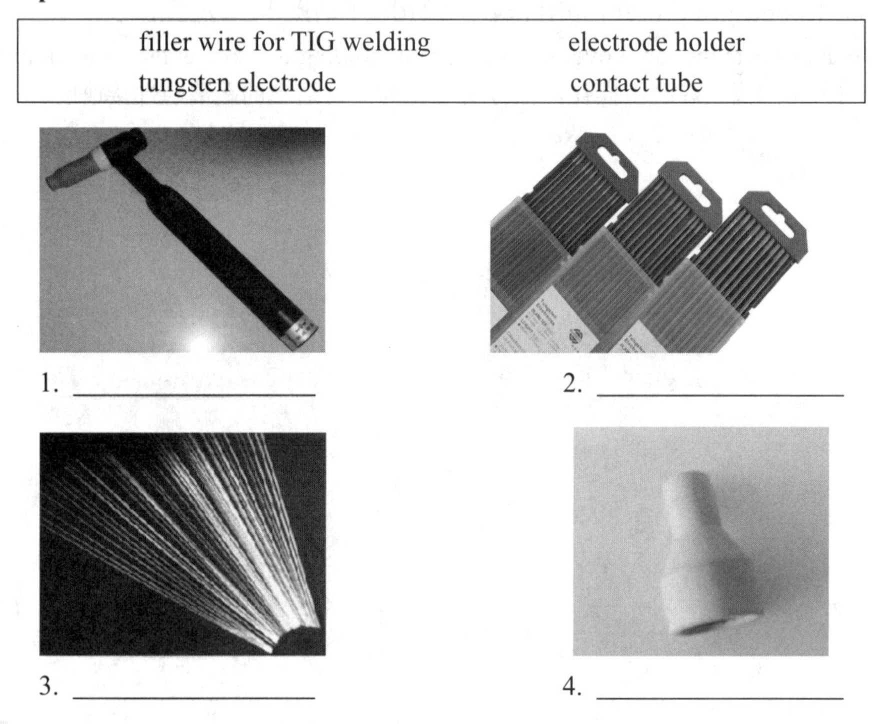

1. ______________ 2. ______________

3. ______________ 4. ______________

Text

1. The definition of gas tungsten arc welding

In the gas tungsten arc welding process (Fig.42.1), a virtually non-consumable tungsten electrode

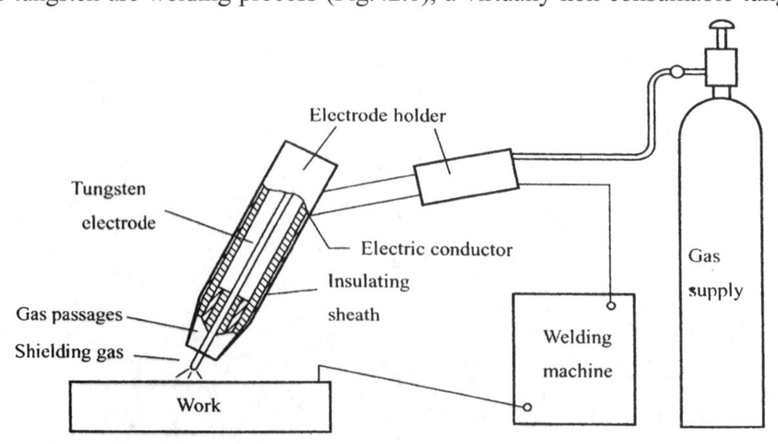

Fig.42.1 Process of gas tungsten arc welding

is used to provide the arc for welding. During the welding cycle a shield of inert gas expels the air from the welding area and prevents oxidation of the electrode, weld puddle, and surrounding heat-affected zone.

2. The characteristics of gas tungsten arc welding

In TIG welding [Fig.42.2(a)], the electrode is used only to create the arc. It is not consumed in the weld. In this way it differs from the regular Shielded Metal Arc Process [Fig.42.2(b)], where the stick electrode is consumed in the weld. For joints where additional weld metal is needed, a filler wire is fed into the puddle in a manner similar to welding with the oxy-acetylene flame process.

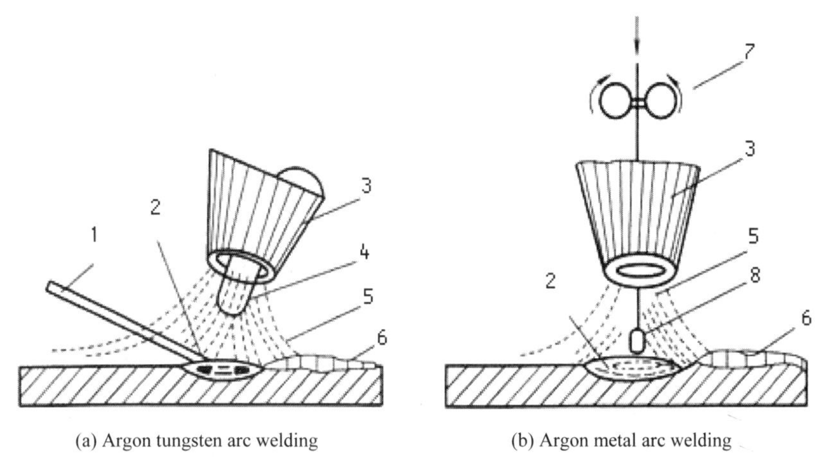

(a) Argon tungsten arc welding (b) Argon metal arc welding

Fig.42.2 Kinds of Argon shielded arc welding

1—filler wire; 2—molten pool; 3—nozzle; 4—tungsten electrode; 5—shielded gas; 6—weld; 7—feed rolls; 8—welding wire

Knowledge extention

Understand the welding methods in Table 42.1.

Table 42.1 Welding methods

English	Chinese
Helium shielded arc welding	氦弧焊
Argon shielded arc welding	氩弧焊
Argon tungsten arc welding	钨极氩弧焊
Argon shielded arc welding-pulse arc	脉冲氩弧焊
Mixed gas welding	混合气体保护焊
Inert gas shielded arc welding	惰性气体保护焊
Metal inert-gas welding(MIG welding)	熔化极惰性气体保护焊

Exercises

1. Look at the picture (Fig.42.3) and translate the terms into Chinese.

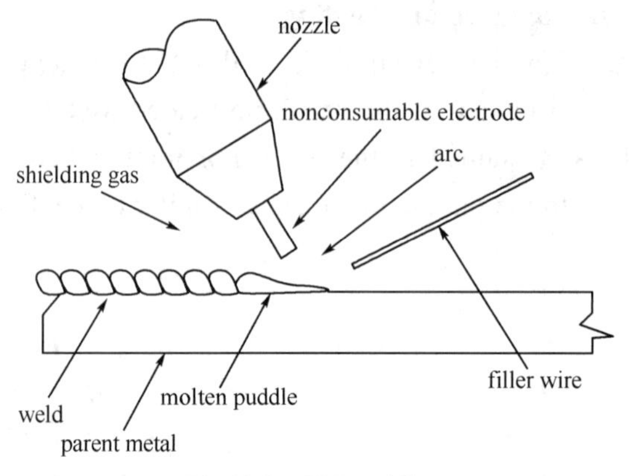

Fig.42.3 TIG welding

parent metal ________________; arc ________________________;

molten puddle ______________; nonconsumable electrode ____________;

filler wire ________________; nozzle ______________________;

weld ____________________; shielding gas ____________________.

2. Match the English with the Chinese. Draw lines.

(1) gas tungsten arc welding　　a. 非熔化电极（焊条）

(2) non-consumable electrode　　b. 氧化

(3) shielding gas　　c. 母材

(4) oxidation　　d. 钨极气体保护电弧焊

(5) parent metal　　e. 保护气体

3. Answer the following questions:

(1) What's the definition of gas tungsten arc welding?

(2) What are the characteristics of gas tungsten arc welding?

4. Translate the sentences into Chinese.

(1) In the gas tungsten arc welding process, a virtually non-consumable tungsten electrode is used to provide the arc for welding.

(2) In TIG welding, the electrode is used only to create the arc. It is not consumed in the weld.

(3) For joints where additional weld metal is needed, a filler wire is fed into the puddle in a manner similar to welding with the oxy-acetylene flame process.

Words and phrases

Gas Tungsten Arc Welding ['tʌŋstən] 钨极气体保护电弧焊

contact tube ['kɒntækt] [tju:b] 导电管/导电嘴

parent metal ['peərənt] 母材

filler wire ['fɪlə(r)] ['waɪə(r)] 填充焊丝

nozzle ['nɔzəl] 喷嘴

non-consumable [kən'sju:məbl] 非熔化的
molten puddle ['məʊltən] ['pʌdl] 熔池
provide the arc [prə'vaɪd] 提供电弧
expel...from [iks'pelz] 把……排除/驱逐
prevent [prɪ'vent] 阻止/防止
surrounding [sə'raʊndɪŋ] 周围/边的
tungsten electrode 钨极焊条
insulating sheath ['insjuleitɪŋ] [ʃi:θ] 绝缘套
electric conductor [ɪ'lektrɪk] [kən'dʌktə(r)] 导电嘴
arc [ɑ:k] 电弧
virtually ['vɜ:tʃuəli] 实际上
be used to 用来
welding area ['eərɪə] 焊接区
oxidation [ˌɒksɪ'deɪʃn] 氧化
weld puddle 熔池
gas passage 气体通道

Lesson 43 Resistance Welding

Look and judge

Judge the following true or false (Table.43.1).

Table. 43.1 The actions during welding

Actions	True or False?
1. Place the wire feeder on an adequately insulated floor.	
2. You can loop cables around your body when the power source is ON.	
3. Do any work on the machine, switch on the power source.	
4. Don't immediately replace any undimensioned cables.	
5. Immerse the tungsten electrode in liquid in oder to cool it.	
6. During welding, don't touch the wire reel.	

Text

1. The principle of resistance welding

Resistance welding (Fig.43.1) is done by passing an electric current through two pieces of metal pressed together. The pieces coalesce at the surfaces of contact because more resistance and heat are concentrated there. The heat is localized where needed, the action is rapid, no filler metal is needed. It is one of the oldest of the electric welding processes in use by industry today.

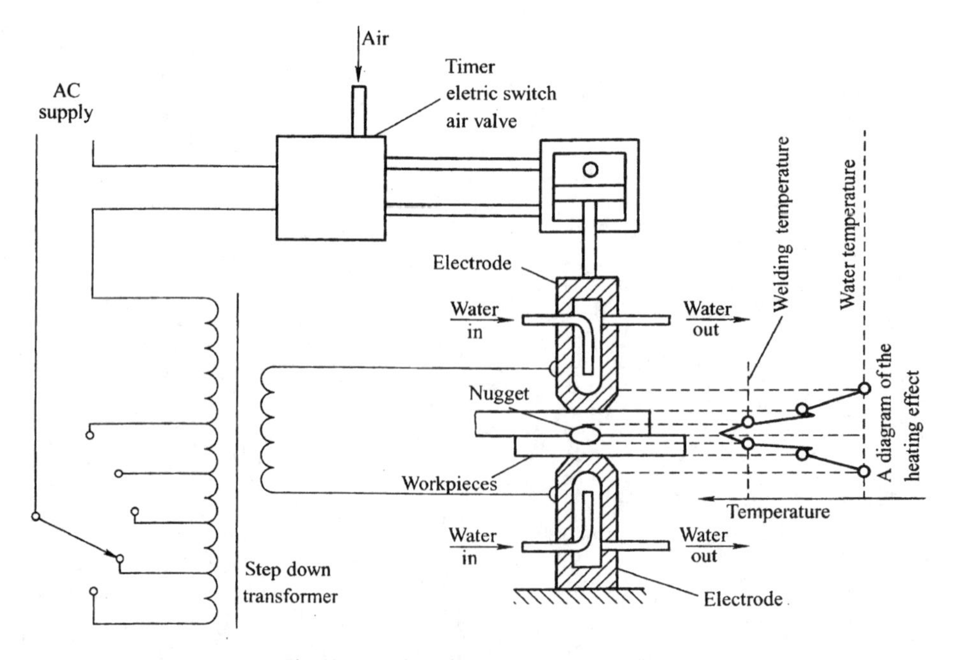

Fig.43.1 Principle of resistance welding

2. The main advantages of resistance welding

This welding operation requires little skill and can easily be automated, and these advantages make the process suitable for large-quantity production. All the common metals and dissimilar metals can be resistance welded although special precautions are necessary for some. The parent metal is normally not harmed, and none is lost. Many difficult shapes and sections can be processed.

Knowledge extention

Joints of resistance welding (Fig.43.2).

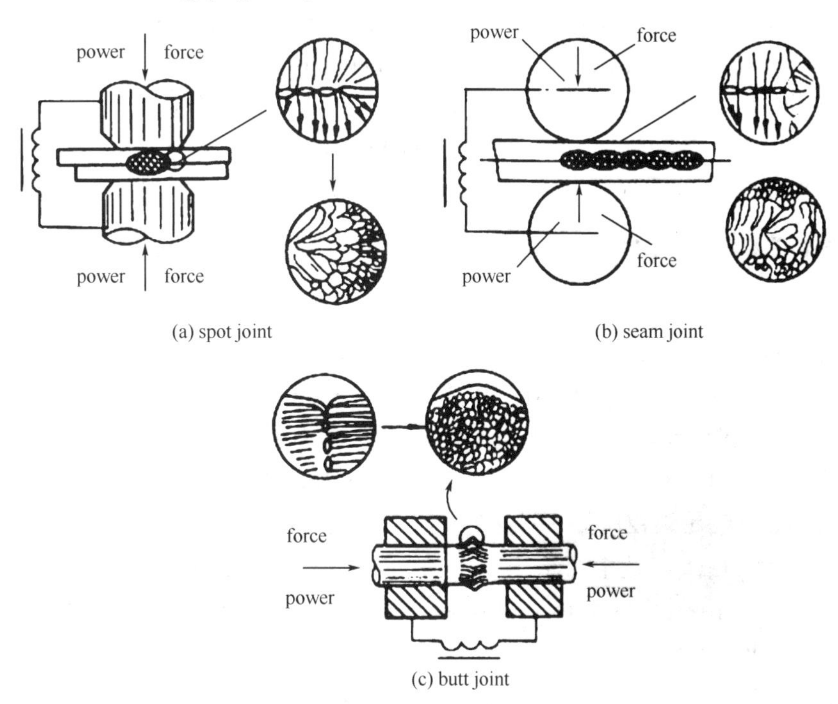

Fig.43.2 Joints of resistance welding

Exercises

1. Match the English with the Chinese. Draw lines.

(1) wire feeder	a. 电缆
(2) cable	b. 电源
(3) power source	c. 接合，焊在一起
(4) tungsten electrode	d. 送丝机
(5) wire reel	e. 焊丝盘
(6) coalesce	f. 大批量生产
(7) filler metal	g. 优势
(8) advantage	h. 填充金属
(9) large-quantity production	i. 钨极

2. Fill in the blanks with the proper words in the text.

(1) The system of resistance welding include ____________________, transformer, timer, electric ____________, air ____________ and water etc.

(2) There are three kinds of joints in resistance welding: ________ joint, ________ joint and ____________ joint.

3. Answer the following questions.

(1) What are the principle of resistance welding?

(2) What are the advantages of resistance welding?

4. Translate the sentences into Chinese.

(1) This welding operation requires little skill and can easily be automated.

(2) The parent metal is normally not harmed, and none is lost.

(3) Many difficult shapes and sections can be processed.

(4) Place the wire feeder on an adequately insulated floor.

(5) During welding, don't touch the wire pool.

Words and phrases

resistance [rɪ'zɪstəns] 电阻
pass ... through 通过，经过
electric current [ɪ'lektrɪk] ['kʌrənt] 电流
two pieces of ['pi:sɪz] 两片，两块
contact ['kɒntækt] 接触
at the surface of ['sə:fisiz] 在……表面
AC supply [sə'plaɪ] 交流电源
electric switch [ɪ'lektrɪk] [swɪtʃ] 电开关
electrode [ɪ'lektrəʊd] 电极
nugget ['nʌgɪt] 熔核
temperature ['temprətʃə(r)] 温度
heating effect ['daɪəgræm] [ɪ'fekt] 热影响
wire feeder ['waɪə(r)] ['fi:də(r)] 送丝机
adequately ['ædɪkwətlɪ] 充分地，足够地
insulated floor ['ɪnsjuleɪtɪd] [flɔ:(r)] 绝缘地面
power source ['paʊə(r)] [sɔ:s] 电源
immediately [ɪ'mi:diətli] 立即，马上
press together [pres] [tə'geðə(r)] 挤压在一起
step down transformer [træns'fɔ:mə(r)] 降压变压器
roll [rəʊl] 旋转，辊子
wire reel 焊丝盘
metal ['metl] 金属
coalesce [ˌkəʊə'les] 联合，接合
air [eə(r)] 空气
concentrate ['kɒnsntreɪt] 集中
timer ['taɪmə(r)] 时间继电器
air valve [vælv] 空气阀门
water in / out 进水/出水
workpiece ['wɜ:kˌpi:s] 工件
diagram ['daɪəgræm] 图表
loop ... around 环绕，缠绕
switch [swɪtʃ] 开关
undimensioned 未注明尺寸的
immerse [ɪ'mɜ:s] 浸入
cool 冷却
liquid ['lɪkwɪd] 液体

Lesson 44 Spot and Butt Welding and Other Welding Methods

Look and learn

Learn the four common joints of gas welding (Fig.44.1).

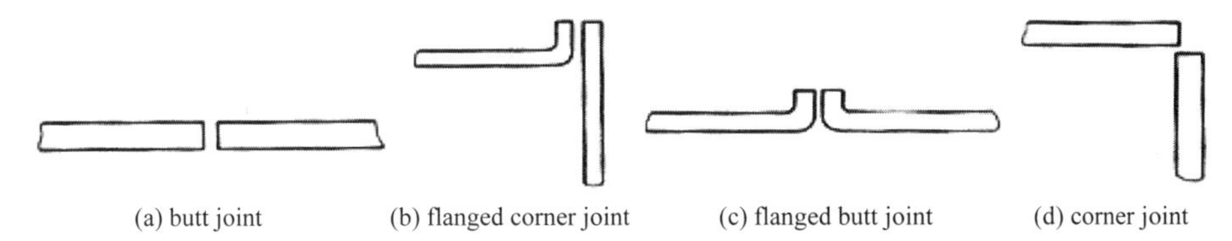

(a) butt joint (b) flanged corner joint (c) flanged butt joint (d) corner joint

Fig.44.1 Common joints of gas welding

Text

1. Spot welding

Spot welding is the most common form of resistance welding (Fig.44.2) and the simplest. It is accomplished when current is caused to flow through the electrode tips and the separate pieces of metal to be joined. The resistance of the base metal to electric current flow causes localized heating in the joint and the weld is made shown in Fig.44.3.

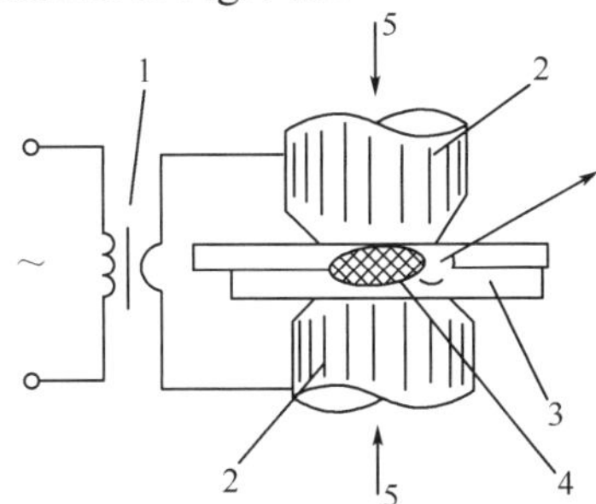

Fig.44.2 Principle of spot-resistance welding

1—transformer; 2—electrode; 3—workpiece; 4—nugget; 5—power up and apply pressure

Force
Horn
Electrodes
Joint
Nugget(fused metal)
Parent metal
Force

(a) Typical spot welding circuit (b) Spot-welded joint (c) Enlarged cross section through a spot weld (d) Indication of the way the current spreads out when passed through several sheets

Fig.44.3 Diagram of spot welding

2. Butt welding

Upset-butt welding (Fig.44.4) consists of pressing two pieces of metal together and passing a current through them. For flash-butt welding (Fig.44.5), two pieces to be flashed welded are clamped with their ends not quite touching. Most commercial metals may be flash welded.

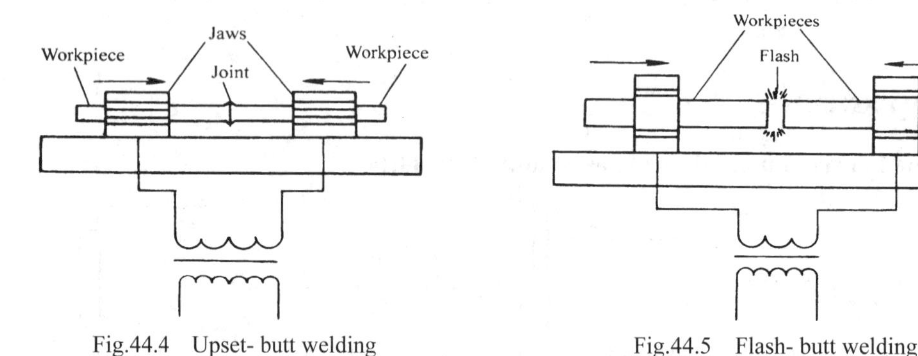

Fig.44.4 Upset- butt welding   Fig.44.5 Flash- butt welding

Knowledge extention

1. Electron beam welding

Energy may be supplied for welding and cutting by directing a concentrated beam of electrons to bombard the work in the manner shown in the picture. The beam is created in a high vacuum. If the work is done in a vacuum of around 0.1 micron (ц), no electrodes, gases, or filler metal need contaminate it, and pure welds can be made illustrated in Fig.44.6.

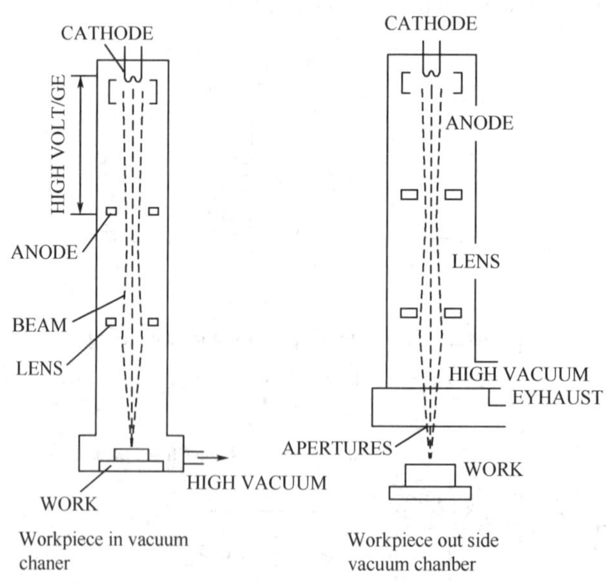

Fig.44.6 Electron beam welding

2. Special welding methods

The special welding methods more difficult to understand are **induction welding** and **laser welding** (Fig.44.7). Others are **ultrasonic welding**, **cold welding**, **friction welding**, **brazing** and **plasma arc welding** etc.

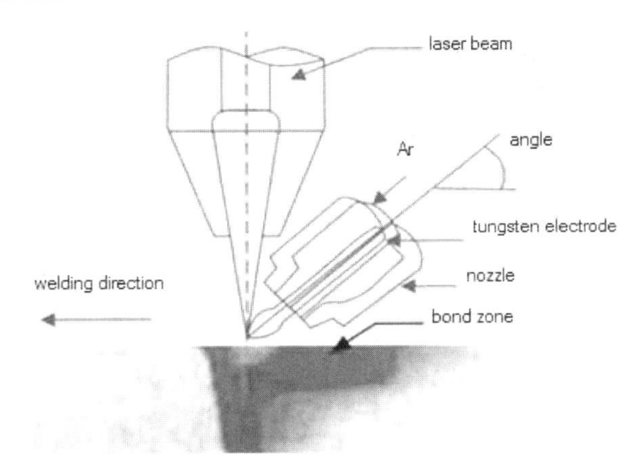

Fig.44.7 Diagram of laser welding

Exercises

1. Match the English with the Chinese. Draw lines.

(1) force a. 卷边角焊
(2) parent metal b. 压紧
(3) nugget c. 母材
(4) flanged corner joint d. 电子束
(5) electric current e. 焊核
(6) press f. 闪光对焊
(7) flash-butt welding g. 真空
(8) electron beam h. 力
(9) vacuum i. 电流

2. Judge the following true or false.

(1) The joints of gas welding are divided into butt joint, flanged corner joint , flanged butt joint and corner joint. ( )

(2) Spot welding is the most common form of fusion welding and the simplest. ( )

(3) Butt welding includes upset-butt welding and flash-butt welding. ( )

(4) During the electron beam welding, the electron beam is created in the air. ( )

(5) The gas involved in laser welding is oxygen. ( )

3. Fill in the blanks with the proper words in the text.

(1) Spot welding is accomplished when _______ is caused to flow through the _______ tips.

(2) Upset-butt welding consists of pressing ___________ together and passing a current through them.

(3) For flash-butt welding, two pieces to be flashed welded are ___________ with their ends not quite _____________. Most commercial _____________ may be flash welded.

(4) _____________ may be supplied for welding and cutting by directing _____________ to _______________ the work in the manner shown in the picture.

(5) The special welding methods more difficult to understand are _____________ and laser welding. Others are ___________, cold welding, ___________, brazing and ___________ etc.

4. Translate the following reasons into Chinese.

导致焊接时没有保护气体的原因可能是：

(1) The gas cylinder is empty.

(2) The gas pressure regulator is faulty.

(3) The gas hose is damaged.

(4) The welding torch is defective.

(5) The gas solenoid valve is defective.

导致焊接质量差的原因可能是：

(1) Incorrect welding parameters.

(2) Poor grounding connection.

(3) Not enough shielding gas, or none at all.

(4) Welding torch is leaking.

(5) Wrong contact tube, or contact tube is worn out.

Words and phrases

most common form ['kɒmən] form 最普遍的形式

nugget [□nʌgɪt] 熔核

the simplest ['sɪmplɪst] 最简单

parent metal ['peərənt] 母材

electron beam welding [ɪ'lektrɒn] [bi:m] 电子束焊

high voltage ['vəʊltɪdʒ] 高电压

high vacuum ['vækjʊəm] 高度真空

exhaust [ɪg'zɔ:st] 排出

ultrasonic [ˌʌltrə'sɒnɪk] 超声的

friction ['frɪkʃn] 摩擦

plasma arc welding ['plæzmə] 等离子弧焊

gas cylinder ['sɪlɪndə(r)] 气瓶

incorrect [ˌɪnkə'rekt] 不正确的

be damaged ['dæmɪdʒd] 损坏

defective [dɪ'fektɪv] 有瑕疵/缺陷的

parameter [pə'ræmɪtə] 参数

none at all [nʌn] 一点儿也不

force [fɔ:s] 力

horn [hɔ:n] 角，喇叭

joint [dʒɔɪnt] 接头

cathode ['kæθəʊd] 阴极

lens [lenz] 透镜

anode ['ænəʊd] 阳极

work (workpiece) 工件

aperture ['æpətʃə(r)] 孔

laser ['leɪzə(r)] 镭射，激光

brazing [breɪzɪŋ] 钎焊

etc. [ˌet'setərə] 等等

empty ['emptɪ] 空的

faulty ['fɔ:ltɪ] 错误，毛病

gas hose [həʊz] 气体软管

welding torch [tɔ:tʃ] 焊炬，焊枪

leak [li:k] 泄漏

worn out [wɔ:n] 破旧，磨损

gas pressure regulator ['preʃə(r)] ['regjuleɪtə(r)] 气压校准器

gas solenoid valve ['sɒlənɔɪd] [vælv] 燃气电磁阀

poor grounding connection [pʊə(r)] ['graʊndɪŋ] [kə'nekʃn] 不良接地

enough shielding gas [ɪ'nʌf] [ʃi:ldɪŋ] 足够的保护气体

wrong contact tube [rɒŋ] ['kɒntækt] [tju:b] 导电铜管/导电嘴损坏

Part Ⅲ
Welding Inspection

Lesson 45 Defects

Look and select

Look at the pictures and select the correct terms from the box.

slag	porosity	lack of fusion	undercut	pit
crater	crack	lack of penetration	overlap	burn through

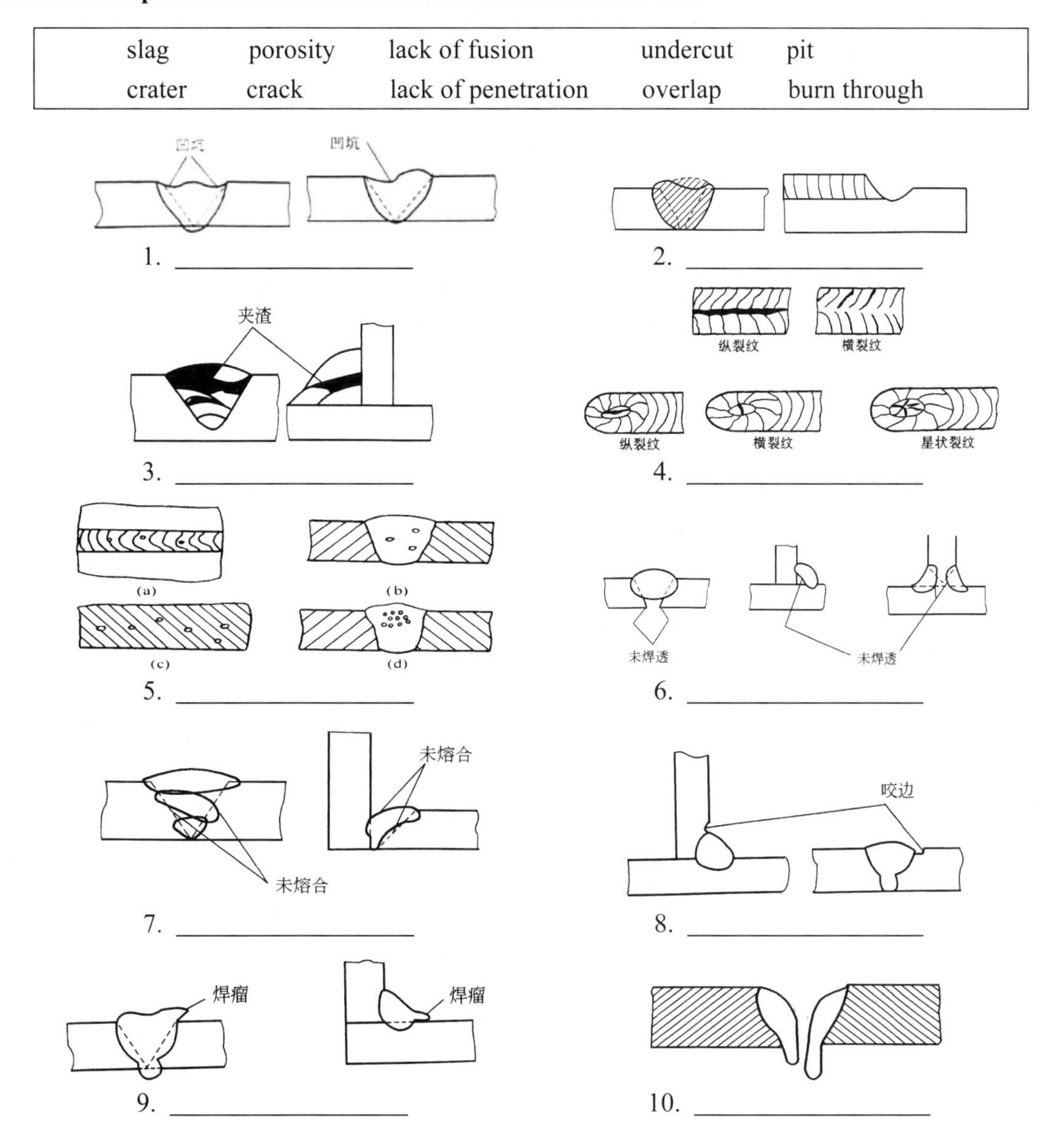

Text

1. Nondestructive testing (NDT)

NDT includes visual inspection, radiography testing (X-rays, Gamma rays and Neutron rays), ultrasonic testing, magnetic-particle inspection, liquid penetrant inspection, acoustic emission inspection, eddy-current testing and TOFT testing etc.

2. Defects

Some welding defects (Table 45.1) that can be checked visually: insufficient throat, excessive undercut, overlap, undercut etc. The internal defects and variations that can be checked out by radiography testing: porosity, crack, lack of fusion, inclusion, geometry variation, corrosion and density variations etc.

Table 45.1 Welding defects

English	Chinese	English	Chinese
Hot crack	热裂纹	Cold crack	冷裂纹
misalignment	错边	Burn-through	烧穿
Tungsten inclusion	夹钨	Undercut	咬边
Lack of penetration	未焊透	Lack of fusion	未熔合
Crater crack	弧坑裂纹	Pit	凹坑
Excessive penetration	下塌	Overlap	焊瘤
Incompletely filled groove	未焊满	Arc scratch	电弧擦伤

Knowledge extention

Quality assurance

The quality of a product may be stated in terms of a measure of the degree to which it conforms to specifications and standards of workmanship. The quality assurance function is charged with the responsibility of maintaining product quality consistent with those requirements and involves the following four basic activities:

(1) Quality spccification

(2) Inspection

(3) Quality analysis

(4) Quality control

Exercises

1. Match the English with the Chinese. Draw lines.

(1) slag　　a. 未熔合

(2) porosity　　b. 夹渣

(3) lack of fusion　　c. 凹坑

(4) undercut　　d. 焊瘤

(5) pit　　e. 咬边

(6) crater f. 气孔
(7) crack g. 未焊透
(8) lack of penetration h. 弧坑
(9) overlap i. 烧穿
(10) burn through j. 裂纹

2. Fill in the blanks with the proper words in the text.

NDT includes _________, __________ (X-rays, Gamma rays and Neutron rays), ultrasonic testing, __________, liquid penetrant inspection, acoustic emission inspection, ___________, TOFT testing etc.

3. Answer the following questions.

(1) What are the welding defects mentioned in the text?

(2) Which four aspects does the quality assurance include?

Words and phrases

defect ['di:fekt] 缺陷
check [tʃek] 检查
overlap [ˌəʊvə'læp] 焊瘤
internal [in'tə:nəl] 内部的，里面的
be checked out by [tʃekt] 由……检查出来
lack of fusion [læk] ['fju:ʒn] 未熔合
inclusion [ɪn'klu:ʒn] 夹,包含
density variation ['densətɪ] 密度变异
misalignment [ˌmɪsə'laɪnmənt] 错边
excessive [ɪk'sesɪv] 过度/分的
cold [kəʊld] 冷
arc [ɑ:k] 电弧
specification [ˌspesɪfɪ'keɪʃn] 规范，规格
inspection [ɪn'spekʃn] 检验
porosity [pɔ:'rɒsətɪ] 气孔
visually ['vɪʒʊəlɪ] 视觉地，直观地
undercut 咬边
etc. [ˌet'setərə] 等等
variation [ˌveəri'eɪʃn] 变异
crack [kræk] 裂纹
tungsten ['tʌŋstən] 钨
corrosion [kə'rəʊʒn] 腐蚀
hot crack [kræk] 热裂纹
crater ['kreɪtə(r)] 弧坑
burn [bɜ:n] 燃烧
scratch [skrætʃ] 擦伤
quality ['kwɒlətɪ] 质量
analysis [ə'næləsɪs] 分析
insufficient throat [ˌɪnsə'fɪʃnt] [θrəʊt] 焊缝厚度不足
excessive undercut [ɪk'sesɪv] [ˌʌndə'kʌt] 过咬边
radiography testing [ˌreɪdɪ'ɒgrəfɪ] ['testɪŋ] 射线测试/探伤
geometry variation [dʒɪ'ɒmətrɪ] [ˌveəri'eɪʃn] 几何变异

Lesson 46 RT

Look and select

Look at the pictures and select the correct terms from the box.

RT meter	RT machine	lead gate	film viewer (illuminator)

1. ________________

2. ________________

3. ________________

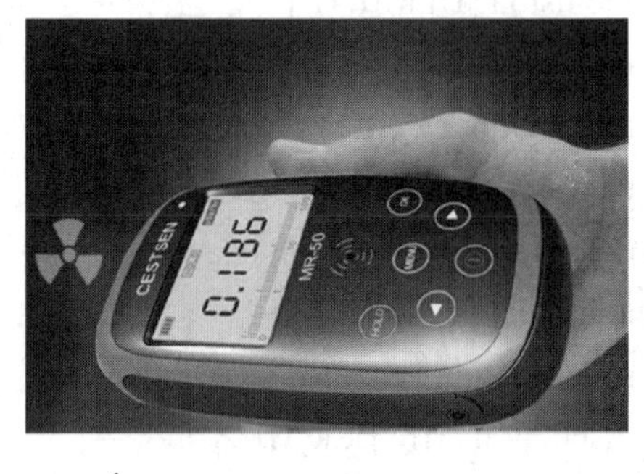

4. ________________

Text

The definition of radiographic examination

Radiographic examination (Fig.46.1) is a nondestructive examination method that uses invisible X-ray, or gamma radiation to examine the interior of materials. Radiographic examination gives a permanent film record of defects that is relatively easy to interpret. Although this is a slow and expensive method of nondestructive examination, it is a positive method for detecting porosity, inclusions, cracks, and voids in the interior of castings, welds, and other structures, shown in Fig.46.2.

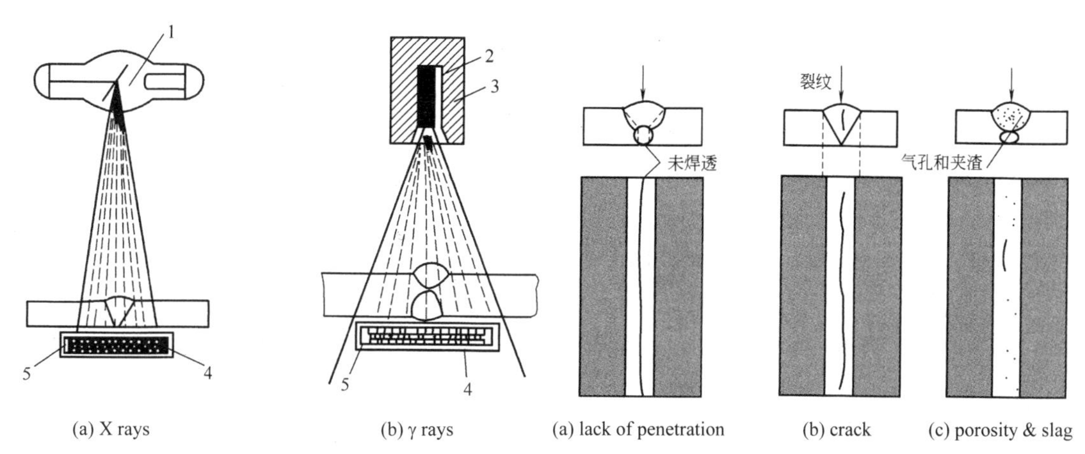

(a) X rays (b) γ rays

Fig.46.1 Diagram of X rays and γ rays examination

1—X rays tube; 2—γ rays source; 3—lead box;

4—film; 5—film holder

(a) lack of penetration (b) crack (c) porosity & slag

Fig.46.2 Defects in the films

Knowledge extention

The principle of radiographic examination

X-rays are produced by electrons hitting a tungsten target inside an X-ray tube (Fig.46.3). In addition to the X-ray tube, the apparatus consist of a high-voltage generator with necessary controls. Gamma rays are produced by radioactive decay of certain radioisotopes. The radioisotopes normally used are cobalt-60, iridium-192, thulium-170, and cesium-137. The isotopes are contained in a lead or spent uranium vault or capsule to provide safe handling.

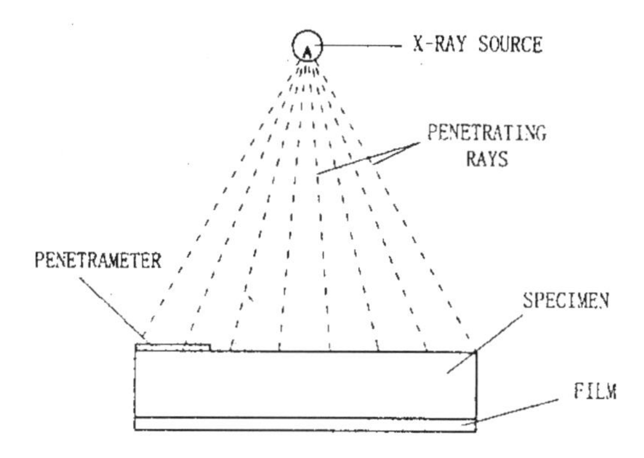

Fig.46.3 The principle of radiographic examination

Exercises

1. Match the English with the Chinese. Draw lines.

(1) film viewer a. 射线探测仪

(2) lead gate b. 辐射

(3) RT meter c. 观片灯

(4) radiation
(5) radiographic examination
(6) X rays
(7) gamma rays
(8) neutron radiography

d. X 射线
e. 伽马射线
f. 中子射线探伤
g. 铅门
h. 射线探伤

2. Fill in the blanks with the proper words in the text.

(1) RT is a positive method for detecting _______, _______, _______, and voids in the interior of castings, _______, and other structures.

(2) X-rays are produced by _______ hitting a tungsten target inside an _______. In addition to the X-ray tube, the apparatus consist of a ___________ with necessary controls.

3. Answer the following questions:

(1) What is radiographic examination?

(2) Describe the principles of radiographic examination.

4. Translate the sentence about neutron radiography.

Neutron radiography is a form of nondestructive examination that uses a specific type of particulate radiation, called neutrons, to from a radiographic image of a test piece.

Words and phrases

RT (Radiographic examination) ['reɪdɪəʊ 'græfɪk] [ɪgˌzæmɪ'neɪʃn] 射线检测/探伤
lead gate [li:d] [geɪt] 铅门
method ['meθəd] 方法
invisible [ɪn'vɪzəbl] 看不见的，无形的
relatively ['relətɪvli] 相比较地，相对地
interpret [ɪn'tɜ:prɪt] 解释，翻译
positive ['pɒzətɪv] 明确的，肯定的
detect [dɪ'tekt] 探测
inclusion [ɪn'klu:ʒn] 夹渣，包含
voids [vɔɪdz] 空隙，孔洞
other ['ʌðə(r)] 其他的
principle ['prɪnsəpl] 原理，原则
penetrative rays ['penɪtrətɪv] 射线
penetrameter ['penɪtrəmɪtər] 像质计
specimen 试件，样品，标本
meter ['mi:tə(r)] 仪表/器
interior [ɪn'tɪərɪə(r)] 内部
x-ray [reɪ] X 射线
material [mə'tɪərɪəl] 材料
slow [sləʊ] 缓慢的
expensive [ɪk'spensɪv] 昂贵的
porosity [pɔ:'rɒsətɪ] 气孔
crack [kræk] 裂纹
casting ['kɑ:stɪŋ] 铸造
structure ['strʌktʃə(r)] 结构
X-ray source [sɔ:s] X 射线源
gamma ray 伽马射线
neutron ['nu:ˌtrɔn, 'nju:-] 中子
film [fɪlm] 底片，胶片
film viewer (illuminator) [fɪlm] ['vju:ə(r)] [ɪ'lju:mɪneɪtə] 观片灯
give a permanent film record ['pɜ:mənənt] [fɪlm] ['rekɔ:d] 留下永久性胶片记录
nondestructive ['nɒndɪs'trʌktɪv] 非破坏性的
gamma radiation ['gæmə] [ˌreɪdi'eɪʃn] 伽玛辐射

Lesson 47 UT

Look and select

Look at the pictures and select the correct terms from the box.

transducer	UT staff	specimen	UT meter

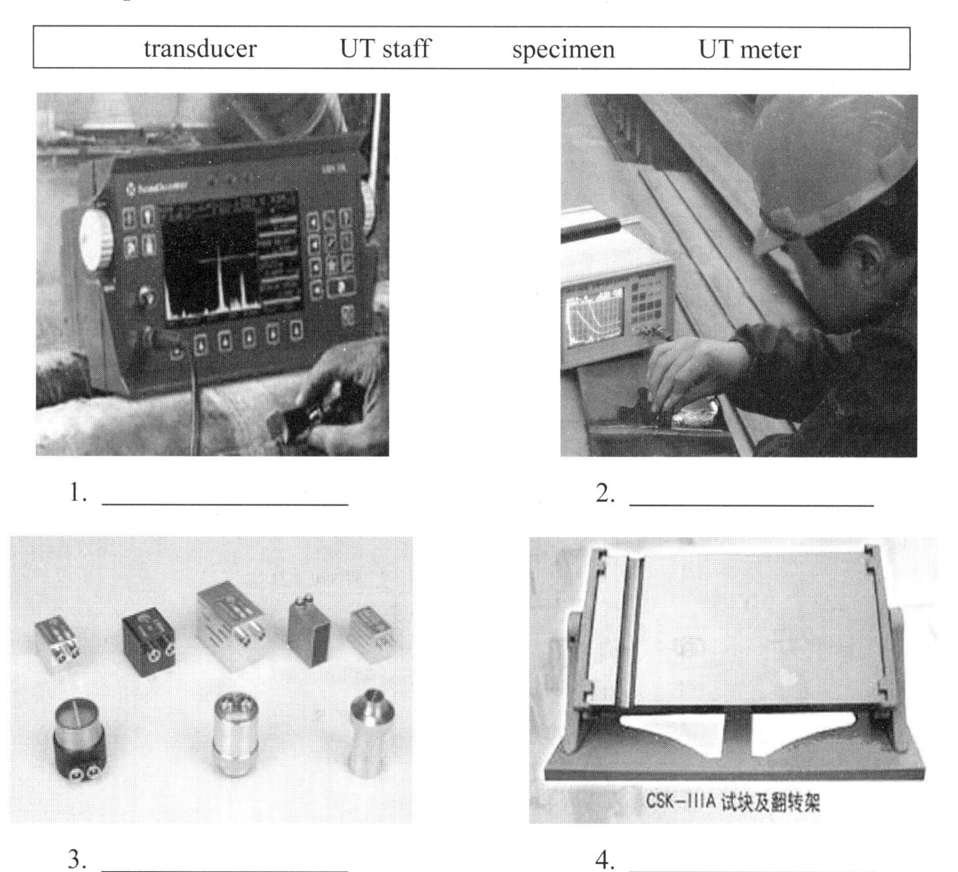

1. ________________ 2. ________________

3. ________________ 4. ________________

Text

1. Definition of ultrasonic examination

Ultrasonic examination (UT) is a nondestructive examination method that uses mechanical vibrations similar to sound waves but of a higher frequency (Fig.47.1). A beam of ultrasonic energy is directed into the specimen to be examined. This beam travels through a material with only a small loss, except when it is interpreted and reflected by a discontinuity or by a change in material.

2. Principle of ultrasonic examination

The transducer is sending out a beam of ultrasonic energy. Some of the energy is reflected by the internal flaw, and the remainder is reflected by the back surface of the specimen (see Fig.47.2).

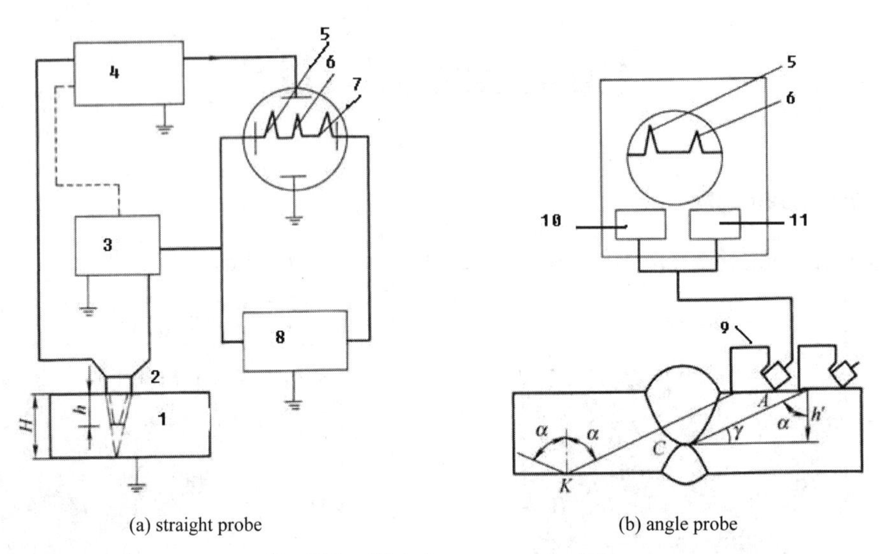

Fig.47.1 High-frequency pulse UT

1—workpiece; 2—straight probe; 3—HF pulse generator; 4—receiving amplifer; 5—start pulse; 6—defect pulse; 7—base pulse; 8—scanning generator; 9—angle probe; 10—emit; 11—receive

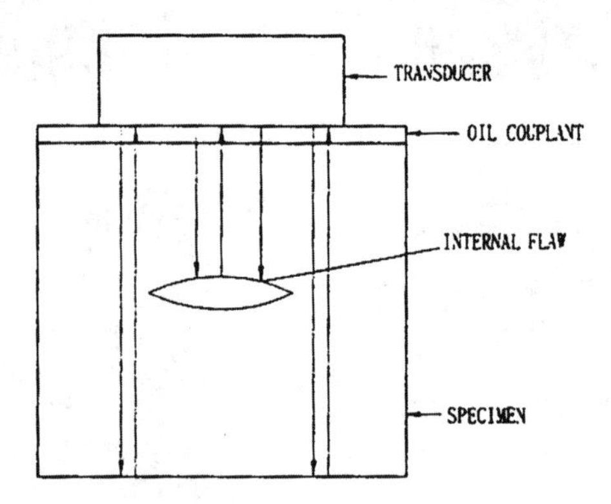

Fig.47.2 Principle of UT

Knowledge extention

1. The following picture (Fig.47.3) shows a typical display as presented on the oscilloscope screen.

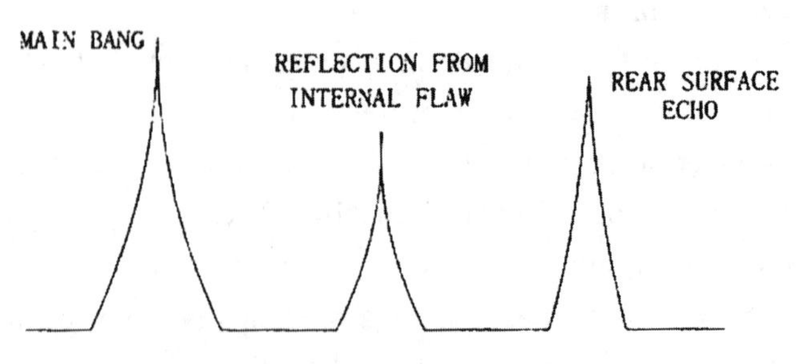

Fig.47.3 Oscillograph display

2. Answer the following question then translate them into Chinese.

What are displayed by a trace on the screen of a cathode-ray oscilloscope?

The initial signal or main bang, the returned echoes from the discontinuities, and the echo of the rear surface of the test material are all displayed by a trace on the screen of a cathode-ray oscilloscope.

Exercises

1. Match the English with the Chinese. Draw lines.

(1) transducer	a. 仪器
(2) staff	b. 试件
(3) meter	c. 耦合剂
(4) specimen	d. 高频脉冲
(5) high-frequency pulse	e. 示波器
(6) oil couplant	f. 人员
(7) oscilloscope	g. 探头

2. Fill in the blanks with the proper words in the text.

(1) There are three pulses in HF pulse UT examination. They are ___________, defect pulse and ______________.

(2) Ultrasonic examination (UT) is a __________ examination method that uses _________ similar to sound waves but of a higher frequency. A beam of ______________________ energy is directed into the specimen to be examined. This travels through a material with only a small loss, except when it is interpreted and reflected by a __________________ or by a change in material.

3. Answer the following questions.

(1) What is Ultrasonic examination?

(2) Describe the principle of ultrasonic examination.

Words and phrases

ultrasonic examination (UT) [ˌʌltrəˈsɒnɪk] [ɪgˌzæmɪˈneɪʃn] 超声波检测

NDT staff [stɑ:f] 无损检测/探伤人员

transducer [trænzˈdju:sə(r)] 探头

mechanical [mɪˈkænɪkəl] 机械的

vibration [vaɪˈbreɪʃn] 震动，摆动

sound wave [saʊnd] [weɪv] 声波

be similar to [ˈsɪmələ(r)] 与……相似

oscilloscope [əˈsɪləskəʊp] 示波器

energy [ˈenədʒɪ] 能量/力

be directed into [dɪˈrektɪd] 直射进……

travel through [ˈtrævəlz] 穿过去

with only a small loss [lɒs] 只有很少损失

material [məˈtɪəriəl] 材料

a beam of ultrasonic energy 一束超声波

end out [end] 发出

be reflected by [rɪˈflektɪd] 被……反射

remainder [rɪˈmeɪndə(r)] 剩余的

back surface [bæk ˈsɜ:fɪs] 背面

oil couplant [ˈku:plɑ:nt] 耦合剂

internal flaw [inˈtə:nəl] [flɔ:] 内部缺陷

main bang [meɪn] [bæŋ] 主脉冲

cathode-ray ['kæθəʊd] [reɪ] 阴极射线　　beam [biːm] 束
higher frequency ['haɪə(r)] ['friːkwənsɪ] 较高频率
specimen to be examined ['spesɪmən] [ɪg'zæmɪnd] 要检测的的试件
discontinuity [ˌdɪsˌkɒntɪ'njuːətɪ] 不连续
change in material [tʃeɪndʒ] [mə'tɪəriəl] 材料中的变化

Lesson 48 MT

Look and select

Look at the pictures and select the correct terms from the box.

standard	magnetic particles	magnet	MT

1. ________________

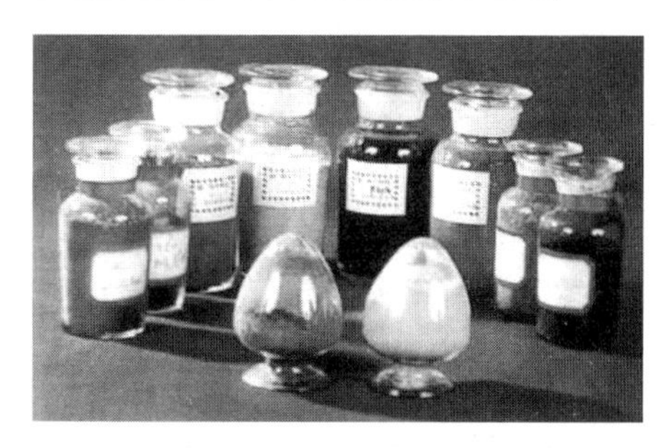

2. ________________

3. ________________

ICS 19.100
J 04

GB

中华人民共和国国家标准

GB/T 15822.3—2005/ISO 9934-3:2002

无损检测 磁粉检测
第3部分:设备

Non-destructive testing—Magnetic particle testing—
Part 3: Equipment

4. ________________

Text

1. Magnetic particle examination

It is a nondestructive method (Fig.48.1) of detecting cracks, porosity, seams, inclusions, lack of fusion, and other discontinuities in ferromagnetic materials. Surface discontinuities and shallow subsurface discontinuities (Fig.48.2) can be detected by using this method. There is no restriction as to the size and shape of the parts to be inspected; only ferromagnetic materials can be examined by this method.

2. When a crack is present, what are set up at its edge?

If a magnet is bent and the two poles are joined so as to form a closed ring, no external poles exist and hence it will have no attraction for magnetic materials. This is the basic principle of magnetic particle inspection. When a crack is present, north and south magnetic poles are set up at the edge of the crack. The magnetic particles will be attracted to the poles that are the edges of the crack or discontinuity.

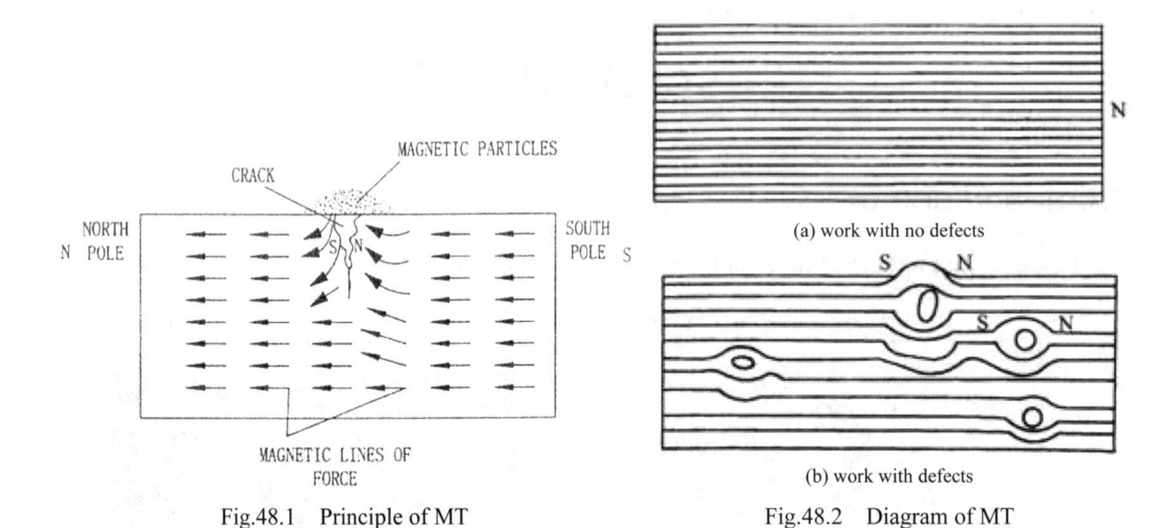

Fig.48.1 Principle of MT

Fig.48.2 Diagram of MT

Knowledge extention

The steps by which magnetic particle examination consist of.

This examination method consists of establishing a magnetic field in the test object, applying magnetic particles (Fig.48.3) to the surface of the test object, and examining the surface for accumulations of the particles that are the indications of defects.

The Working principle of MT machine is as follow shown in Fig.48.4.

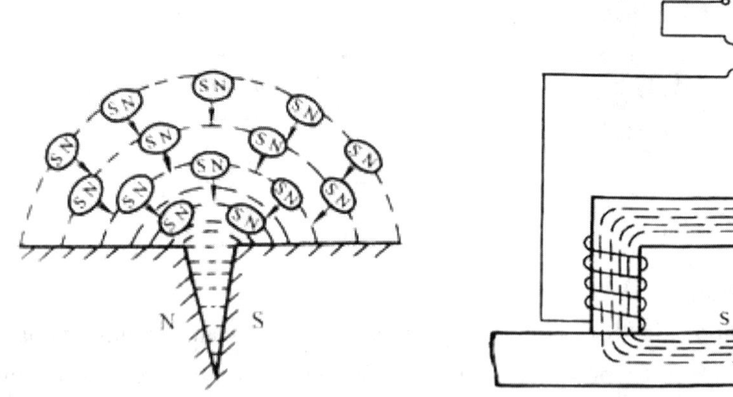

Fig.48.3 Particles' accumulation at the defects

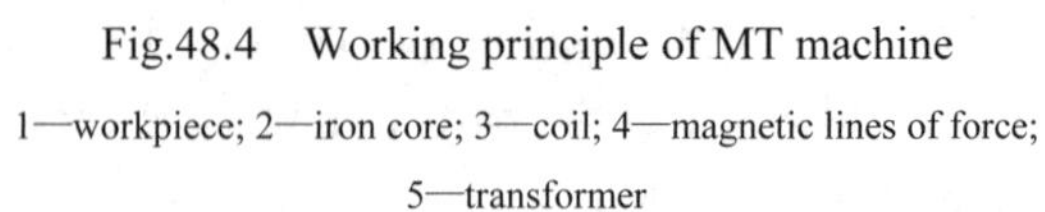

Fig.48.4 Working principle of MT machine

1—workpiece; 2—iron core; 3—coil; 4—magnetic lines of force; 5—transformer

Exercises

1. Match the English with the Chinese. Draw lines.

(1) magnet　　　　a. 颗粒

(2) particle　　　　b. 南极

(3) north pole　　　c. 磁铁

(4) south pole d. 北极
(5) magnetic lines of force e. 堆积
(6) accumulation f. 线圈
(7) transformer g. 磁力线
(8) iron core h. 变压器
(9) coil i. 铁芯

2. Judge the following true or false.

(1) Magnets will attract magnetic material only where the lines of force enter the magnet at the poles. ( )

(2) Alternating current is the most desirable type of current for subsurface discontinuities. ( )

(3) All of steel weldments in the aircraft industry are examined by the magnetic particle method. ( )

3. Fill in the blanks with the proper words in the text.

(1) It is a nondestructive method of detecting ______, ______, seams, ______, lack of fusion, and other ____________ in ferromagnetic materials. Surface discontinuities and ____________ can be detected by using this method.

(2) There is no restriction as to the ________ and ________ of the parts to be inspected; only ____________ materials can be examined by this method.

4. Answer the following questions.

(1) What is magnetic particle examination?

(2) Describe the principle of magnetic particle examination.

(3) What are the steps of MT ?

Words and phrases

magnetic particle [mæg'netɪk] ['pa:tɪklz] 磁粉
MT (magnetic particle testing)磁粉探伤
detect [dɪ'tekt] 检测
porosity [pɔ:'rɒsətɪ] 气孔
lack of fusion [læk] ['fju:ʒn] 未熔合
restriction [rɪ'strɪkʃn] 限制
the size and shape [saɪz] [ʃeɪp] 尺寸和形状
be examined by [ɪg'zæmɪnd] 用……测试
north pole [nɔ:θ] [pəʊl] 北极
standard ['stændəd] 标准
magnet ['mægnət] 磁铁
crack [kræk] 裂纹
seam [si:m] 焊缝
inclusion [ɪn'klu:ʒənz] 夹渣
as to 就，关于
surface 表面
be detected by 用……检测
south pole [saʊθ] 南极
the parts to be inspected [pa:ts] [in'spektid] 被检测的部分
magnetic lines of force [mæg'netɪk] [laɪnz] [fɔ:s] 磁力线
ferromagnetic materials [ˌferəʊmæg'netɪk] [mə'tɪərɪəlz] 铁磁性的材料
shallow subsurface discontinuities ['ʃæləʊ] ['sʌb'sɜ:fɪs] 近表面的缺陷（不连续）

Lesson 49 PT

Look and select

Look at the pictures and select words from the box.

steel plate	developer	penetrant	barrel

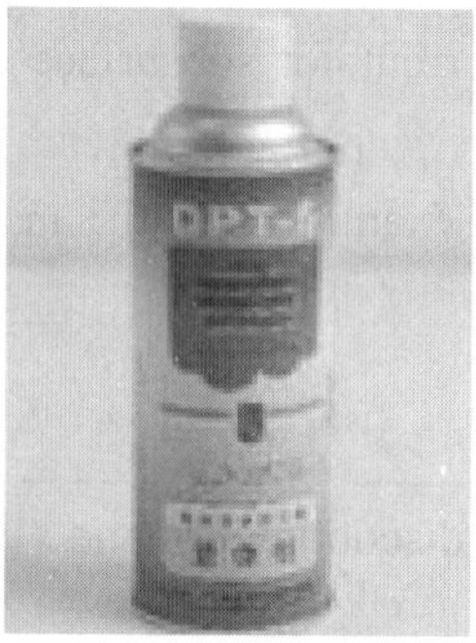

1. ________________

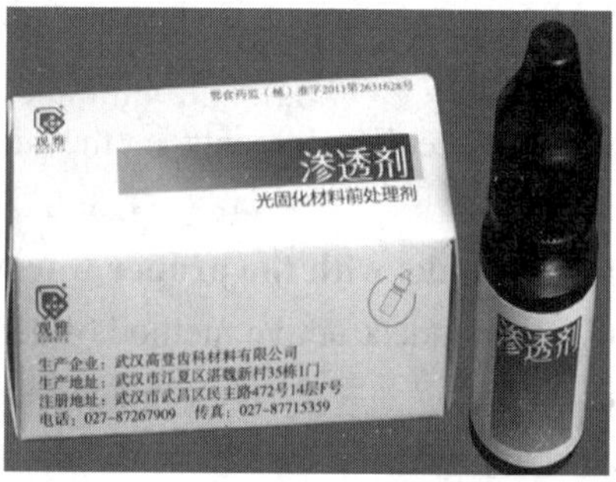

2. ________________

3. ________________

4. ________________

Text

1. Process of liquid-penetrant examination

A liquid penetrant is applied to the surface of the part to be inspected (Fig.49.1). The penetrant remains on the surface and seeps into any surface opening. The penetrant is drawn into the surface opening by capillary action. The parts may be in any position when tested. After sufficient penetration time has elapsed, the surface is cleaned and excess penetrant is removed.

2. Liquid- penetrant examination

It is a highly sensitive, nondestructive method for detecting minute discontinuities such as cracks, pores, and porosity, which are open to the surface of the material being inspected. This method may be applied to many materials, such as ferrous and nonferrous metals, glass, and plastics.

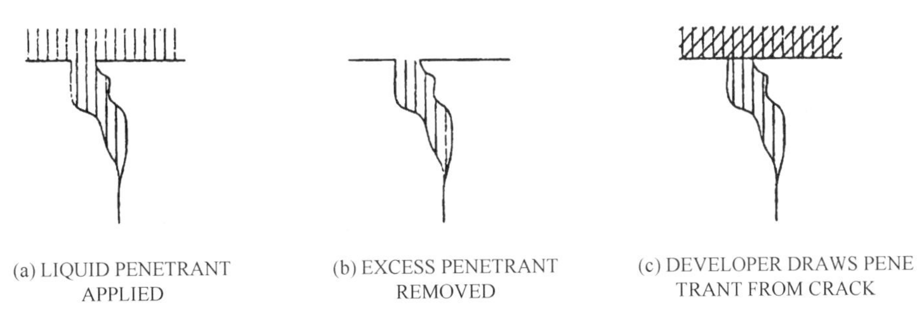

Fig.49.1 Process of penetrant examination

Knowledge extention

1. Five operations for PT using the water-washable system

These five essential operations are shown schematically for the water-washable system in Fig.49.2. The operations are similar for the other liquid-penetrant system.

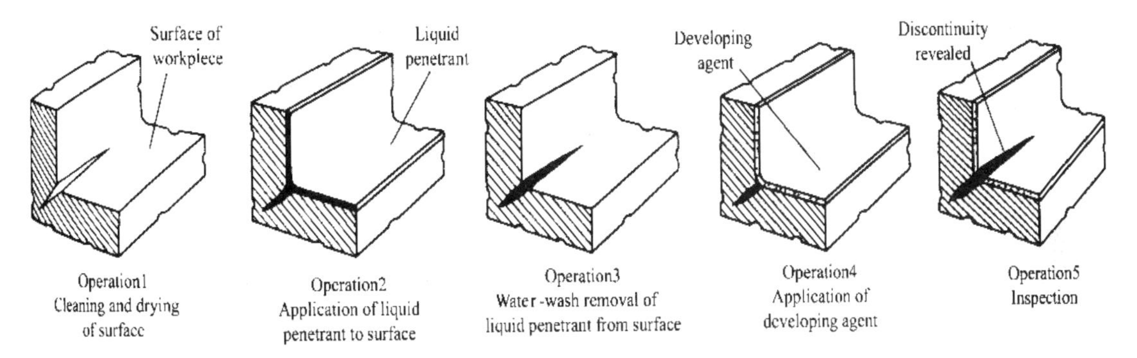

Fig.49.2 Five operations for PT using the water-washable system

2. Fluorescent inspection (Fig.49.3)

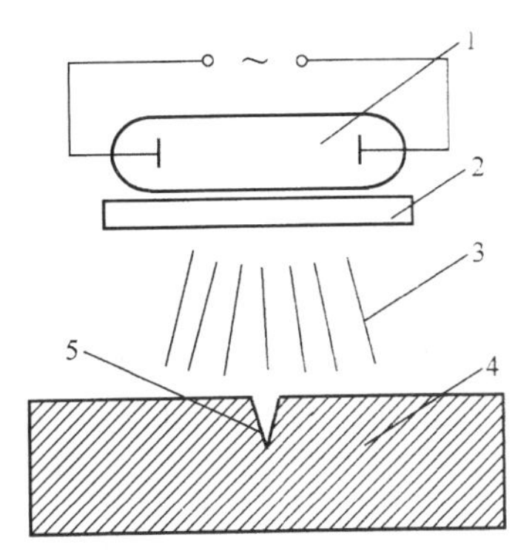

Fig.49.3 Fluorescent inspection

1—ultraviolet light source; 2—filter plate; 3—ultraviolet rays; 4—workpiece to be inspected; 5—defect filled with fluorescent powder

Exercises

1. Match the English with the Chinese. Draw lines.

A.

(1) developer	a. 毛细管作用
(2) penetrant	b. 灵敏的
(3) sensitive	c. 多余的
(4) capillary action	d. 应用，施加
(5) remove	e. 非破坏性
(6) excess	f. 荧光检测
(7) application	g. 渗透剂
(8) nondestructive	h. 显影剂
(9) fluorescent inspection	i. 去除

B.

(1) cleaning and drying of surface	a. 清洁并使表面干燥
(2) aplication of liquid penetrant to surface	b. 把液体渗透剂从表面洗去
(3) water-wash removal of liquid penetrant from surface	c. 将液体渗透剂施加于表面
(4) application of developing agent	d. 检验
(5) inspection	e. 施加显影剂于表面

2. Fill in the blanks with the proper words in the text.

(1) PT is the short form of ______________________. Its process is divided into three stages: (a) ______________, (b) excess penetrant removed and (c) ________________.

(2) It is a highly _________, _________method for detecting minute discontinuities such as ________, ________, and ________, which are open to the surface of the material being inspected.

3. Answer the following questions.

(1) What is liquid-penetrant examination?

(2) Describe the principle of liquid-penetrant examination.

(3) What are the five steps of PT ?

Words and phrases

liquid-penetrant examination ['lɪkwɪd] ['penətrənt] [ɪgˌzæmɪ'neɪʃn]（液体）渗透检测

steel plat [sti:l] [pleɪt] 钢板

penetrant ['penətrənt] 渗透剂

highly sensitive ['haɪli] ['sensətɪv] 高灵敏的

nondestructive ['nɒndɪs'trʌktɪv] 非破坏性的

discontinuity ['disˌkɔnti'nju(:)itiz] 非连续性

minute ['mɪnɪt] 最小

developer [dɪ'veləpə(r)] 显影剂

barrel ['bærəl] 桶，筒体

method ['meθəd] 方法

detecting [dɪ'tektɪŋ] 检测

pore ['pɔ:z] 孔

crack [kræk] 裂纹

porosity [pɔ:'rɒsətɪ] 气孔
the material being inspected 待检材料
be applied to [ə'plaɪd] 应用于，施加到
glass [glɑ:s] 玻璃
draw...from 把……从……里面吸出
the part to be inspected [in'spektid] 待检部件
seep into [si:p] 渗进
be in any position [pə'zɪʃn] 在任何位置
sufficient [sə'fɪʃnt] 足够的
excess [ɪk'ses] 多余
operation [ˌɒpə'reɪʃn] 操作
be similar ['sɪmələ(r)] 相似
application [ˌæplɪ'keɪʃn] 施加，应用
be open to the surface 表面开口
ferrous ['ferəs] 铁的
nonferrous [nɒn'ferəs] 非铁的
plastics ['plæstɪks] 塑料
principle ['prɪnsəpl] 原理
remain [rɪ'meɪn] 保留
opening ['əʊpnɪŋ] 开口
be drawn into [drɔ:n] 渗透进
be cleaned [kli:nd] 清理，清洁
be removed [rɪ'mu:vd] 清除
figure ['figə] 图
dry [draɪ] 干燥
revealed [rɪ'vi:ld] 显示
developing agent [dɪ'veləpɪŋ] ['eɪdʒənt] 显像剂
water-washable system ['wɒʃəbl] ['sɪstəm] 水洗型渗透探伤
be shown schematically [ʃəʊn] [ski:'mætɪklɪ] 如图所示

Lesson 50 ET

Look and select

Look at the pictures and select the terms from the box.

probe	magnetic field	ET tester	magnetizing coil

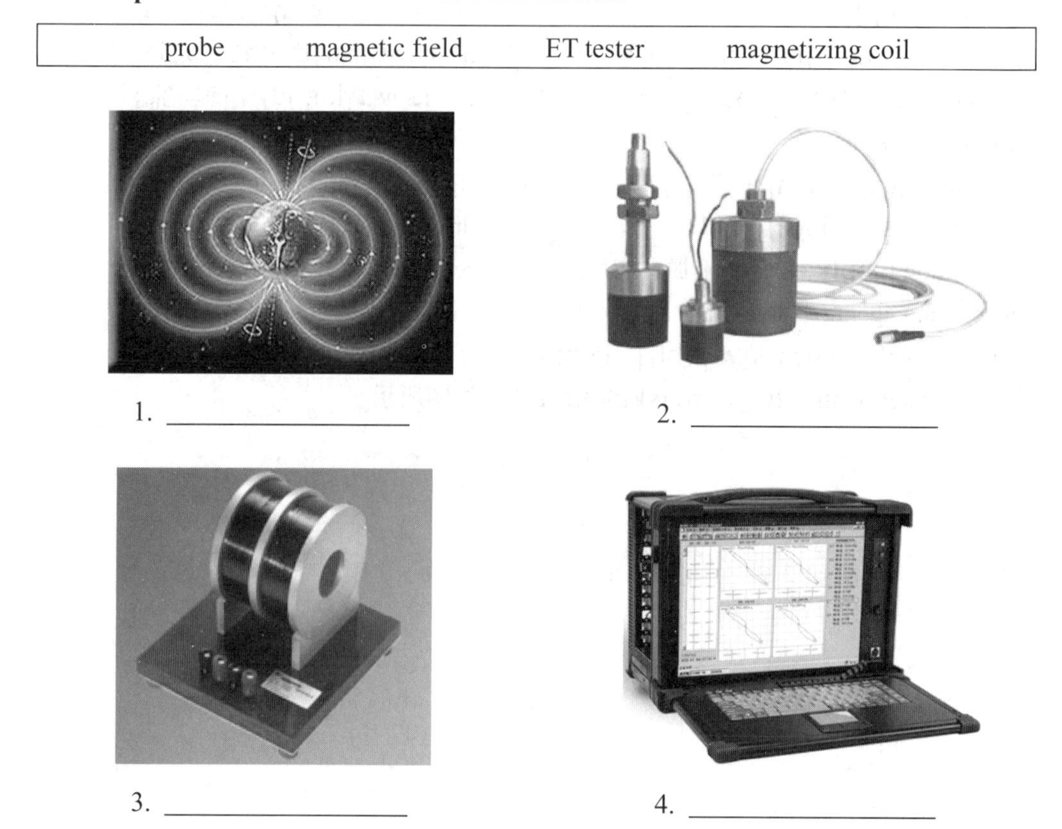

1. ________________ 2. ________________

3. ________________ 4. ________________

Text

1 . The conception of eddy-current testing

When electrically conductive material is subjected to an alternating magnetic field, small circulating electric currents are generated in the the material (Fig.50.1). These so-called eddy currents (涡流) are affected by variations in conductivity (电导率), magnetic permeability (磁导率), mass, and homogeneity (均匀性) of the host material. Conditions that affect these characteristics can be sensed by measuring the eddy-current response of the part.

The eddy currents induced into the part interact with the magnetic field of the exciting coil, thereby influencing the impedance, which is the total opposition to the flow of current from the combined effect of resistance, inductance, and capacitance of the coil, illustrated in Fig.50.2.

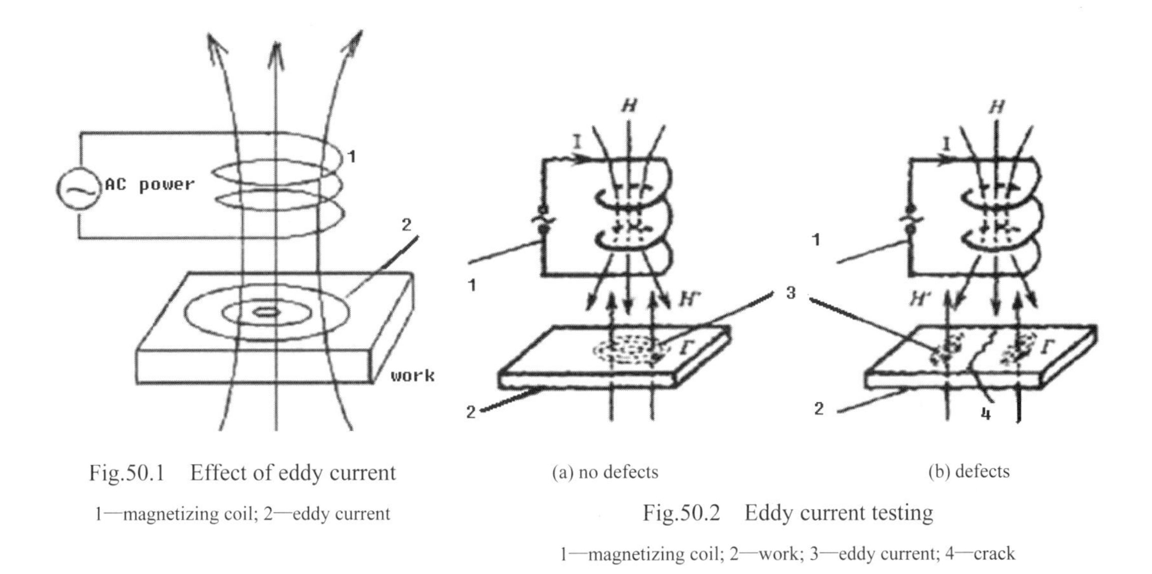

Fig.50.1 Effect of eddy current

1—magnetizing coil; 2—eddy current

(a) no defects (b) defects

Fig.50.2 Eddy current testing

1—magnetizing coil; 2—work; 3—eddy current; 4—crack

2. NASA's report

According to a NASA (Fig.50.3) report, eddy-current testing is not as sensitive to small, open flaw as in liquid penetrant. However, it does not require cleanup operations and is generally faster. Compared with magnetic particle testing, eddy-current tests are not as sensitive as to small flaws but they work equally as well on ferromagnetic and nonferromagnetic material.

Fig.50.3 Logo of NASA

Knowledge extention

Principle of ET

Magnetization of the workpiece may vary depending upon the application, but for optimum indication the direction of the magnetic fields should be nearly right angles to the fault, as shown in Fig.50.4.

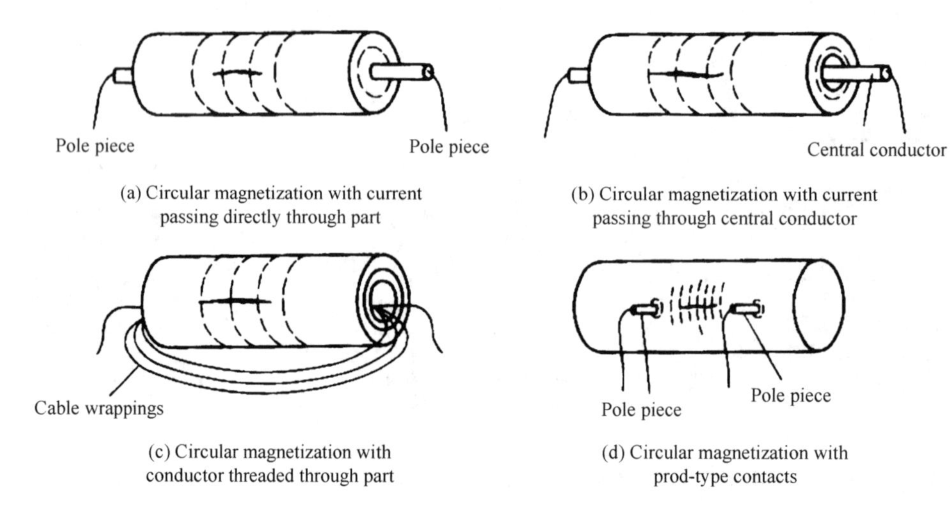

Fig.50.4 Principle of eddy current

Exercises

1. Translate the English terms in the picture (Fig.50.5) into Chinese.

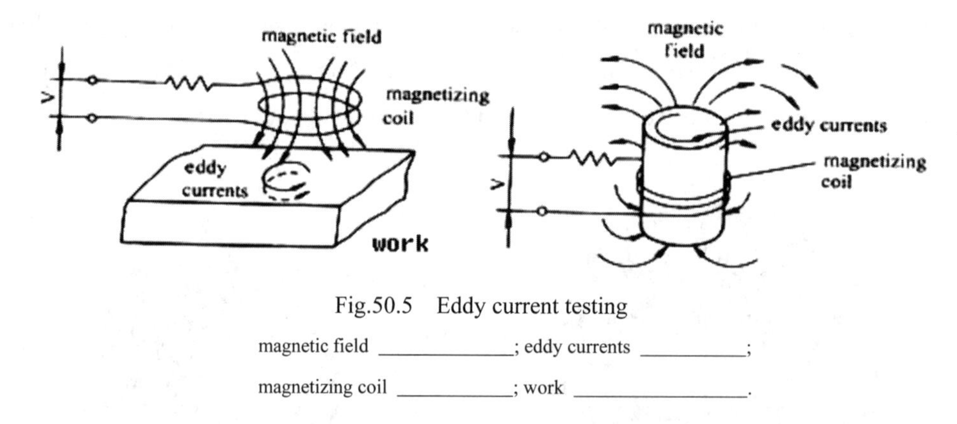

Fig.50.5 Eddy current testing

magnetic field ____________; eddy currents __________;

magnetizing coil __________; work ________________.

2. Match the English with the Chinese. Draw lines.

(1) ET	a. 美国国家航空航天局
(2) probe	b. 裂纹
(3) defect	c. 涡流检测
(4) crack	d. 电导率
(5) AC power	e. 探头
(6) conductivity	f. 缺陷
(7) NASA	g. 交流电源

3. Answer the following questions.

(1) What is the meaning of eddy-current?

(2) What is the principle of eddy-current testing?

Words and phrases

ET (eddy-curent testing) 涡流探伤（检测）

magnetic field [mæg'netɪk] [fi:ld] 磁场

magnetizing coil [kɔɪl] 励磁线圈

be subjected to ['sʌbdʒektid] 受作用于

be affected by [ə'fektɪd] 受……影响

be generated in [d'ʒenəreɪtɪd] 产生于

eddy current ['edɪ] 涡流

magnetic permeability [ˌpɜ:mɪə'bɪlətɪ] 磁导率

mass [mæs] 质量

homogeneity [ˌhɒmədʒə'ni:əti] 均匀性

affect [ə'fekt] 影响

host material [həʊst] [mə'tɪəriəl] 基材

condition [kən'dɪʃnz] 条件

characteristic [ˌkærɪktə'rɪstɪk] 特性

measure ['meʒə(r)] 测量

cable wrapping ['keɪbl] ['ræpɪŋ] 绕圈

NASA 美国航空航天局

electrically conductive material [kən'dʌktɪv] 导电性材料

circulating electric currents ['sɜ:kjʊleɪtɪŋ] 环形电流

alternating current ['ɔ:ltəneɪtɪŋ] ['kʌrənt] 交流电

can be sensed by ['sensd] 通过……感知/应到

circular magnetization ['sɜ:kjələ(r)] [ˌmægnətɪ'zeɪʃən] 周向磁化

central conductor ['sentrəl] [kən'dʌktə(r)] 中心导体

Electrical and Instrument Part

Lesson 51 Basic Knowledge of Electrotechnics

Look and select

Look at the pictures and select the correct terms from the box.

oscilloscope	signal generator	multimeter
ammeter	mega ohmmeter	bridge

1. ______________

2. ______________

3. ______________

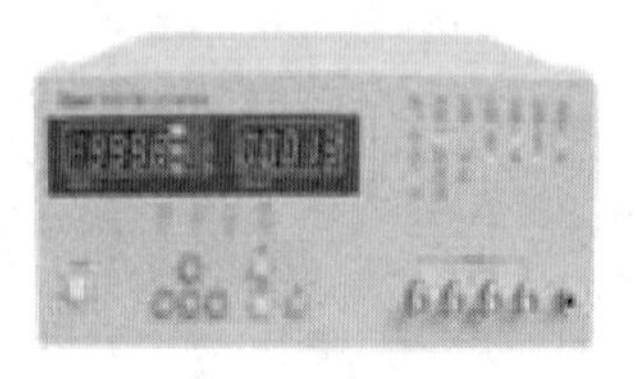

4. ______________

5. ______________

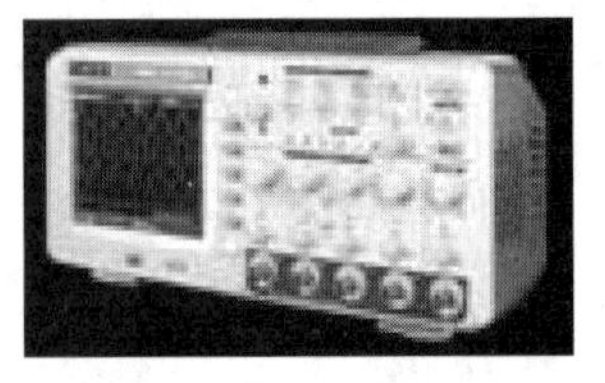

6. ______________

Text

Electrotechnics includes the manufacture and maintenance of electrical machines and instruments. What the electricians care most are five parameters: current, voltage, inductance, capacitance and resistance. In electrical measurement, we measure these parameters to find out the trouble. See Fig.51.1 as following:

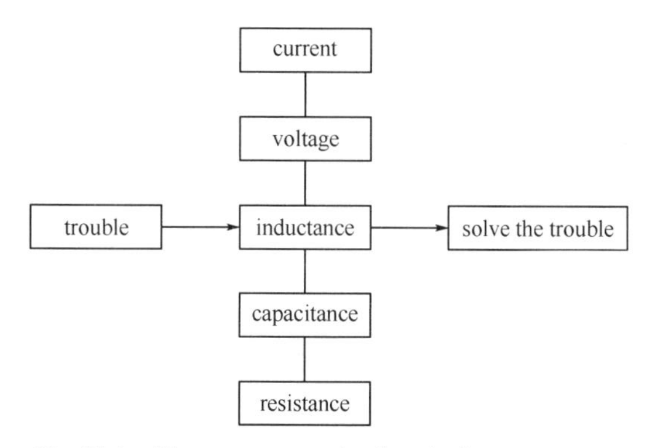

Fig.51.1 Five parameters in electrical measurement

Knowledge extention

My name is Liu Xiaobing. I'm 18 years old. I am a student in Hunan Industrial Technician College (Central South Industrial School). My major is Electrical and Instrument Engineering. I like my major very much. I want to become a good Electrical and Instrument Engineer in the future.

Exercises

1. Fill in the blanks (Table 51.1) with the proper words in the text.

Table 51.1 Five parameters in electrical measurement

项目	英文名称	符号表示	单位	单位的中译文
电压				
电流				
电阻				
电容				
电感				

2. Answer the following questions according to the hints given before.

(1) What is Electrotechnics?

(2) What's your name?

(3) What's your major?

(4) What's your job?

(5) Try to make a simple self-introduction by filling in the blanks.

My name is_名字_. I come from_家乡_. I'm_年龄_years old. I am a student in_学校_ (Central South Industrial School). My hobby is_爱好_. My major is_专业_. I like my major very much. I want to become_职业_in the future.

3. Match the English with the Chinese. Draw lines.

(1) oscilloscope a. 万用表

(2) signal generator b. 安培表

(3) multimeter c. 电桥

(4) ammeter d. 示波器

(5) mega ohmmeter e. 信号发生器

(6) bridge f. 兆欧表

4. Understand the following.

Electrotechnics and electronics are not the same (see Fig.51.2). The former is always about high-voltage or a large current field, and the latter is about low-voltage and a small current field. Here, we just discuss electrotechnics.

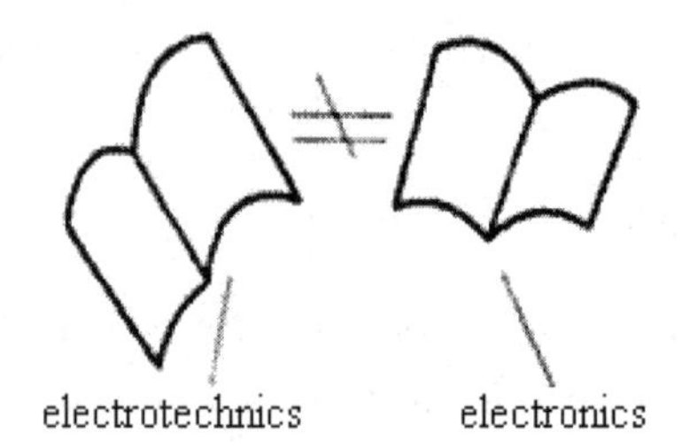

Fig.51.2 Electrotechnics and electronics

New and phrases

oscilloscope [ə'sɪləskəʊp] 示波器

measurement ['meʒəmənt] 测量

current ['kʌrənt] 电流

voltage ['vəʊltɪdʒ] 电

bridge [brɪdʒ] 桥，电桥

electrotechnics [ɪ'lektrə'teknɪks] 电工学

electronics [ɪˌlek'trɒnɪks] 电子学

instrument ['ɪnstrəmənt] 仪表

engineer [ˌendʒɪ'nɪə(r)] 工程师

electrical [ɪ'lektrɪkl] 电气的

multimeter ['mʌltɪmi:tə] 万用表

ammeter ['æmi:tə(r)] 安培表

inductance [ɪn'dʌktəns] 电感

capacitance [kə'pæsɪtəns] 电容

major ['meɪdʒə(r)] 专业

industrial [ɪn'dʌstriəl] 工业的

engineering [ˌendʒɪ'nɪərɪŋ] 工程

technician [tek'nɪʃn] 技师

signal generator ['sɪgnəl] ['dʒenəreɪtə(r)] 信号发生器

mega ohmmeter ['megə] ['əʊmmi:tə(r)] 兆欧表

Lesson 52 Circuits

Look and select

Look at the pictures and select the correct terms from the box.

plier	knife	soldering iron
wire stripper	screwdriver	desoldering tool

1. ____________ 2. ____________ 3. ____________

4. ____________ 5. ____________ 6. ____________

Text

An electric circuit often consists of four basic parts (Fig.52.1): the power supply such as a battery, the conductor or wire, the control device such as a switch, and the load or electric appliance.

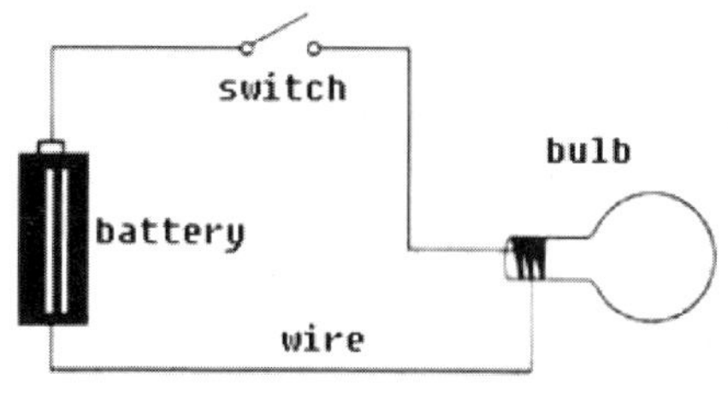

Fig.52.1 Circuit of flashlight

A circuit usually has three states: closed circuit or loop, open circuit and short circuit.

There are three kinds of circuits: the series circuit, parallel circuit and series-parallel circuit. They are shown in Fig.52.2.

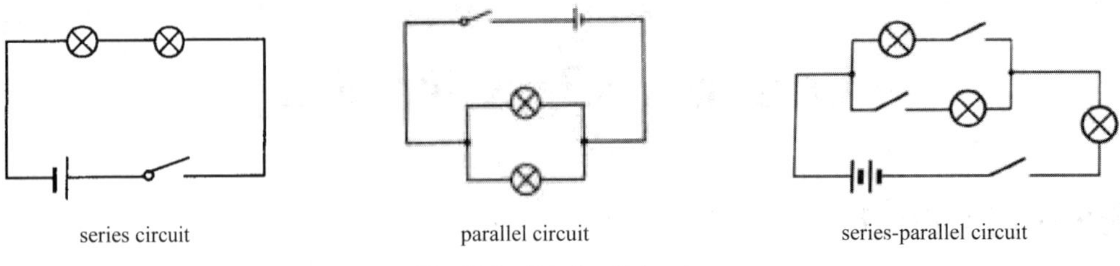

Fig.52.2 Kinds of circuits

Knowledge extention

Current will flow only if a circuit is complete. Switches turn circuits on and off by making them complete or incomplete (Fig.52.3).

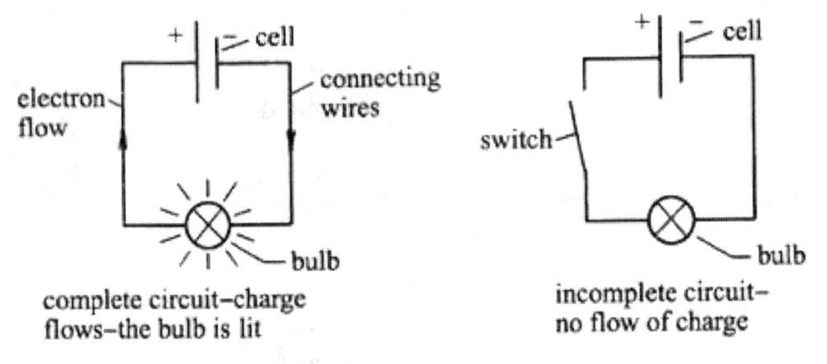

Fig.52.3 Complete and incomplete circuits

Exercises

1. Fill in the blanks with the proper letters.

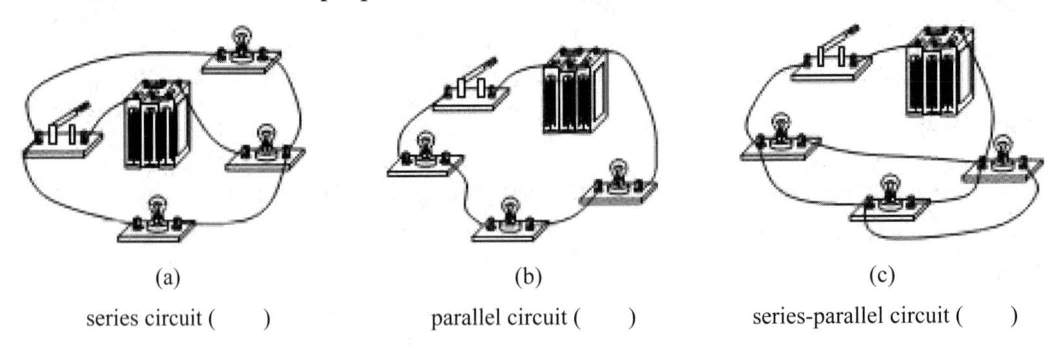

(a) series circuit (   )

(b) parallel circuit (   )

(c) series-parallel circuit (   )

2. Fill in the blanks with the proper words in the text.

(1) An electric circuit often consists of four basic parts: the power supply such as a __________, __________, the control device such as a __________, and __________.

(2) A circuit usually has three states: __________ , __________ and short circuit.

3. Translate the following into Chinese.

(1) In a series circuit (Fig.52.4), the current has only one path to follow. There are no branches. One switch will turn the whole circuit on or off.

(2) In a parallel circuit (Fig.52.5), there are several paths the current might follow. Switches can be used to turn the whole or just part of the circuit on or off.

When S is closed, the bulbs light up.

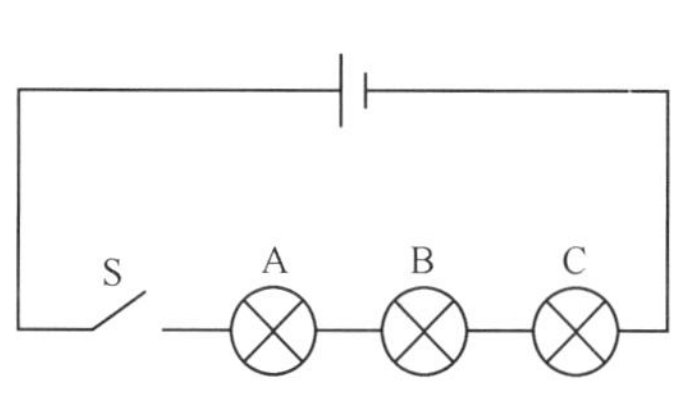

Fig.52.4 Series circuit

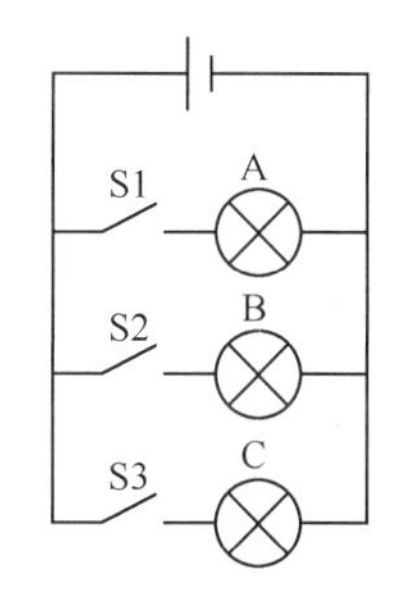

Fig.52.5 Parallel circuit

4. Match the English with the Chinese. Draw lines.

A

(1) plier	a. 电工刀
(2) knife	b. 电烙铁
(3) soldering iron	c. 吸锡器
(4) wire stripper	d. 旋具（起子）
(5) screwdriver	e. 剥线钳
(6) desoldering tool	f. 钳子

B

(1) complete	a. 灯泡
(2) incomplete	b. 开关
(3) bulb	c. 电子
(4) switch	d. 完整，闭合
(5) electron	e. 流动
(6) flow	f. 不完整，断开

Words and phrases

circuit ['sɜːkɪt] 电路
knife [naɪf] 电工刀
wire stripper 剥线钳
desoldering tool 吸锡器
consist of 由……组成
battery ['bætəri] 电池
wire ['waɪə(r)] 导线，电线
switch [swɪtʃ] 开关
electric appliance 电气设备
closed circuit 闭合电路
short circuit 短路
parallel circuit 并联电路
complete 完整的，闭合的
cell [sel] 电池
connecting wire 连接导线
charge flow 电荷流动
path [pɑːθ] 路径，小道
turn on/off 打开/关闭
plier ['plaɪə] 虎钳
soldering iron 烙铁
screwdriver 旋具，螺丝刀
basic part 基本部分
power supply 电源
conductor [kən'dʌktə(r)] 导体
control device 控制仪器/设备
load [ləʊd] 负荷，负载
state [steɪt] 状态
loop [luːp] 回路
series circuit 串联电路
current 电流
incomplete 不完整的，断开的
electron flow 电子流动
bulb [bʌlb] 灯泡
be lit 点亮的，燃着的
branch [brɑːntʃ] 支路

Lesson 53 Electromagnetism

Look and select

Look at the pictures and select the correct terms from the box.

wire	magnetic field	motor	generator

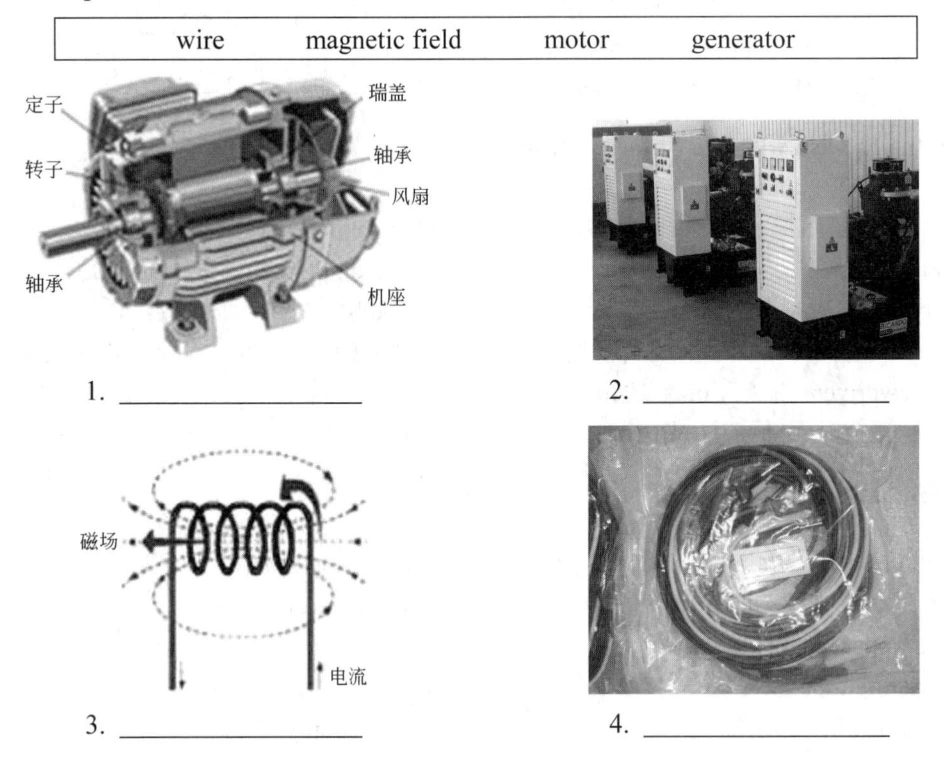

1. ________________ 2. ________________

3. ________________ 4. ________________

Text

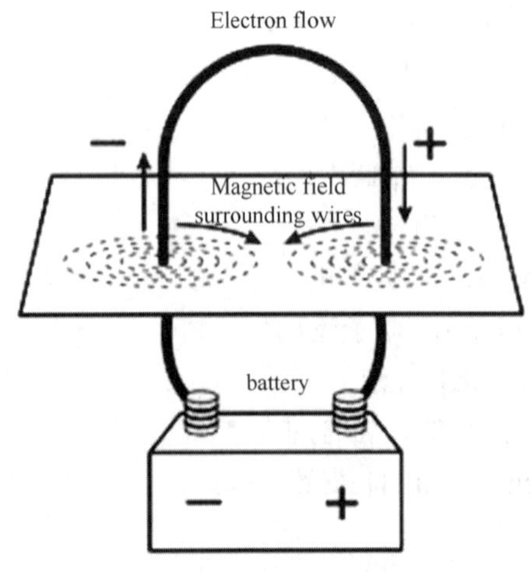

Fig.53.1 Current and magnetic field

Electromagnetism is the foundation of electric motors and generators. It is mainly about the relationship between three terms: the current in wire, the magnetic field and the force (Fig.53.1).

The left-hand rule: The direction of the magnetic field must go from the palm to the back of the palm, and the other four fingers must point to the current direction in the wire, so the force will have a trend to let the wire move in the thumb's direction (Fig.53.2).

The right-hand rule: The moving of the wire goes firstly. The direction of magnetic field must go from the palm to the back of the palm, and the thumb must point to the moving direction. By doing so, the current in the wire is in the other four fingers' direction (Fig.53.2).

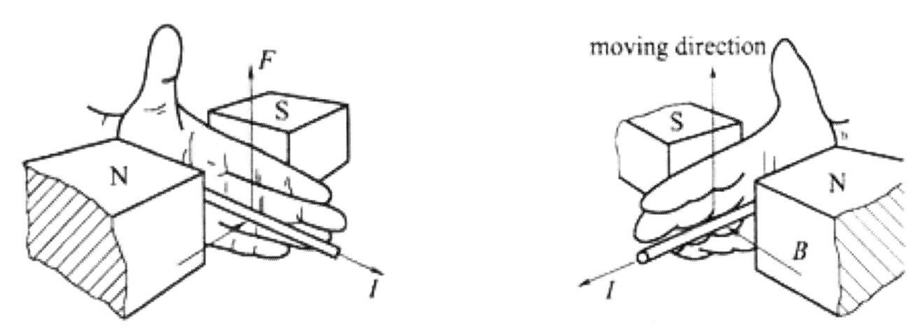

Fig.53.2 Left hand and right hand rules

Knowledge extention

Electricity and magnetism

At the beginning of the 19th century, a series of discoveries in the field of electromagnetism started off a new technology revolution. In 1819, a Danish scientist, Hans Christian Oersted (Fig.53.3) discovered by chance that when a compass was placed near an electrical wire, the needle would point in the direction perpendicular to the wire (Fig.53.4). After this, he discovered the relationship between electricity and magnetism.

Fig.53.3 Hans Christian Oersted

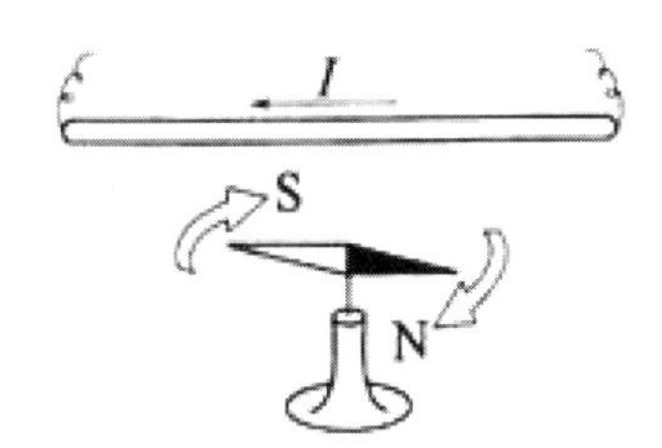

Fig.53.4 Compass and electrical wire

Exercises

1. Answer the following questions.

(1) How significant electromagnetism is in physics?

(2) What are the three terms of electromagnetism?

(3) How does the left-hand rule describe?

(4) How does the right-hand rule describe?

2. Match the English with the Chinese. Draw lines.

(1) electromagnetism	a. 手掌
(2) magnetic field	b. 手指
(3) palm	c. 电磁学

(4) finger d. 磁场
(5) force e. 基础
(6) direction f. 力
(7) foundation g. 方向

3. True or false.

(1) Electromagnetism is mainly about the relationship between three terms: the current in wire, the magnetic field and the force. (　　)

(2) As for the left-hand rule, the thumb points to the current direction in the wire, so the other four fingers will have a trend to let the wire move in the thumb's direction. (　　)

(3) As for the right-hand rule, the other four fingers must point to the moving direction. By doing so, the current in the wire is in the thumb direction. (　　)

4. Fill in the blanks (Table 53.1) with the phrases in the Knowledge extention.

Table 53.1 Electricity and magnetism

Field	
Time	
Country	
Name	
Discovery	
Relationship	

Words and phrases

electromagnetism [ɪˌlektrəʊ'mægnətɪzəm] 电磁学
magnetic [mæg'netɪk] 磁性的
generator ['dʒenəreɪtə(r)] 发电机
foundation [faun'deiʃən] 基础
relationship [rɪ'leɪʃnʃɪp] 关系
direction [dɪ'rɛkʃən, daɪ-] 方向
palm [pɑ:m] 手掌
finger ['fɪŋgə(r)] 手指
trend [trend] 趋势
thumb [θʌm] 大拇指
wire ['waɪə(r)] 电线，导线
field [fi:ld] 场地，田野
motor ['məʊtə(r)] 电机，马达
electric [ɪ'lektrɪk] 电的
term [tɜ:m] 术语，名词
force [fɔ:s] 力，力量
back [bæk] 背，后面
point to 指向
move [mu:v] 移动
be mainly about 主要关于

Lesson 54 Resistance

Look and select

Look at the pictures and select the correct terms from the box.

RCD	MCB	circuit	electrical symbol

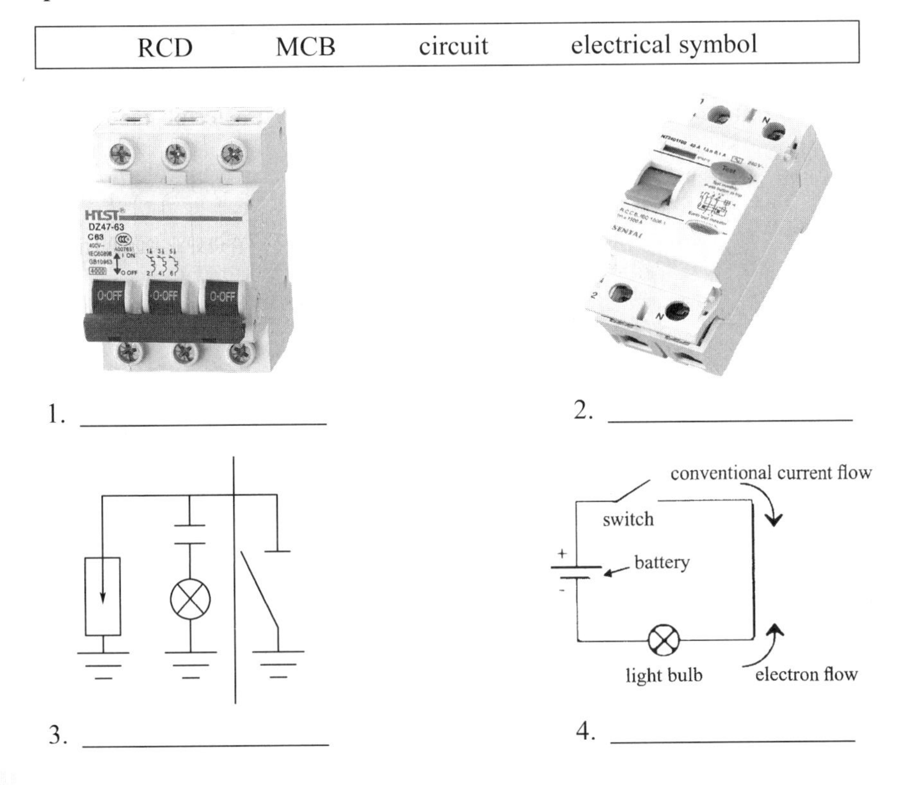

Text

Resistance impedes the flow of current. It is represented by the symbol R and is measured in unit of ohm.

Resistance components are as follows: heating elements, resistors, variouble resistors, thermistors, light-dependent resistors (LDRs) and diodes.

heating element　resistor　variable resistor　thermistor　LDR　diode

If the resistors are connected in series, their total resistance R_T is given by:

$$R_T = R_1 + R_2 + R_3 + R_4 + \cdots + R_N$$

If the resistors are connected in parallel, the resistance R_T can be found using the equation:

$$\frac{1}{R_T} = \frac{1}{R_1} + \frac{1}{R_2} + \frac{1}{R_3} + \cdots + \frac{1}{R_n}$$

Knowledge extention

Resistance and current

Resistance is the opposition to the flow of current and represented by the letter symbol R. The unit of resistance is the ohm, expressed by using (Ω). Larger amount of resistance are commonly expressed in kilo-ohm (kΩ) and in mega-ohm (MΩ).

The resistance of a piece of wire depends upon the length of the wire, the cross-section area of the wire, the temperature of the wire and the material from which the wire is made.

To control the size of the current flowing in circuit we use resistor (Fig.54.1). The resistors have values measured in ohms (Ω). A resistor of 100 Ω is a much greater obstacle of current than a resistor of 10Ω.

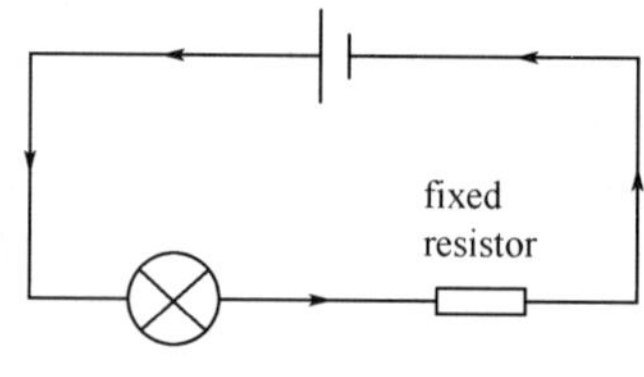

Fig.54.1 Circuit

Exercises

1. Match the English with the Chinese. Draw lines.

(1) heating element　　a. 可变电阻
(2) resistor　　b. 热敏电阻
(3) variable resistor　　c. 加热元件
(4) thermistor　　d. 电阻器
(5) light-dependent resistor　　e. 二极管
(6) diode　　f. 光敏电阻
(7) RCD　　g. 小型断路器
(8) MCB　　h. 漏电保护器

2. Answer the following questions:

(1) What is the importance of resistance?

(2) What are the components of resistance?

(3) What would happen to the total resistance R_T if the resistors are connected in series or in parallel?

3. Select and fill in the blanks about symbols with the letters.

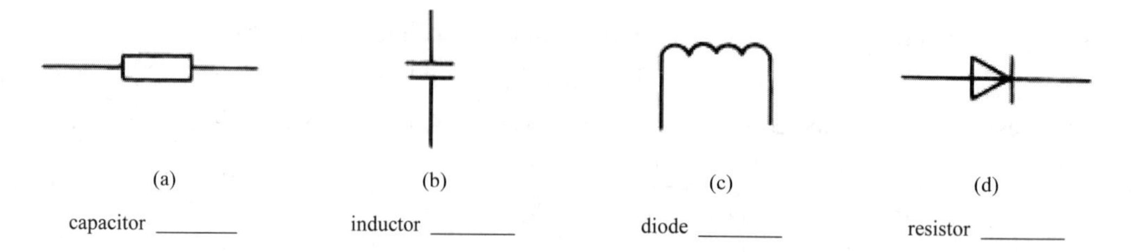

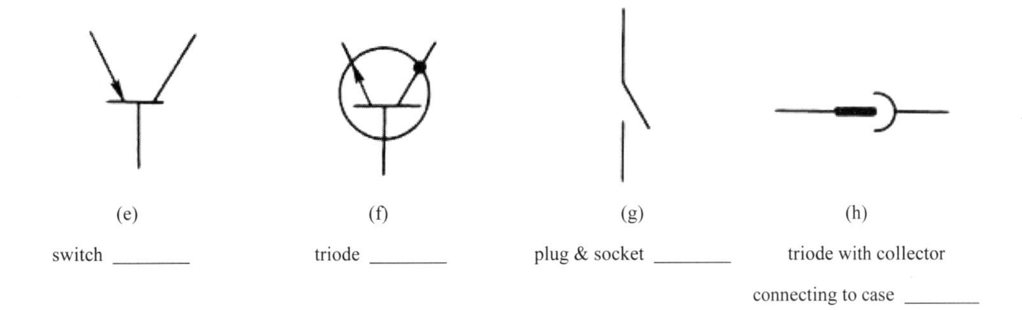

Words and phrases

RCD (residual current device) 漏电保护器
MCB (miniature circuit breaker) 小型断路器
light-dependent resistor (LDR) 光敏电阻
symbol ['sɪmbəl] 符号，象征
measure ['meʒə(r)] 测量
component [kəm'pəʊnənt] 元件，组件
element ['elɪmənt] 元件，元素
variable ['veəriəbl] 可变/可调的
diode ['daɪəʊd] 二极管
series ['sɪəri:z] 串联
use [ju:s] 使用，利用
be connected in 以……连接
as follow 如下
be found 得到，发现
letter 字母
commonly 普通地，一般地
kilo ['ki:ləʊ] 千，千克
a piece of 一根，一条
length [lɛŋkθ] 长度
temperature ['temprətʃə(r)] 温度
be made from 制成
size [saɪz] 规格，尺寸
obstacle ['ɒbstəkl] 障碍，阻碍
impede [ɪm'pi:d] 阻止，阻抗
unit ['ju:nɪt] 单位，单元，装置
heat [hi:t] 加热，热
resistor [rɪ'zɪstə(r)] 电阻（器）
thermistor [θɜ:'mɪstə] 热敏电阻
total ['təʊtl] 总计，总的
parallel ['pærəlel] 并联
equation [ɪ'kweɪʒn] 公式，等式
be given by 给出，得出
be represented by 用……表示/代表
opposition [ˌɑpə'zɪʃən] 相反
larger 较大
be expressed in 用……表示
mega ['megə] 兆
depends upon 取决于……
cross-section area 横截面
material [mə'tɪəriəl] 材料
control [kən'trəʊl] 控制
value ['vælju:] 价值，数值
fixed resistor 不变/固定电阻

Lesson 55 Current, Inductors and Capacitors

Look and select

Look at the pictures and select the correct terms from the box.

soldering gun	safety sign	power line	engineer

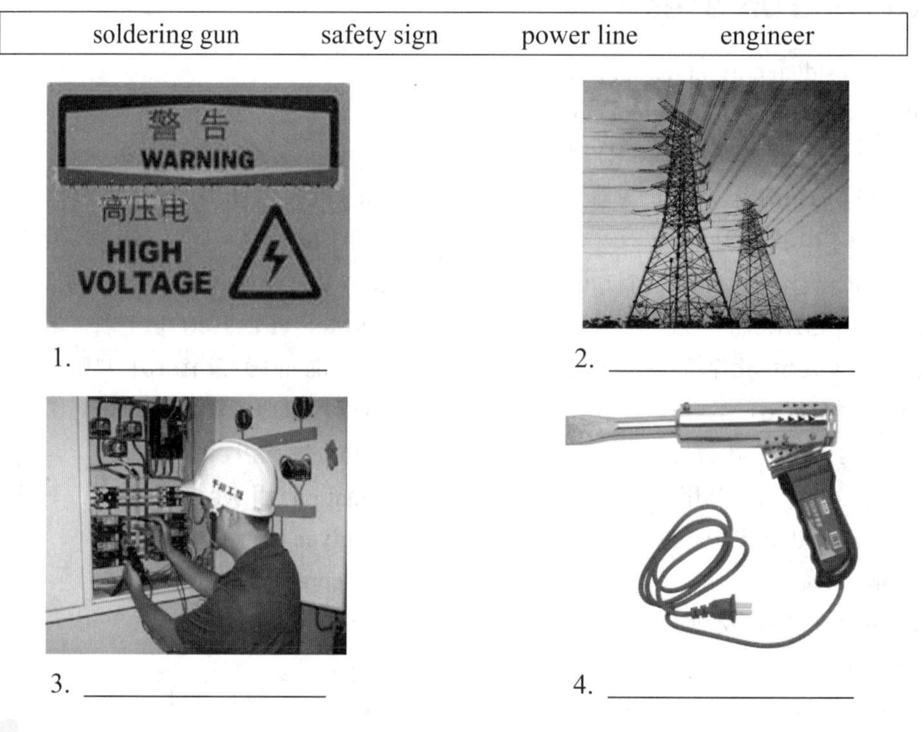

1. ________ 2. ________

3. ________ 4. ________

Text

Current

An electric current is the flow of electrons. The electrons move from minus to plus, the electric current flows from the positive to the negative in a completed circuit (Fig.55.1). The unit of current is ampere.

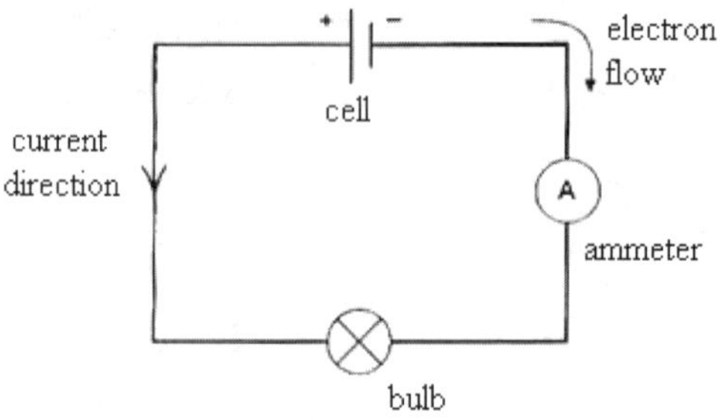

Fig.55.1 Current direction and electron flow

Inductors

Inductor is a device to resist change in current flow. It is simply a coil of wire with or without a magnetic core. There are two types of inductors, air core and iron core (Fig.55.2). Its value is expressed in Henry (or milli-henry).

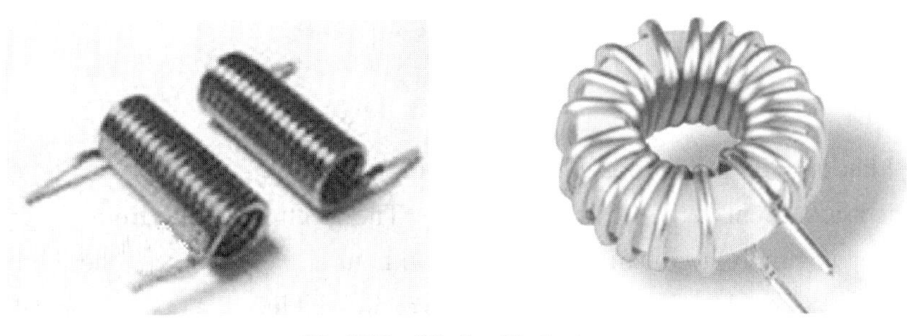

Fig.55.2 Kinds of inductors

Capacitors

Capacitor is a device that resists change in voltage. It stores charge. Its value is expressed in Farads (or micro-farads). There are many types, such as the ceramic capacitor, the paper capacitor, the plastic film capacitor, the metal silicon dioxide capacitor and the electrolytic capacitor (Fig.55.3).

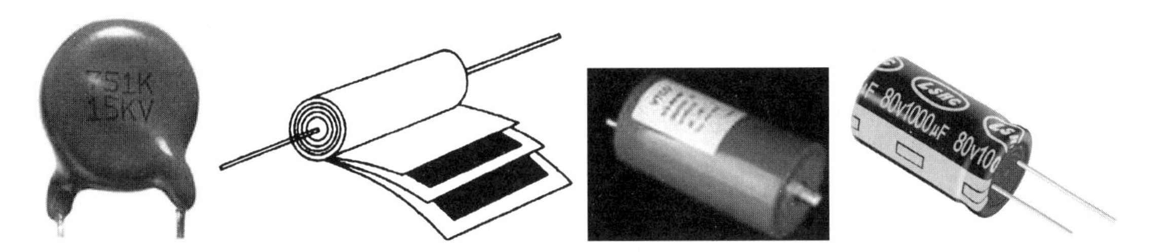

Fig.55.3 Kinds of capacitors

Knowledge extention

Total capacitor can be calculated by adding each of the capacitances connected in parallel. The formula is as follows:

$$C_T=C_1+C_2+C_3+\cdots+C_n$$

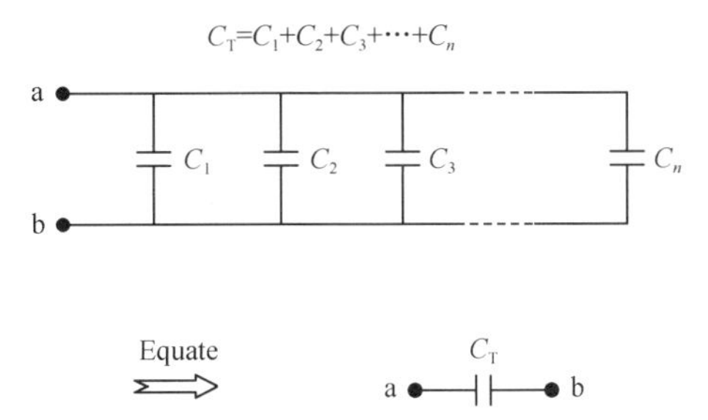

Total capacitance can be calculated from capacitances connected in series by the following formula:

$$\frac{1}{C_T}=\frac{1}{C_1}+\frac{1}{C_2}+\frac{1}{C_3}+\cdots+\frac{1}{C_n}$$

Exercises

1. Fill in the blanks with the proper words in the text.

(1) An electric current is the flow of ________. The electrons move from ______ to ______.

(2) Inductor is simply a coil of wire with or without a __________. There are two types of inductors, _________ and _________. Its value is expressed in _________ (or milli-henry).

(3) Capacitor's value is expressed in ________ (or micro-farads). There are many types.

2. Match the English with the Chinese. Draw lines.

(1) ceramic capacitor	a.（塑料）薄膜电容器
(2) paper capacitor	b. 电解电容器
(3) plastic film capacitor	c. 金属二氧化硅电容器
(4) metal silicon dioxide capacitor	d. 纸质电容器
(5) electrolytic capacitor.	e. 陶瓷电容器

3. True or false.

A.

(1) The electric current flows from the positive to the negative in a completed circuit. So do the electrons. ( )

(2) The unit of current is volt. ( )

(3) Inductor is a device to resist change in electron flow. ( )

(4) Its value is expressed in Ampere. ( )

B.

(1) A capacitor is a device to resist change in current. ( )

(2) A capacitor stores charge. ( )

(3) The electrolytic capacitor is one kind of capacitors. ( )

(4) The formulas for calculation of capacitances in series and in parallel are the same. ( )

Words and phrases

soldering gun ['sɒldərɪŋ] [gʌn] 钎焊枪

flow 流动

move 移动

plus [plʌs] 加，正的

negative ['negətɪv] 负的，阴极

unit 单位，单元

inductor [ɪn'dʌktə] 电感

safety sign 安全标志

electron [ɪ'lektrɒn] 电子

minus ['maɪnəs] 减，负的

positive ['pɑzɪtɪv] 正的，阳极

completed circuit 闭合电路

ampere ['æmpeə(r)] 安培

device [dɪ'vaɪs] 仪器，设备

resist [rɪ'zɪst] 阻止
coil [kɔɪl] 圈，电圈
core [kɔ:(r)] 芯，核
air core 空芯
value ['vælju] 数值，价值
milli 毫
store [stɔ:(r)] 储存，贮存
farad 法拉
charge [tʃɑ:dʒ] 电荷
magnetic [mæg'netɪk] 磁性的
type [taɪp] 种类
iron core ['aɪən] 铁芯
henry 亨利
capacitor 电容
be expressed in 用……表示
micro ['maɪkrəʊ] 微
ceramic capacitor [sə'ræmɪk] 陶瓷电容器
paper capacitor 纸质电容器
plastic film capacitor（塑料）薄膜电容器
electrolytic capacitor [ɪˌlektrə'lɪtɪk] 电解电容器
metal silicon dioxide capacitor ['metl] ['sɪlɪkən] [daɪ'ɒksaɪd] 金属二氧化硅电容器

Lesson 56 Electrical Measurement

Look and select

Look at the pictures and select the correct terms from the box.

fire alarm	fuse	telecommunication system	switch

1. ________________

2. ________________

3. ________________

4. ________________

Text

When we choose a measuring meter, we must pay attention to the following (Fig.56.1):

(a) Choose the right measuring range.

(b) Choose the appropriate measurement accuracy.

(c) Don't take down the meter at random.

(d) Have a rectification after a long-time use.

(e) Differentiate different types of measurement and then choose the right meter.

(f) Consider the current magnitude. Only if the current is under the meter rating, we can use the meter.

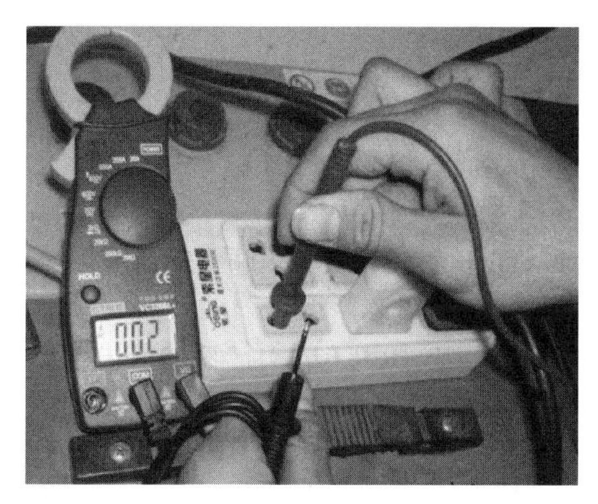

Fig.56.1 Electrical measurement

Knowledge extention

The electrical measuring tools are divided into two classes: one is called the absolute instrument, the other is called the secondary instrument. The absolute instrument can give the value directly with no comparison. The secondary instrument must be calibrated with the help of the absolute instruments such as ammeter, voltmeter and wattmeter etc.

Exercises

1. Match the English with the Chinese. Draw lines.

(1) fire alarm	a. 开关
(2) fuse	b. 量程
(3) switch	c. 熔丝，保险丝
(4) measuring range	d. 调校
(5) rectification	e. 火警
(6) measurement	f. 额定值
(7) current magnitude	g. 测量
(8) rating	h. 电流大小

2. True or false.

(1) When we choose a measuring meter, we should shoose the right measuring range and the right measurement accuracy. (   )

(2) The meter can be taken down at random. (   )

(3) After a long-time use of the meters, they should be rectified. (   )

(4) All the types of measurement are the same. (   )

(5) When we can use the meter, we should consider the current magnitude first, and make sure the current is under the meter rating. (   )

3. Understand the tip.

These pictures show different signs of different meters. “m” means “milli-”, “u” means “micro-”, and “k” means “kilo”.

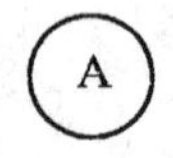

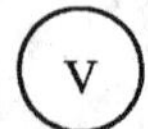

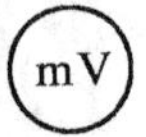

Words and phrases

telecommunication system [ˌtelɪkəˌmju:nɪ'keɪʃn] ['sɪstəm] 电讯/电讯系统

fire alarm ['faɪə(r)] [ə'lɑ:m] 火警

choose [tʃu:z] 选择

pay attention to 注意

appropriate [ə'prəʊprɪət] 适当的，合适的

measuring range 测量范围，量程

after a long-time use 长时间使用之后

at random ['rændəm] 随意，随机

rectification [ˌrektɪfɪ'keɪʃn] 调校，修整

differentiate [ˌdɪfə'renʃɪeɪt] 区别，区分

be under the meter rating 在仪器额定值之内

magnitude ['mægnɪtju:d] 幅度，大小

sign 标志

fuse [fju:z] 保险丝，熔丝

measuring meter 测量仪器

following 以下，如下

right 正确的

accuracy ['ækjərəsi] 精度

take down 拆卸

consider [kən'sɪdə(r)] 考虑

only if 只有当……

tip 提示，建议

show 展示,显示

different 不同的

mean 意味着，意思是

Lesson 57 Measurement of Voltage and Current

Look and select

Look at the pictures and select the correct terms from the box.

multimeter	ammeter	voltmeter	mV meter

1. ________________

2. ________________

3. ________________

4. ________________

Text

Measurement of voltage

We can use a voltmeter or a multimeter to measure voltages. A simple circuit of measuring voltage is shown in Fig.57.1. What we must take notice of is the correct connection of the voltmeter. The anode must be connected to the anode of the battery, and the cathode to the cathode of the battery. Besides, we must also take the range of measurement into consideration. Before taking a measurement, we must check for a right measuring range. If we use a smaller range to measure a higher voltage, the voltmeter will be burned out.

Measurement of current

We need to measure the current in a loop circuit with the help of multimeter or ammeter. Dislike the voltage measurement, the meter here should be connected into the loop circuit, as a part of it (Fig.57.2). The principle is that the current is from the red terminal to the black. So the red (anode) should be connected to the anode of the battery.

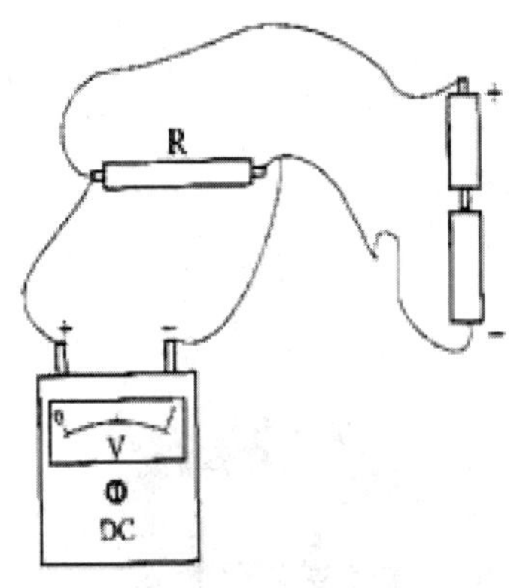

Fig.57.1 Measurement of voltage

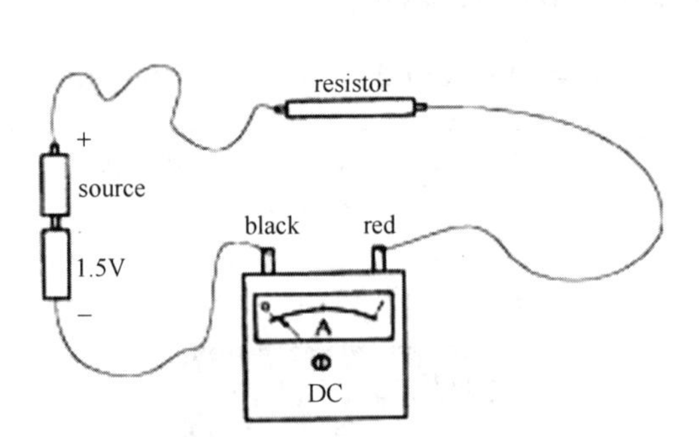

Fig.57.2 Measurement of current

Knowledge extention

Tips: Attention! We can't use an mV meter to measure a 220V voltage directly (Fig.57.3). It will lead to the damage of the meter.

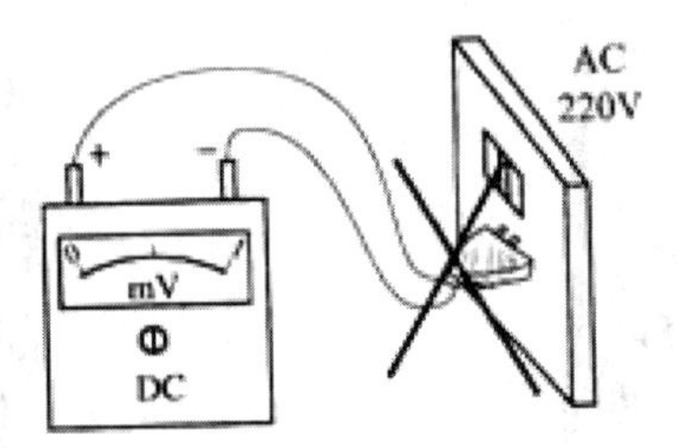

Fig.57.3 Incorrect measurement

Exercises

1. Fill in the blanks with the proper words in the text.

(1) What we must take notice of is the correct _______ of the voltmeter. The ______ must be connected to the ________ of the battery, and the ________ to the ________ of the battery.

(2) The _________ of current measuring is that the current is from _________ to the black. So the red (anode) should be connected to _________.

2. Match the English with the Chinese. Draw lines.

(1) multimeter	a. 电流表
(2) ammeter	b. 电压表
(3) voltmeter	c. 万用表
(4) mV meter	d. 毫伏表

3. Answer the following questions.

(1) What should we do before taking a measurement? Why?

(2) What is the difference between voltage measurement and current measurement?

Words and phrases

ammeter ['æmi:tə(r)] 安培表
voltmeter ['vəʊltmi:tə(r)] 电压表
take notice of 注意
anode ['ænˌəʊd] 阳极
battery ['bætəri] 电池
besides 除此之外
range of measurement 量程
check [tʃek] 检查
burn out 烧坏
dislike 与……不一样
be connected into 连接进……
principle ['prɪnsəpl] 原理
attention 注意，留心
probe [prəʊb] 探针
mV meter 毫伏表
multimeter ['mʌltɪmi:tə] 万用表
connection [kə'nekʃn] 连接
be connected to 连接到……
cathode ['kæθəʊd] 阴极
take...into consideration 把……考虑进来
take a measurement 测量
right measuring range 正确的量程
with the help of 借助于...帮助
voltage measurement 电压测量
part 部分
terminal ['tɜ:mɪnl] 端子，端部
damage ['dæmɪdʒ] 损坏

Lesson 58 Measurement of Resistance

Look and select

Look at the pictures and select the correct terms from the box.

dial gauge	outside micrometer	vernier caliper	electrical pen

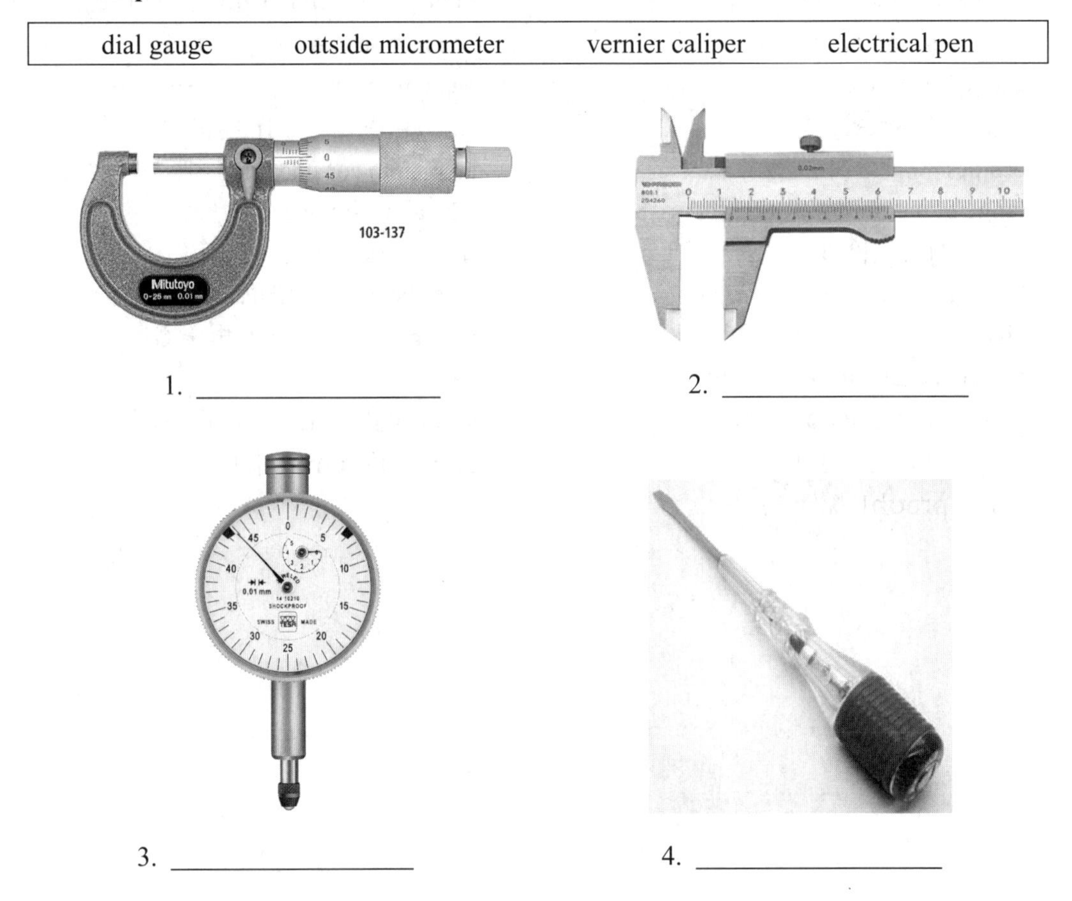

1. ________________ 2. ________________

3. ________________ 4. ________________

Text

Measurement of resistance

When we want to measure a resistance, we must be sure of the size of the resistor at first. If it's a small resistor, we can use a multimeter or an ohmmeter to measure it. And if it's a very large one, such as of many millions of ohms, we must use a megger. The method of measurement is very simple here. We just need to connect the two probes to the two terminals of the resistor. We can have a sense of the measurement from Fig.58.1.

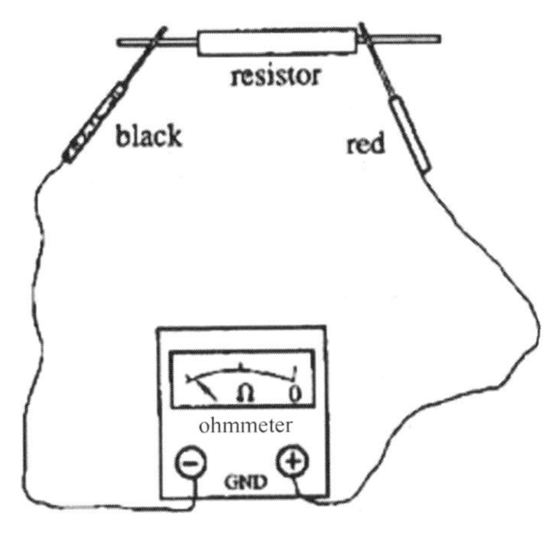

Fig.58.1 Measurement of resistance

Knowledge extention

1. Steel ruler (Fig.58.2)

Fig.58.2 Steel rulers

2. Dial gauge (Fig.58.3)

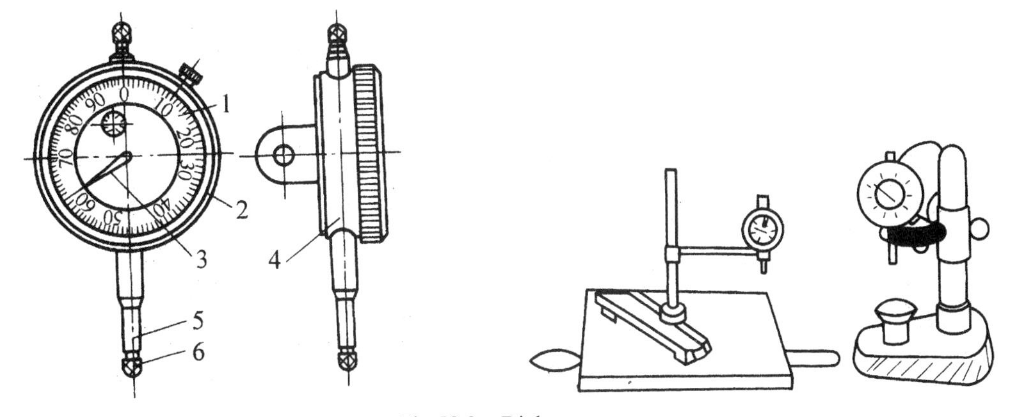

Fig.58.3 Dial gauges

1—dial plate; 2—bezel; 3—pointer; 4—body; 5—metering rod; 6—metering head

3. Outside micrometer (Fig.58.4)

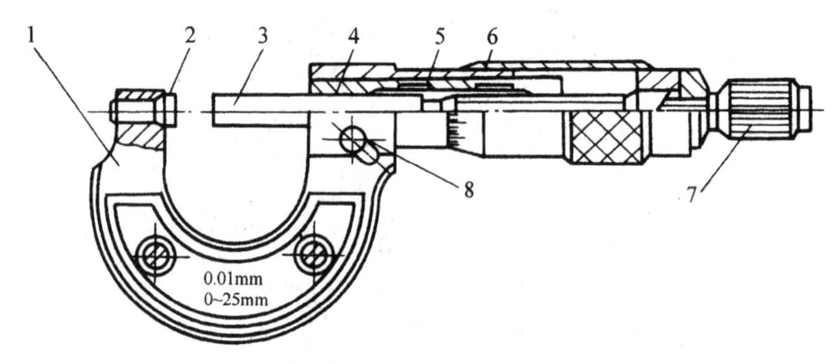

Fig.58.4 Outside micrometer

1—ruler frame; 2—anvil; 3—micrometric screw; 4—threaded sleeve; 5—fixed sleeve; 6—microdrum; 7—force measuring device; 8—locking device

4. Vernier caliper (Fig.58.5)

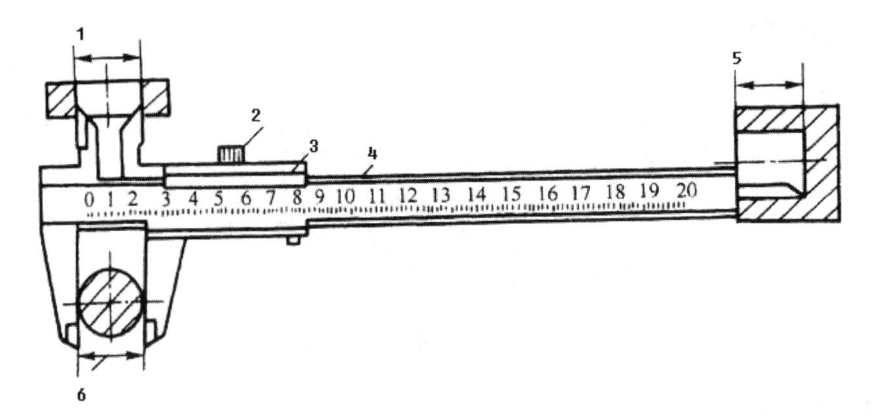

Fig.58.5 Vernier caliper

1—measure inner surface; 2—braking screw; 3—auxiliary ruler; 4—main ruler; 5—measure depth; 6—measure outer surface

Exercises

1. Match the English with the Chinese. Draw lines.

A.

(1) steel ruler	a. 外径千分尺
(2) dial gauge	b. 钢直尺
(3) outside micrometer	c. 游标卡尺
(4) vernier caliper	d. 百分表

B.

(1) ruler frame	a. 套管
(2) sleeve	b. 指针
(3) pointer	c. 尺架
(4) screw	d. 螺钉，螺纹

2. True or false.

(1) When we want to measure a resistance, we must be sure of the size of the resistor at first. (　　)

(2) If it's a small resistor, we can use a megger to measure it. (　　)

(3) If it's a very large one, such as of many millions of ohms, we can use a multimeter or an ohmmeter to measure it. (　　)

(4) We just need to connect the two probes to the two terminals of the resistor: the red end connected to the positive; the black to the negative. (　　)

Words and phrases

megger ['megə] 兆欧表
have a sense of 感受，感觉
outside micrometer [maɪ'krɒmɪtə(r)] 外径千分尺
vernier caliper ['vɜ:nɪə] ['kælɪpə] 游标卡尺
bezel ['bezl] 表圈
outer surface ['sɜ:fɪs] 外表面
metering head [hed] 测量头
micrometric screw [skru:] 测微螺杆
threaded sleeve 螺纹轴套
force measuring device 测力装置
fixed sleeve [fɪkst] [sli:v] 固定套管
braking screw 制动螺钉
main ruler 主尺
probe [prəʊb] 探针
dial gauge ['daɪəl] [geɪdʒ] 百分表
dialplate ['daiəl pleit] 表盘
electrical pen 电工笔
pointer ['pɔɪntə(r)] 指针
metering rod [rɒd] 量杆
ruler frame [freɪm] 尺架
anvil ['ænvɪl] 测砧
inner surface 内表面
microdrum [maɪk'rɒdrʌm] 微分筒
locking device 锁紧装置
auxiliary ruler [ɔ:g'zɪliəri] 辅尺
depth [depθ] 深度

Lesson 59 Ohm's Law

Look and select

Look at the pictures and select the correct terms from the box.

PCB	LED	resistor	charger

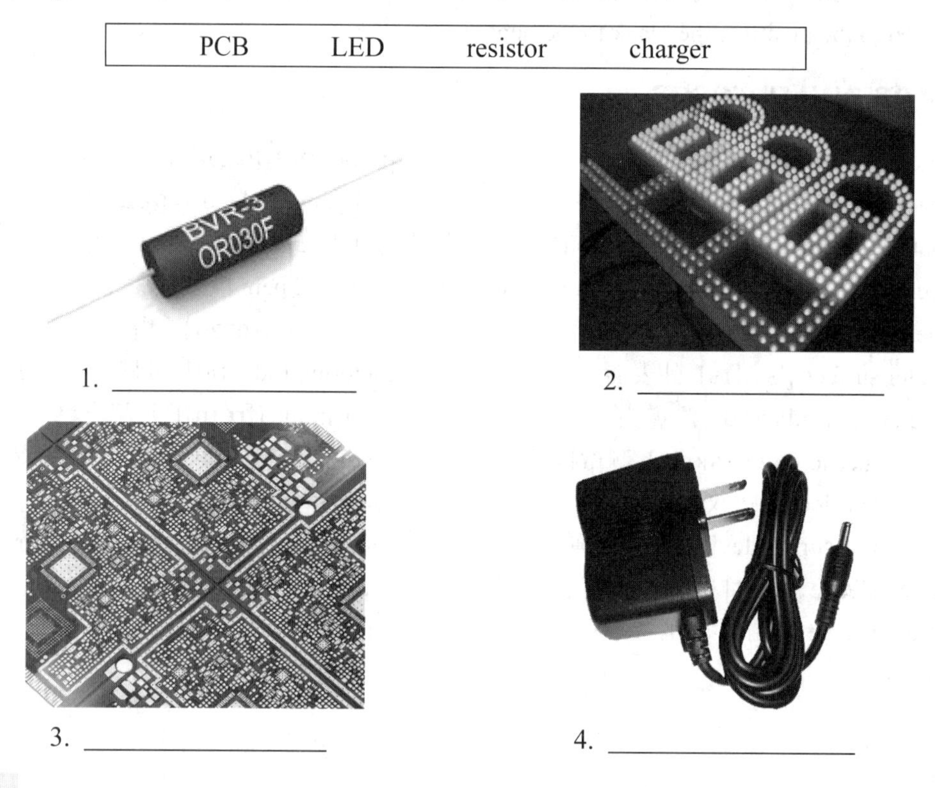

1. ________________ 2. ________________

3. ________________ 4. ________________

Text

Fig.59.1 Ohm

Ohm's law

In 1825, a German physicist called Ohm (Fig.59.1) discovered that a simple relationship exists between the three quantities: resistance, current and voltage. The law can be expressed in the following formula:

$$I = U / R$$

Of which, I means current, U means voltage and R means resistance. If the voltage remains the same, the greater the resistance, the smaller the current. From Ohm's law, we also know that the voltage across the ends of a wire is equal to the current timing the resistance.

Knowledge extention

Tips:

Ohm's Law is not only used in DC circuits (Fig.59.2), but also in AC circuits. You must be aware that they are different. You can use the formula in the text directly in DC circuits, but not in AC circuits. In AC circuits, we should take the phase into consideration. It's more difficult.

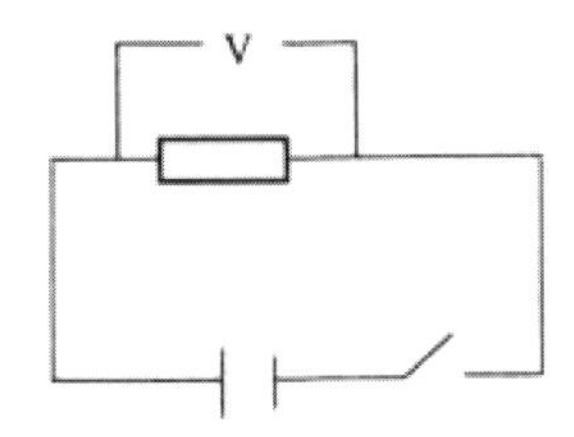

Fig.59.2 DC circuit

Exercises

1. Fill in the blanks.

(1) PCB is the short form of ______________________________. LED is the short form of ____________________.

(2) The law can be expressed in the following formula:

$$I = U / R$$

Of which, I means ____________, U means ____________ and R means ____________.

2. Answer the following questions:

(1) Who discovered the Ohm's law? When?

(2) What are the three quantities involved in Ohm's law?

(3) In the formular of $I = U / R$, what does I mean?

(4) If the voltage is the same, what is the relationship between current and resistance?

(5) What can we know from Ohm's law?

3. Explain the example.

The voltage applied across the conductor (Fig.59.3) is 6V, and the current flowing through the conductor is 0.5A, we can find the resistance of the conductor is 12Ω ($R= U / I = 6/0.5 = 12$).

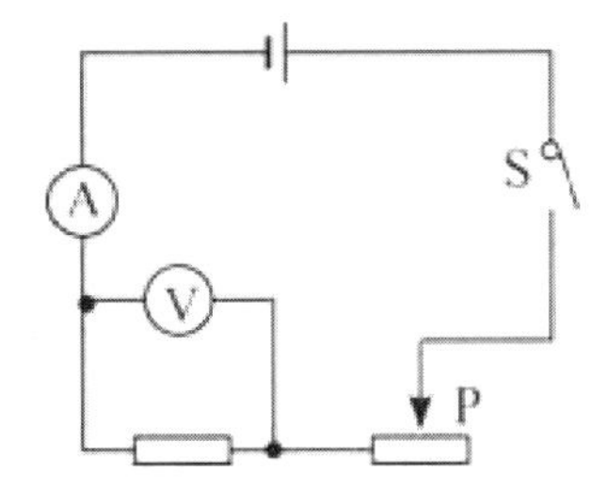

Fig.59.3 Ohm's Law

Words and phrases

PCB (printed circuit board) [p'rɪntɪd] [bɔ:d] 印刷电路板
LED (light-emitting diode) [ɪ'mɪt] 发光二极管
resistor [rɪ'zɪstə(r)] 电阻
German ['dʒɜ:mən] 德国人
discover [dis'kʌvə] 发现
relationship [rɪ'leɪʃnʃɪp] 关系
between 在……两者之间
law [lɔ:] 法则，法规
following ['fɒləʊɪŋ] 以下，如下
of which 其中
remain [rɪ'meɪn] 保持，保留
greater [greɪtə] 较大，越大
across [ə'krɔs] 通过。穿过
not only … but also 不仅……而且
AC circuit 交流电路
different 不同的
take … into consideration 把……考虑进来
be applied across the conductor 施加到导体上
charger ['tʃɑ:dʒə(r)] 充电器
physicist ['fɪzɪsɪst] 物理学家
simple ['sɪmpl] 简单的
exist [ɪg'zɪst] 存在
quantity ['kwɒntəti] 量
be expressed in 在……表示
formula ['fɔ:mjələ] 公式
mean [mi:n] 意思是
same [seɪm] 同样的
smaller [s'mɔ:lər] 较小，越小
wire 电线，导线
be equal to 等于
be aware that 意识到
directly 直接地
DC circuit 直流电路

Lesson 60 Diodes

Look and select

Look at the pictures and select the correct terms from the box.

rectifier	diode	photo-diode	light-emitting diode

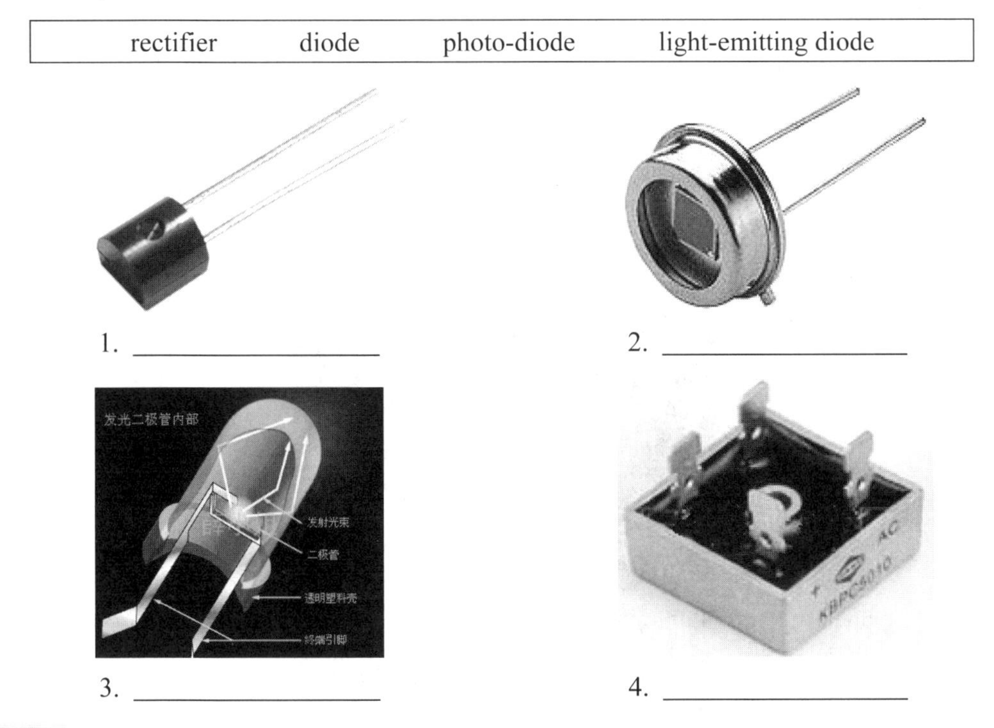

1. ________________
2. ________________
3. ________________
4. ________________

Text

Diode

A diode is a semiconductor that allows current to flow in only one direction. It is made from a P-N junction with two leads, an anode and a cathode (Fig.60.1).

Fig.60.1 Diode

A diode is forward biased when the voltage of its anode is higher than the voltage of its cathode, it will cause current to flow. A diode causes no current to flow when it is reversed biased, i.e. the voltage of its anode is lower than the voltage of its cathode.

Usually a diode can be used as a rectifier because of the above characteristic. The value of a rectifier is expressed in ampere. Furthermore there are numerous types of diodes, such as Zener diodes, photo-diodes, light-emitting diodes (Fig.60.2) etc.

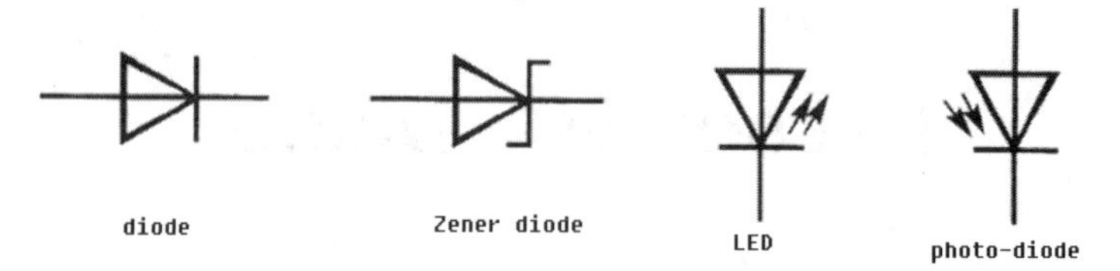

Fig.60.2 Many types of diodes

Knowledge extention

A **light-emitting diode (LED)** looks like a normal semiconductor diode which has low resistance in the forward direction and high resistance in the reversed direction.

Electrically LEDs are P-N junction diodes which emit light when conducting. They can be used as indicator lamps (Fig.60.3) to show the present working conditions, for instance, power on, power off, and standby, etc. LEDs can be obtained in a wide range of colors, red, yellow, green and so on.

Fig.60.3 LED

Exercises

1. Match the English with the Chinese. Draw lines.

A.

(1) rectifier	a. 二极管
(2) diode	b. 光电二极管
(3) photo-diode	c. 发光二极管
(4) light-emitting diode	d. 稳压/齐纳二极管
(5) Zener diode	e. 整流器

B.

(1) semiconductor	a. 指示灯
(2) indicator lamp	b. 半导体
(3) working condition	c. 电源开
(4) power on	d. 电源关

(5) power off　　e. 备用，待命
(6) standby　　f. 工作状态

2. Fill in the blanks with the words in the text.

A diode is forward biased when the voltage of its anode is ________ than the voltage of its ________, it will cause current to flow. A diode causes no current to flow when it is _______, i.e. the voltage of its ________ is lower than the voltage of its _________.

3. Try to answer the questions.

(1) What is a diode?
(2) What is it made from?
(3) Could you list some types of diodes?
(4) Explain the forward biased and reverse biased of diode.

4. True or false.

(1) A diode is a semiconductor. ( )
(2) A diode allows current to flow in any direction. ( )
(3) A diode is made from a P-N junction. ( )
(4) Zener diodes are rectifiers. ( )

Words and phrases

rectifier ['rektɪfaɪə] 整流器
light-emitting diode [ɪ'mɪtɪŋ] 发光二极管
semiconductor [ˌsemikən'dʌktə(r)] 半导体
direction [də'rekʃn] 方向
cause [kɔ:z] 引起，造成
be used as 被用于
forward biased ['fɔ:wəd] ['baɪəst] 正向偏置
reversed biased [rɪ'vɜ:st] 反向偏置
characteristic [ˌkærəktə'rɪstɪk] 特性，特征
furthermore ['fə:ðəˌmɔ:] 而且，更进一步
numerous ['nju:mərəs] 无数，许多
Zener diode 稳压/齐纳二极管
diode ['daɪəʊd] 二极管
photo-diode 光电二极管
allow [ə'laʊ] 允许
be made from 由……制成/组成
usually 通常地
i.e. [ˌaɪ 'i:] 即
lead [lid] 导线
junction ['dʒʌŋkʃn] 结
value ['vælju:] 值，数值
be expressed in 用……表示
type [taɪp] 类型
etc [ˌet 'setərə] 等等

Lesson 61 Transistor

Look and select

Look at the pictures and select the correct terms from the box.

stereo	transistor	radio	amplifier

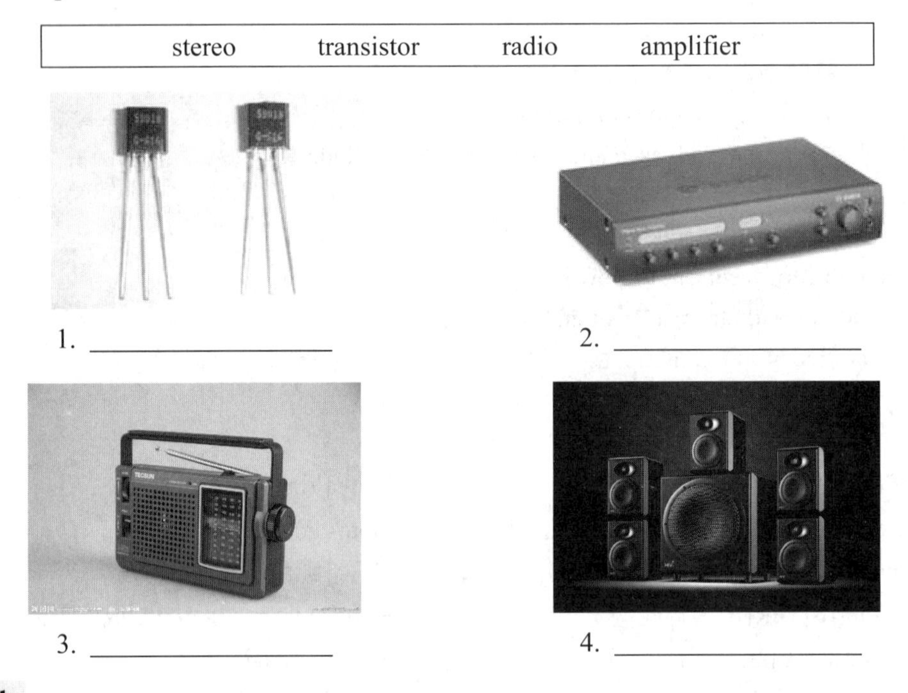

1. ________________ 2. ________________

3. ________________ 4. ________________

Text

Transistor

One of the newest developments in electricity is the transistor. A transistor is called a semiconductor because under one set of conditions it will not allow current to pass through, but under other conditions it will. This characteristic is important in controlling current in radios and other industrial devices, such as television and hearing aids etc.

In application, triodes are usually called transistors. Transistors are made from three layers of semiconductors, i.e. NPN or PNP (two types of transistors). A transistor has at least three electrodes, they are the emitter, the collector and the base (see Fig.61.1).

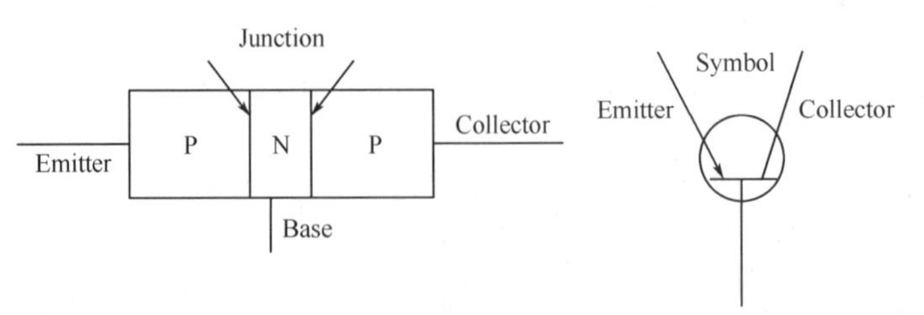

Fig.61.1 Transistor

Knowledge extention

If a small current is applied to the base, a large current can be achieved at the emitter and the collector. See Fig.61.2. The formulas are as follow:

$$I_C = \alpha I_E \quad \text{or} \quad I_C = \beta I_B$$

Where:

α is the common-base current gain.

β is the common-emitter current gain.

The currents of three electrodes are specified as follows:

$$I_E = I_B + I_C$$

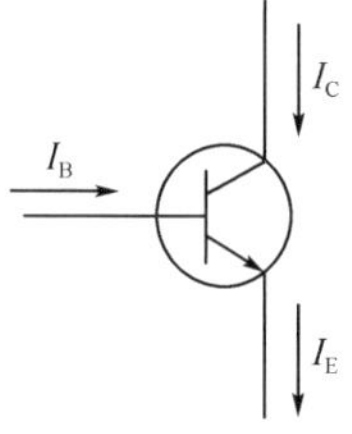

Fig.61.2 Current

Exercises

1. Match the English with the Chinese. Draw lines.

(1) transistor a. 三极管

(2) triode b. 结

(3) semiconductor c. 晶体管

(4) junction d. 发射极

(5) emitter e. 集电极

(6) collector f. 基极

(7) base g. 半导体

2. Fill in the blanks with the words in the text.

(1) A transistor is called a __________ because under one set of conditions it will not allow _______ to pass through, but under other conditions it will.

(2) In application, triodes are usually called __________. Transistors are made from three _______ of semiconductors, i.e. NPN or __________ (two types of transistors).

3. Try to answer the questions.

(1) What are the three electrodes of a transistor?

(2) What are the two types of transistors?

4. True or false.

(1) In application, diodes are often called transistors. ( )

(2) The two types of transistors are PNP and NPN. ( )

(3) The three electrodes of a transistor are the emitter, the collector and the base. ( )

Words and phrases

stereo ['sterɪəʊ] 立体声系统

radio ['reɪdɪəʊ] 无线电，收音机

application [ˌæplɪ'keɪʃn] 应用

layer ['leiə] 层

electrode [ɪ'lektrəʊd] 极，电极，焊条

transistor [træn'zɪstə(r)] 晶体管

amplifier ['æmplɪfaɪə(r)] 放大器

triode ['traɪəʊd] 三极管

at least 至少

emitter [ɪ'mɪtə] 发射极

collector [kə'lektə(r)] 集电极
base [beɪs] 基极
be applied to 施加到，应用到
be achieved at 在……获得
formula ['fɔːmjələ] 公式
as follow 如下
be specified ['spesɪfaɪd] 表述
symbol ['sɪmbl] 符号
common-base current gain ['kɒmən] [geɪn] 共基极放大系数/增益
common-emitter current gain 共发射极放大系数/增益

Lesson 62 Integrated Circuits

Look and select

Look at the pictures and select the correct terms from the box.

PGA	SIP	DIP	ULSIC

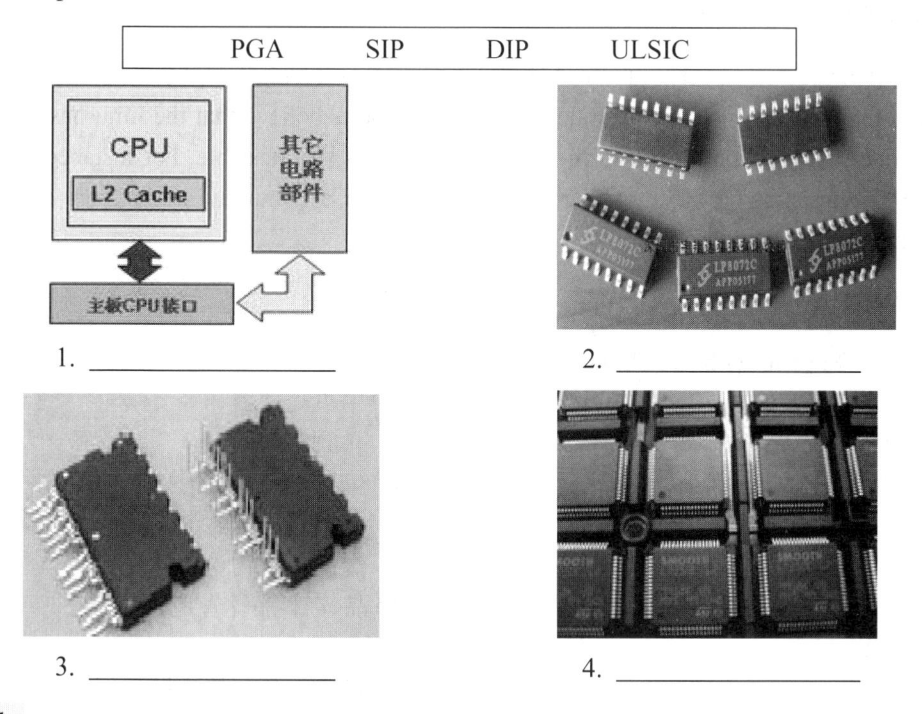

1. ____________ 2. ____________

3. ____________ 4. ____________

Text

The integrated circuit is sometimes called an IC (Fig.62.1). The IC was invented in 1958 and it looks like a tiny chip of metal. It combines transistors, diodes, resistors, and small capacitors on a single IC chip, usually silicon.

Fig.62.1 Integrated circuit

Integrated circuits are classified in a variety of ways, for instance, analog ICs or digital ICs, while digital ICs are grouped into TTL (transistor-transistor logic) and MOS (metal oxide semiconductor).

Integrated circuits come in various kinds of packages. The most common is the dual in-line package (DIP). Other housing methods are the single in-line package (SIP), pin grid array package (PGA), and so on.

Knowledge extention

The advantage of an integrated circuit is a tremendous saving in space when compared with discrete components, which are separate units. Some experts believe that the limit has nearly been reached today; others think much further miniaturization is still possible. For instance, there were only 2 300 transistors in the first microprocessor chip in a computer while there are now 10^9 transistors in an updated single chip.

Exercises

1. Match the English with the Chinese. Draw lines.

(1) IC　　　　a. 中央处理器
(2) PGA　　　b. 单列式封装
(3) SIP　　　c. 集成电路
(4) DIP　　　d. 阵列式引脚封装
(5) ULSIC　　e. 金属氧化物半导体
(6) CPU　　　f. 晶体管-晶体管逻辑
(7) TTL　　　g. 双列式封装
(8) MOS　　　h. 超大规模集成电路

2. Try to fill in the blanks (Table 62.1) with the words about ICs.

Table 62.1　Integrated circuits

Be invented in 发明年代			
Look like 外形			
Combine 组成			
Be classified in 分类	analog		
	digital		
Packages 封装			

3. True or false.

(1) An integrated circuit combines transistors, diodes, resistors and small capacitors on a silicon chip. (　　)

(2) An integrated circuit may be a single in-line package. (　　)

(3) Integrated circuit can be classified in only one way. (　　)

4. Translate the following sentences into Chinese.

(1) The advantages of an integrated circuit is a tremendous saving in space when compared

with discrete components, which are separate units.

(2) For instance, there were only 2 300 transistors in the first microprocessor chip in a computer.

(3) There are now 10^9 transistors in an updated single chip.

Words and phrases

integrated ['ɪntɪgreɪtɪd] 集成的
circuit ['sɜːkɪt] 电路
be invented in 发明于
look like 看起来像
a tiny chip of ['taɪnɪ] [tʃɪp] 一小片
metal ['metl] 金属
combine [kəm'baɪn] 使……结合/化合
transistor [træn'zɪstə(r)] 晶体管
single ['sɪŋgl] 单个的
silicon ['sɪlɪkən] 硅
be classified in 按……分类
a variety of ways 各种方式
for instance 例如
analog ['ænəlɔːg] 模拟
digital ['dɪdʒɪtl] 数字
be grouped into 分类为……
variety [və'raɪətɪ] 种类
logic ['lɒdʒɪk] 逻辑
package ['pækɪdʒ] 包裹，封装
oxide ['ɒksaɪd] 氧化物
dual ['djuːəl] 双的
housing ['haʊzɪŋ] 外壳
pin [pɪn] 管脚
grid [grɪd] 格栅
tremendous [trɪ'mɛndəs] 极大的
array [ə'reɪ] 排列
microprocessor chip 微处理器芯片
most common 最普遍
is a tremendous saving in 在……大大节省
advantage 优点，优势
space 空间
compare with 与……相比较
discrete components 分散的独立元件
separate units 单独元件/装置
expert 专家
believe 相信
miniaturization [ˌmɪnətʃəraɪ'zeɪʃn] 小型化
microprocessor [ˌmaɪkrəʊ'prəʊsesə(r)] 微处理器
TTL (transistor-transistor logic) 晶体管-晶体管逻辑
MOS (metal oxide semiconductor) 金属氧化物半导体
dual in-line package (DIP) 双列式封装
single in-line package (SIP) 单列式封装
pin grid array package (PGA) 阵列式引脚封装
ULSIC (ultra large scale integrated circuit) 超大规模集成电路
come in various kinds of 以……不同种类出现
limit has nearly been reached today 当今已经几乎达到了极限
much further miniaturization is still possible 更进一步微型化是有可能的

Lesson 63 Multimeters

Look and select

Look at the pictures and select the correct terms from the box.

DMM	VOM	capacitor	ADSL modem

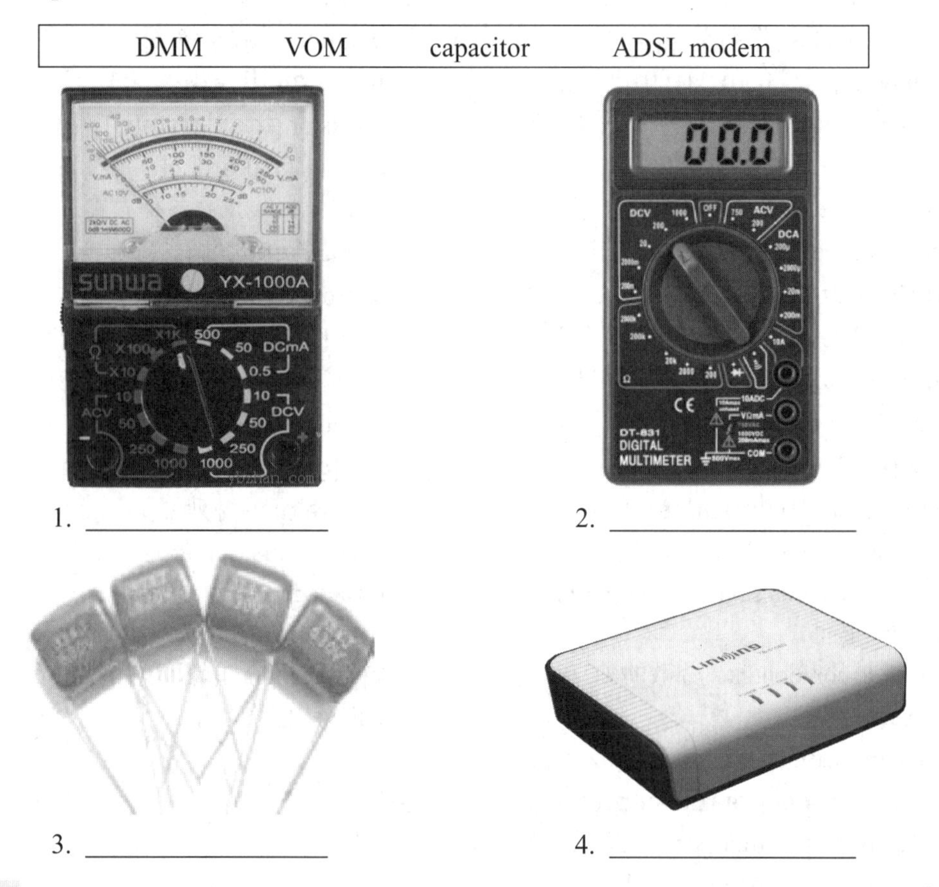

1. ________________ 2. ________________

3. ________________ 4. ________________

Text

Multimeter

A multimeter is a circuit tester. It is also an instrument to measure voltage, current and resistance. Multimeter is named after it's multi-functions.

It is divided into the analog style (VOM: volt-ohm-milliammeter) and the digital style (DMM: digital multimeter).

For the analog multimeter, it has three separate scales aiming at current, voltage and resistance (Fig.63.1). For a digital multimeter, it has only one LCD display screen. For example, the correct connection is shown in Fig.63.2.

Fig.63.1 Analog multimeter

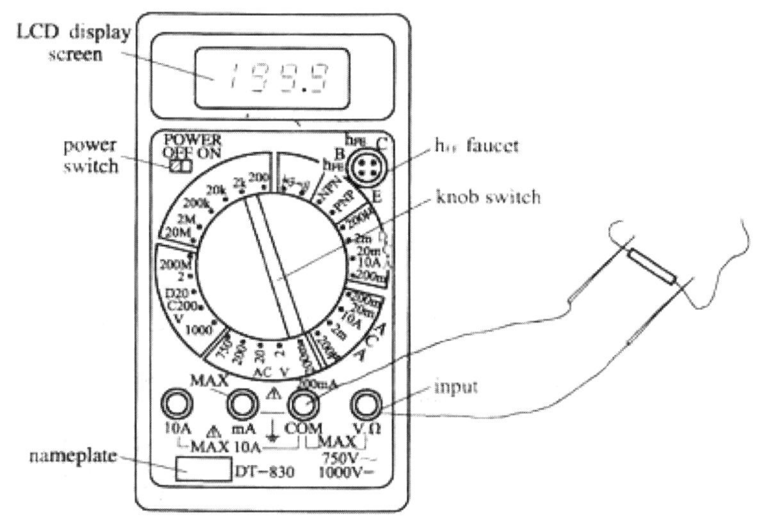

Fig.63.2 Digital multimeter

Knowledge extention

The digital multimeter (DMM) is one of the test equipment most used by technicians today. It is a prime choice in most cases for its accuracy, expanded measurement capabilities, ease of reading, and lighter loading of the circuit under test, compared to the VOM (volt-ohm-milliammeter).

Exercises

1. Fill in the blanks with the words in the text.

(1) A multimeter is a ________ tester. It is also an instrument to measure _________, current and _____________. Multimeter is named after it's _____________.

(2) It is divided into the ____________ style (VOM: volt-ohm-milliammeter) and the digital _____________ (DMM: digital multimeter).

(3) For the analog multimeter, it has three separate ___________ aiming at current, voltage and resistance. For a digital multimeter, it has only one display screen.

2. Match the English with the Chinese. Draw lines.

A.

(1) DMM	a. 伏特-欧姆-毫安表
(2) VOM	b. 数字万用表
(3) ADSL modem	c. 数字的
(4) analog	d. 调制解调器
(5) digital	e. 模拟的
(6) scale	f. 显示屏
(7) display screen	g. 刻度

B.

(1) connection	a. 连接
(2) faucet	b. 旋钮

(3) nameplate　　c. 插孔
(4) knob switch　　d. 铭牌
(5) input　　e. 输出
(6) output　　f. 输入

3. Try to answer the questions.

(1) What is multimeter? Please detail it.

(2) What are the styles of multimeters?

(3) Explain the connection for DMM while measuring the voltage.

(4) Why we call the DMM is a prime choice while comparing to the VOM?

4. Understand the tip.

You can never use this connection to measure a high voltage. You must change the connection from "mA" to "V". Otherwise, the meter will be burnt out (Fig.63.3).

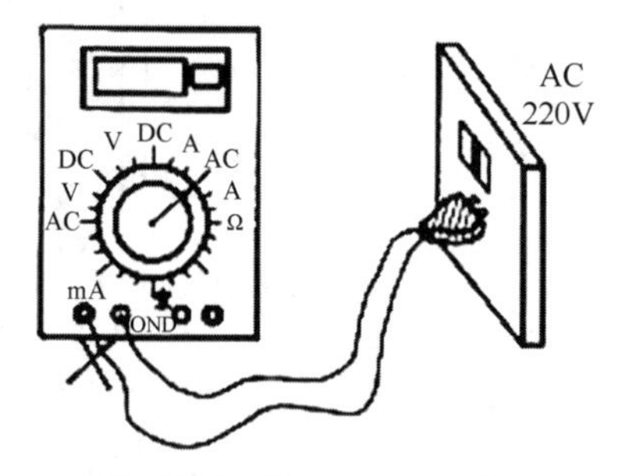

Fig.63.3　Wrong connection

Words and phrases

DMM (digital multimeter) ['dɪdʒɪtl] ['mʌltɪmi:tə] 数字万用表

VOM(volt-ohm-milliammeter) [mɪlɪ'æmɪtə] 伏特-欧姆-毫安表

ADSL modem (asymmetric digital subscriber line) [ˌeɪsɪ'metrɪk] [səb'skraɪbə(r)] 调制解调器

instrument ['ɪnstrəmənt] 仪表

multi-functions ['fʌŋkʃn] 多功能

analog style ['ænəlɔ:g] [staɪl] 模拟式

separate scale ['seprət] [skeɪl] 各自的刻度

aim at [eɪm] 旨在，目标是

be shown in 展示/显示于……

nameplate 铭牌

faucet ['fɔ:sɪt] 插孔

test equipment [ɪ'kwɪpmənt] 测试设备

input ['ɪnpʊt] 输入

prime choice [praɪm] [tʃɔɪs] 首选

accuracy ['ækjərəsi] 精度，准确度

be compared to [kəm'pɛrd] 与……相比

change [tʃeɪndʒ] 变更，改变

be burnt out [bɜ:nt] 烧坏

tester ['testə(r)] 测试仪

be named after 以……命名

be divided into [dɪ'vaɪdɪd] 分为

digital style 数字式

LCD display screen 液晶显示屏

correct connection 正确的连接

power switch 电源开关

knob switch [nɒb] 旋钮开关

most used by 多数被……所用

technician [tek'nɪʃn] 技术员

in most cases 在大多数情况下

ease of reading 易于读数

connection [kə'nekʃn] 连接

otherwise ['ʌðəwaiz] 否则

expanded measurement capabilities [ɪks'pændɪd] [ˌkeɪpə'bɪlɪti:z] 扩展测量能力

lighter loading of the circuit under test ['ləʊdɪŋ] 测量时对电路的轻载

Lesson 64 Oscilloscope

Look and select

Look at the pictures and select the correct terms from the box.

bulb	electron gun	LCD	wave

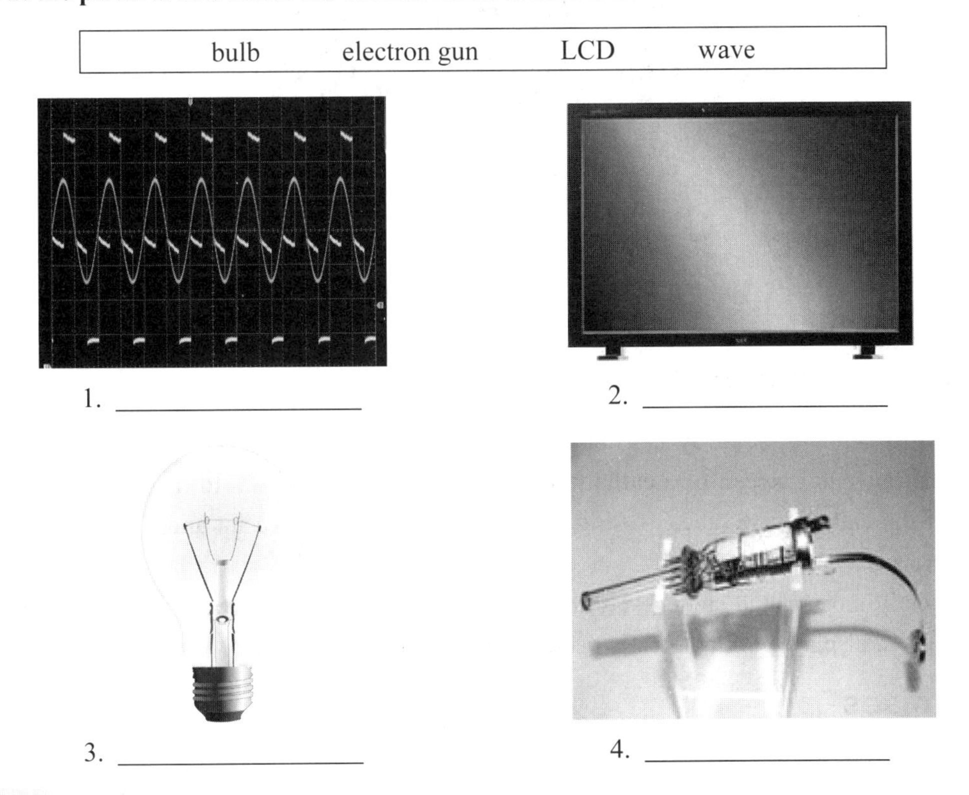

1. ________________
2. ________________
3. ________________
4. ________________

Text

The cathode-ray oscilloscope is a very useful electronic measuring instrument for research and development in the electronic industry, such as designing, trouble shooting, signal monitoring and many other applications where the observation of an electrical wave-form is desired (Fig.64.1).

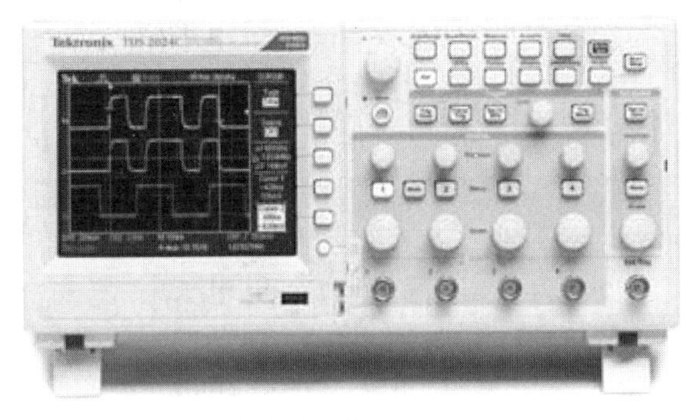

Fig.64.1 Oscilloscope

The heart of the oscilloscope is the cathode-ray tube (Fig.64.2). This consists of the base, neck (an electron gun is included), bulb, and the face-plate (screen). The electron gun consists of a cathode, a control grid, an anode, and two sets of deflection plates.

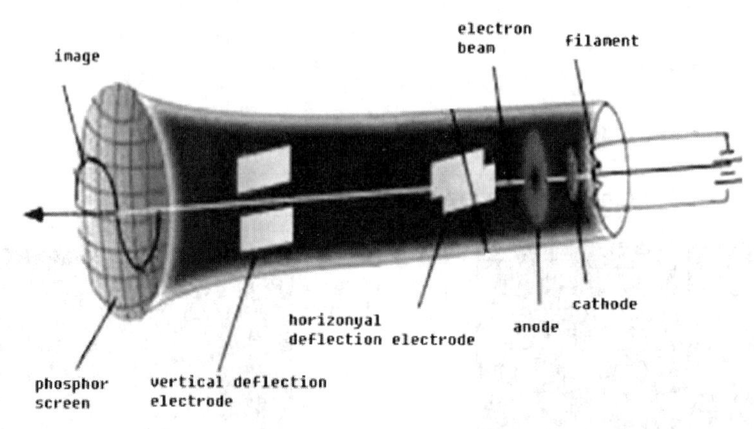

Fig.64.2 Cathode-ray tube

Knowledge extention

Oscilloscopes operate by causing an electron beam to sweep rapidly from left to right across the phosphor screen of a cathode-ray tube. A signal is applied to the vertical deflection plates of the cathode-ray tube, causing the electron beam to move up and down as the signal voltage varies.

Recently most of the cathode-ray tubes used in oscilloscopes have been replaced by LCDs (liquid crystal displays), which greatly decrease the weight and size of the machine.

Exercises

1. Match the English with the Chinese. Draw lines.

A.

(1) cathode-ray tube	a. 管颈
(2) base	b. 灯泡
(3) neck	c. 面板
(4) bulb	d. 管座
(5) face-plate	e. 电子枪
(6) electron gun	f. 阴极射线管

B.

(1) phosphor screen	a. 偏转电极
(2) deflection electrode	b. 荧光屏
(3) image	c. 电子束
(4) electron beam	d. 图像
(5) anode	e. 阴极
(6) cathode	f. 阳极

2. True or false.

(1) The cathode-ray oscilloscope is a useful measuring instrument for trouble shooting. (　　)

(2) The heart of the oscilloscope is the screen. (　　)

(3) The cathode-ray tube consists of the base, neck, bulb, and the face-plate (screen). (　　)

(4) The LCDs can greatly decrease the weight and size of the machine. (　　)

3. Translate the following sentences into Chinese.

(1) The cathode-ray oscilloscope is a very useful electronic measuring instrument for research and development in the electronic industry .

(2) Oscilloscopes operate by causing an electron beam to sweep rapidly from left to right across the phosphor screen of a cathode-ray tube.

Words and phrases

LCD (liquid crystal display) 液晶显示屏
oscilloscope [ə'sɪləskəʊp] 示波器
cathode-ray ['kæθəʊd] [reɪ] 阴极射线
useful ['ju:sfl] 有用的
electronic measuring instrument 电子测量仪器
research [rɪ'sɜ:tʃ] 研究
development [dɪ'veləpmənt] 开发
electronic industry 电子工业
trouble shooting ['trʌbl] ['ʃu:tɪŋ] 故障排斥
design [dɪ'zaɪn] 设计
cathode-ray tube [tju:b] 阴极射线管
heart [hɑ:t] 心脏
base [beɪs] 基础，基地，管座
consist of [kən'sɪst] 由……组成
electron gun [ɪ'lɛkˌtrɑn] [gʌn] 电子枪
neck [nek] 颈，管颈
face-plate 面板
control grid [kən'trəʊl] [grɪd] 控制栅
screen [skri:n] 屏幕
anode ['ænˌəʊd] 阳极
two sets of 两套，两对
deflection plate [dɪ'flekʃn] [pleɪt] 偏转板
application [ˌæplɪ'keɪʃn] 应用
signal monitor ['sɪgnəl] ['mɒnɪtə(r)] 信号监测
Observation of an electrical wave-form is desired. 需要观察电子波形。

Oscilloscopes operate by causing an electron beam to sweep rapidly from left to right across the phosphor screen of a cathode-ray tube. 示波器是以电子束在阴极射线管的屏幕上从左至右快速扫描的方式来工作的。

A signal is applied to the vertical deflection plates of the cathode-ray tube, causing the electron beam to move up and down as the signal voltage varies. 在阴极射线管的垂直偏转板上施加一个信号，电子束因信号电压的变化而上下移动。

Recently most of the cathode-ray tubes used in oscilloscopes have been replaced by LCDs (liquid crystal displays), which greatly decrease the weight and size of the machine. 现代的示波器大都用液晶显示器代替阴极射线管，这大大减小了示波器的重量和体积。

Lesson 65 Transformers (Ⅰ)

Look and select

Look at the pictures and select the correct terms from the box.

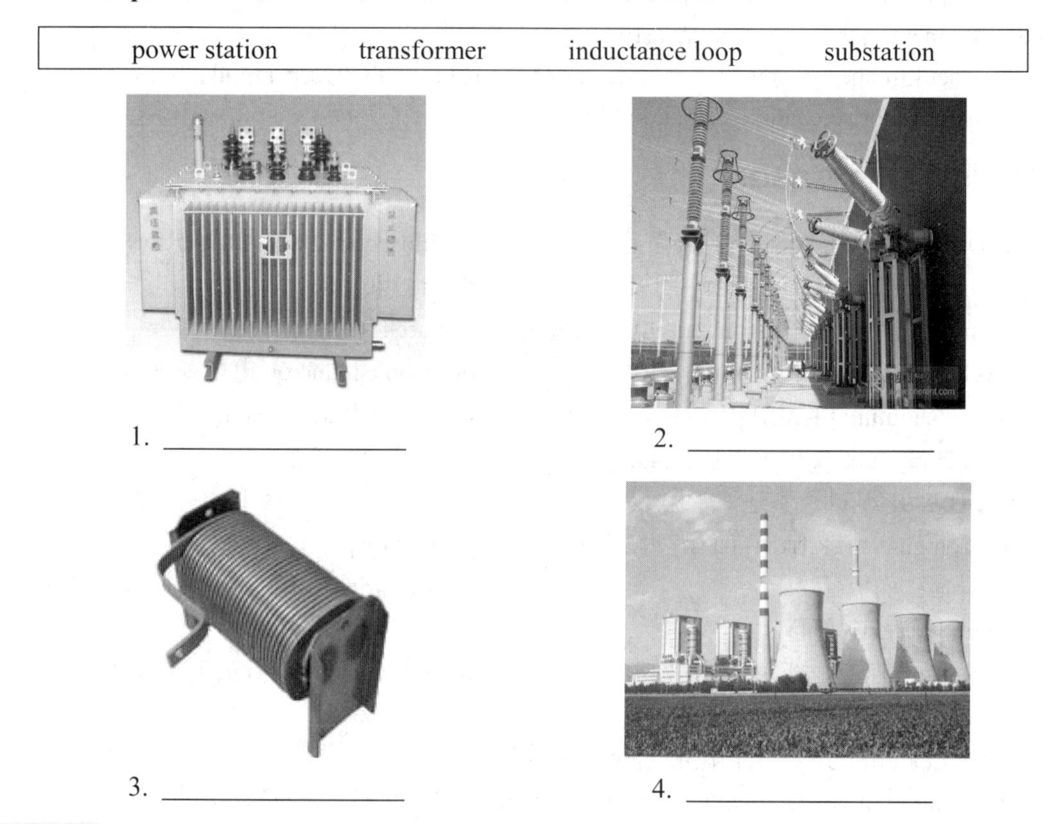

Text

Transformer

In our daily life, we can't live without electricity. But if there are no transformer, we can never get the 220V voltage we need. Furthermore, in an electrical design, a transformer is also used. For a simple transformer, it has the same circuit just as the figures shown in Fig.65.1.

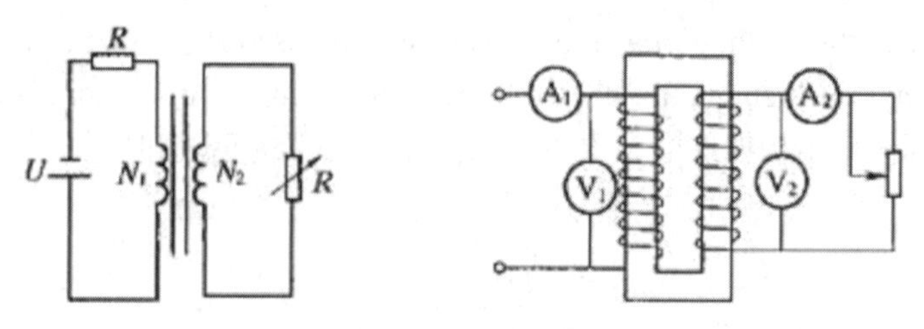

Fig.65.1 Circuit of transformer

Both the input loop circuit and the output loop circuit have only one inductance loop respectively. There are two formulas for this kind of circuit.

$$U_1 / U_2 = N_1 / N_2 , \quad I_1 / I_2 = N_2 / N_1$$

Another kind of circuit can provide two or more output voltages. N_2 / N_1. The voltage and the current are also followed by the formulas above. At the same time, the input power must be equal to the output power, so we can get: $N_1I_1 = N_2I_2 + N_3I_3$.

Knowledge extention

Mains AC is generated in power stations. The layout of a typical fuel-burning station is shown in Fig.65.2. The power is fed into a distribution network called the Grid.

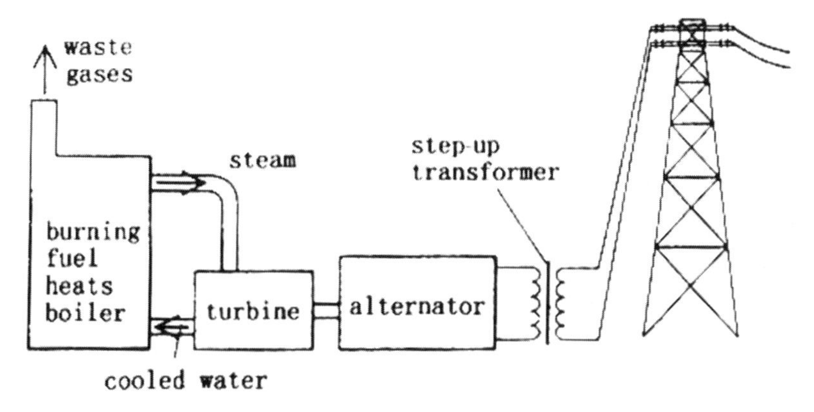

Fig.65.2 Supplying AC for the mains

Power is sent across country through overhead lines at very high voltage (typically 400,000V). The voltage is increased to this level by transformers and reduced again at the far end. As power = voltage × current, transmitting at a higher voltage means a lower current and, therefore, less power wasted as heat because of line resistance.

Exercises

1. Match the English with the Chinese. Draw lines.

A.

(1) transformer a. 发电站

(2) power station b. 变压器

(3) substation c. 电感线圈

(4) inductance loop d. 电网

(5) grid e. 变电站

B.

(1) boiler a. 蒸汽

(2) steam b. 交流发电机

(3) alternator c. 锅炉

(4) cooled water d. 增压变压器

(5) step-up transformer　　e. 冷却水

(6) turbine　　f. 涡轮

2. Answer the following questions.

(1) Why can't we live without electricity in our daily life?

(2) Why the voltage ratio is different from the current ratio?

(3) How to get the formula: $N_1I_1 = N_2I_2 + N_3I_3$?

3. Translate the following phrases into Chinese.

(1) fuel-burning station　distribution network　Grid　increase　reduce　transmit　heat　line resistance

(2) In a transformer, an alternating current in the primary(初级) (input) coil generates a changing magnetic field in the core, which induces an alternating voltage in the secondary (次级)(output) coil. A step-up transformer(增压变压器) increases the voltage. A step-down transformer (减压变压器)reduces it.

Words and phrases

transformer [træns'fɔ:mə(r)] 变压器

inductance loop 电感线圈

in our daily life 在我们日常生活中

furthermore ['fə:ðəˌmɔ:] 而且

simple 简单

figure ['figə] 图

input loop circuit 输入线圈电路

respectively [rɪ'spektɪvli] 分别地

provide [prə'vaɪd] 提供

be followed by 紧跟，遵循……

at the same time 同时

be equal to 等于……

gct 得到

transmit [træns'mɪt] 传输，传送

layout ['leɪaʊt] 布置图，布局，草图

grid [grɪd] 高压输电

overhead line 高架线

far 远的

line resistance 线路电阻

be capable of 能够

power station 发电站

substation ['sʌbsteɪʃn] 变电站

electricity [ɪˌlek'trɪsətɪ] 电

in an electrical design 在电气设计中

the same … just as 正如……一样

be showed in Fig.65.2 所示如图 65.2

output loop circuit 输出线圈电路

formula ['fɔ:mjələ] 公式

output voltage 输出电压

above 上面的，以上的

input power 输入功率/电源

output power 输入功率/电源

typically ['tɪpɪkli] 典型地，有代表性

mains [meɪnz] 电源/电力/电网

practical 实际的，实用的

increase [ɪn'kri:s] 增加，增长

decrease [dɪ'kri:s] 减少

end 末端，尽头

alternating voltage ['ɔ:ltəneɪtɪŋ] 交流电压

fuel-burning station ['fju:əl] ['bɜ:nɪŋ] ['steɪʃn] 热电站

distribution network [ˌdɪstrɪ'bju:ʃn] ['netwɜ:k] 配电网络

Lesson 66 Transformers (Ⅱ)

Look and select

Look at the pictures and select the correct terms from the box.

step-up substation	power plant	step-down substation
distribution power line	consumer	transmission line

1. ____________________

2. ____________________

3. 

4. ____________________

5. ____________________

6. ____________________

Text

1. General

The transformer consists of two parts: iron core and coil. It can be classified according to the following:

a. Structure of iron core: core type and shell type (Fig.66.1).

b. Function: step-up and step-down transformers.

c. Phases: single-phase and three-phase transformers.

d. Mode of cooling and winding insulation: oil immersed type and dry type.

e. Usage: ordinary and special type.

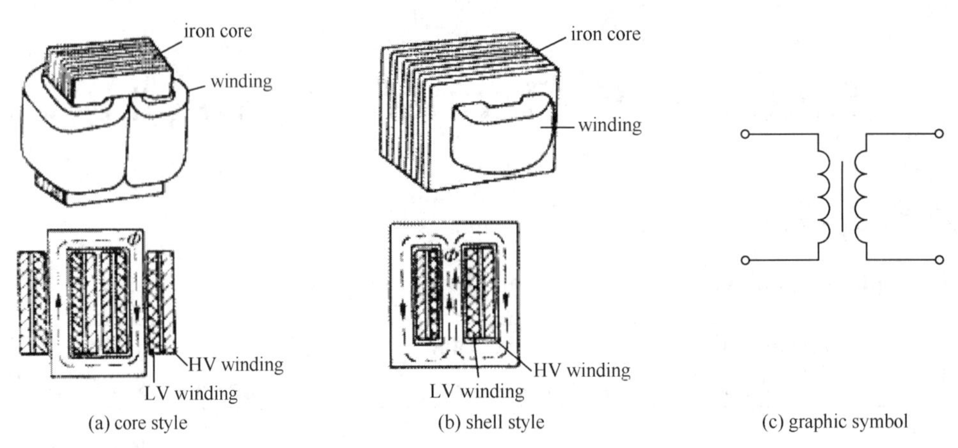

(a) core style (b) shell style (c) graphic symbol

Fig.66.1 Structure and graphic symbol of transformer

2. Nameplate of Transformer (Table 66.1)

Table 66.1 Power transformer

Power Transformer

Standard code ☐

Product code ☐

Model S7-500/10

Capacity 500Kv-a

Rated voltage 10000±5%/400V

Frequency 50Hz

Number of phases 3

Connection group Y,yn0

Impedance voltage 4%

Cooling mode oil coolig

Working condition outside

Weight

Switch location	High voltage		Low voltage	
	Voltage/V	Current/A	Voltage/V	Current/A
1	10500	27.5	400	721
2	10000	28.9		
3	9500	30.4		

☐ Manufacturer ☐ Year ☐ Month Ex-factory No. ☐

Knowledge extention

The basic parts of oil immersed transformer (Fig.66.2) include thermometer, moisture meter, oil conservator, oil meter, safety airway, gas relay, HV bushing, LV bushing, fuel tank, oil discharge valve, trolley, radiator etc.

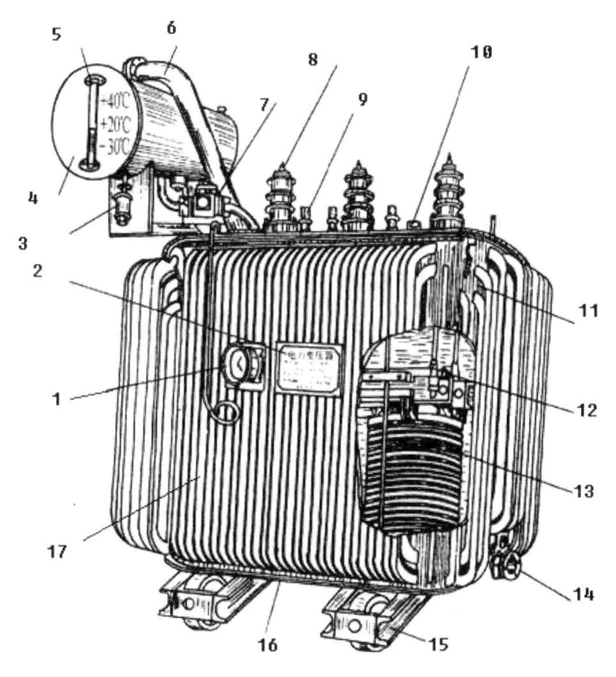

Fig.66.2 Oil immersed transformer

1—thermometer; 2—nameplate; 3—moisture meter; 4—oil conservator; 5—oil meter; 6—safety airway; 7—gas relay; 8—HV bushing; 9—LV bushing; 10—tap switch; 11—fuel tank; 12—iron core; 13—coil & insulation; 14—oil discharge valve; 15—trolley; 16—grounding bolt; 17—radiator

Exercises

1. Match the English with the Chinese. Draw lines.

A.

(1) step-up substation	a. 电厂
(2) power plant	b. 增压变电站
(3) step-down substation	c. 配电线路
(4) distribution power line	d. 输电线路
(5) consumer	e. 高压及低压
(6) transmission line	f. 用户
(7) HV and LV	g. 降压变电站

B.

(1) structure	a. 三极管
(2) function	b. 结构
(3) phase	c. 功能
(4) mode of cooling	d. 相位
(5) winding insulation	e. 使用
(6) usage	f. 图形符号
(7) graphic symbol	g. 绕组绝缘

2. Fill in the blanks with the words in the text.

(1) The transformer consists of two parts: ________ and ________. It can be classified according to the ________, function, ________, mode of cooling and __________ and usage.

(2) There are many types of transformer: ________________ and three-phase ; core type and _______ type; step-up and _________ type; oil _________ type and dry type; ________ and special type.

3. Translate the words into English or Chinese.

(1) 型号　容量　额定电压　频率　相数　连接组　阻抗电压
冷却方式　油冷　使用条件　户外　重量

(2) thermometer　nameplate　moisture meter　oil conservator　oil meter　gas relay
bushing　tap switch　fuel tank　iron core　trolley　oil discharge valve
grounding bolt　radiator

Words and phrases

distribution power line [ˌdɪstrɪ'bju:ʃn] 配电线路
step-up substation ['sʌbsteɪʃn] 增压变电站
step-down substation 降压变电站
consumer [kən'sju:mə(r)] 用户
transmission line [træns'mɪʃn] 输电线
power plant 发电厂
general ['dʒenrəl] 总述
according to [ə'kɔ:dɪŋ] 按照……
core type 芯型
shell type 壳型
function ['fʌŋkʃn] 功能
single-phase ['sɪŋgl] [feɪz] 单相
winding insulation ['waɪndɪŋ] [ˌɪnsju'leɪʃn] 绕组绝缘
mode [məʊd] 方式
oil immersed [ɪ'mɜ:st] 油浸式
dry [draɪ] 干燥
usage ['ju:sɪdʒ] 用途
ordinary ['ɔ:dnri] 普通的
special ['speʃl] 特殊的
model ['mɒdl] 型号
rated voltage ['reɪtɪd] 额定电压
capacity [kə'pæsəti] 容量
connection group [kə'nekʃn] [gru:p] 连接组
frequency ['fri:kwənsi] 频率
impedance voltage [ɪm'pi:dns] 阻抗电压
condition [kən'dɪʃn] 条件
thermometer [θə'mɒmɪtə(r)] 温度计
weight [weɪt] 重量
ex-factory 出厂
radiator ['reɪdieɪtə(r)] 散热器
moisture meter ['mɔɪstʃə(r)] 吸湿器
oil meter 油表
oil conservator [kən'sɜ:vətə(r)] 储油柜
fuel tank ['fju:əl] [tæŋk] 油箱
safety airway ['eəweɪ] 安全气道
gas relay ['ri:leɪ] 气体继电器
HV bushing ['bʊʃɪŋ] 高压套管
trolley ['trɒli] 小车
manufacturer [ˌmænju'fæktʃərə(r)] 制造商，厂家
oil discharge valve [dɪs'tʃɑ:dʒ] [vælv] 放油阀门

Lesson 67 Electric Motors (Ⅰ)

Look and select

Look at the pictures and select the correct terms from the box.

motor	washing machine	refrigerator	servo motor

1. 

2. ____________________

3. 

4. ____________________

Text

Motor

Electric motors are used to give power to many machines that we use in everyday life, such as washing machines, refrigerators, DVD players and so on.

They broadly include: AC motors, DC motors, step motors and servo motors etc.

The main parts of DC motor are as follows (Fig.67.1): 1—electric supply; 2—commutator; 3—armature; 4—field windings; 5—drive shaft; 6—brush; 7—bearing; 8—motor frame.

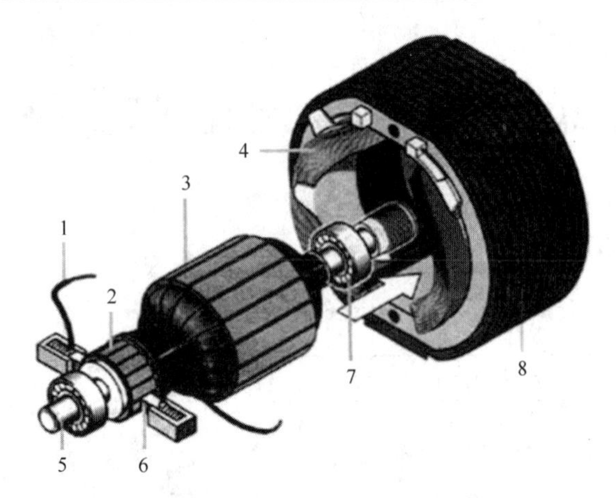

Fig.67.1 Main parts of DC motor

Knowledge extention

The electromagnetic-force law:

If we want to know the details, we must use electromagnetic-force law. When a DC source is connected to a brush circuit, there is current through the electric wire, and then the current must be in the magnetic field. By the Law, the electrical wire will fall under a force that makes the wire move circularly. Fig.67.2 just shows different positions of the wire, and different current orientations, but the force has the same function to make the wire rotate.

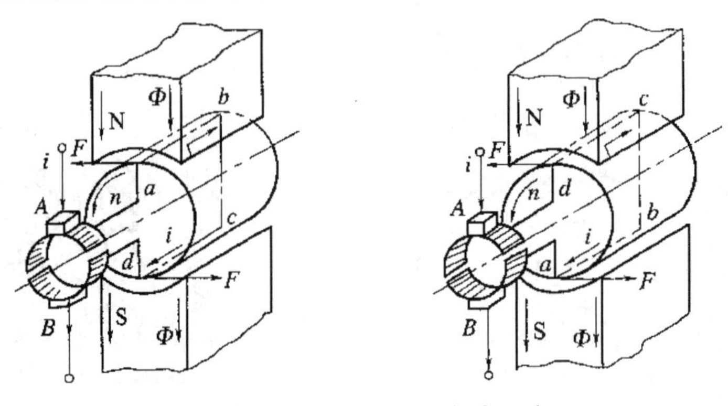

Fig.67.2 The electromagnetic-force law

Exercises

1. Match the English with the Chinese. Draw lines.

A.

(1) AC motor	a. 直流电机
(2) DC motor	b. 交流电机
(3) step motor	c. 伺服电机
(4) servo motor	d. 步进式电机
(5) commutator	e. 电枢，转子
(6) armature	f. 换向

B.

(1) electric supply	a. 励磁绕组
(2) field winding	b. 电源
(3) drive shaft	c. 电刷
(4) brush	d. 驱动轴
(5) bearing	e. 电动机架
(6) motor frame	f. 轴承

2. Fill in the blanks with the words in the text.

(1) Electric motors are used to give ________ to many machines that we use in everyday life, such as __________, refrigerators, __________ and so on.

(2) When a DC source is connected to a __________ circuit, there is current through the electric _________, and then the current must be in the ___________.

(3) By the Law (___________ law), the electrical wire will __________ under a force that makes the wire ___________ circularly.

3. Translate the following passages into English.

(1) Servo motors used in radio controlled models (cars, planes, etc.) are very useful in many kinds of small robotic experiments because they are small, compact and quite cheap.

(2) A servo motor itself has a built-in motor, a gearbox, a position feedback mechanism and controlling electronics.

Words and phrases

washing machine ['wɒʃɪŋ] [mə'ʃi:n] 洗衣机
refrigerator [rɪ'frɪdʒəreɪtə(r)] 冰箱
give power ['paʊə(r)] 提供动力/电源
everyday life ['evrideɪ] [laɪf] 日常生活
broadly ['brɔ:dlɪ] 大致，一般
step motor [step] 步进电动机
main part [meɪn] [pɑ:t] 主要部件/组成
electric supply [sə'plaɪ] 电源/力供应
brush [brʌʃ] 电刷
drive shaft [draɪv] [ʃɑ:ft] 驱动轴
motor frame [freɪm] 电动机架
detail ['di:teɪl] 详情，详细
be connected to [kə'nektɪd] 与……连接
magnetic field [mæg'netɪk] [fi:ld] 磁场
circularly ['səkjələlɪ] 循环地，圆地
position [pə'zɪʃn] 位置
motor ['məʊtə(r)] 马达，电动机
electric [ɪ'lektrɪk] 电的
be used to 用来/于……
and so on 等等
include [ɪn'klu:d] 包含/包括
servo motor ['sɜ:vəʊ] 伺服电动机
as follow ['fɒləʊz] 如下
commutator ['kɒmjuteɪtə(r)] 换向
function ['fʌŋkʃn] 功能
armature ['ɑ:mətʃə(r)] 电枢，转子
bearing ['beərɪŋ] 轴承
source [sɔ:s] 源，电源
brush circuit ['sɜ:kɪt] 电刷线路
rotate [rəʊ'teɪt] 旋转
fall [fɔ:l] 掉下，落下，下降
orientation [ˌɔ:riən'teɪʃn] 方向
electromagnetic-force law [ɪˌlektrəʊmæg'netɪk] [fɔ:s] [lɔ] 电磁定律
field winding [fi:ld] ['waɪndɪŋz] 励磁绕组

Lesson 68 Electric Motors(Ⅱ)

Look and translate

Look at the picture (Fig.68.1) and translate the terms into Chinese.

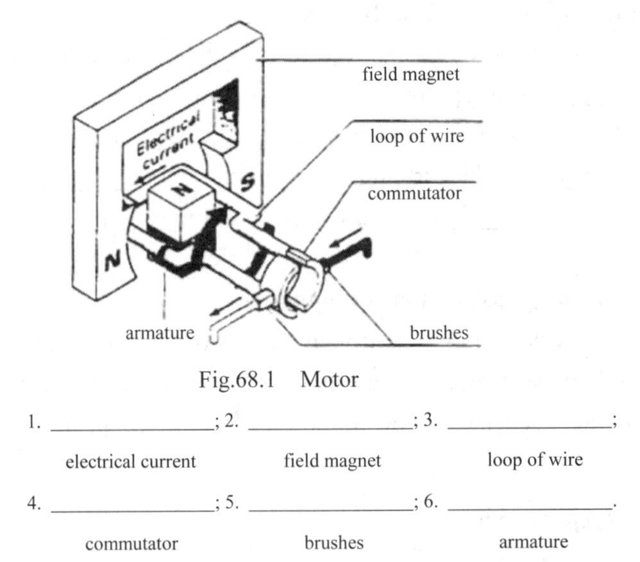

Fig.68.1 Motor

1. ____________; 2. ____________; 3. ____________;
electrical current field magnet loop of wire

4. ____________; 5. ____________; 6. ____________.
commutator brushes armature

Text

Nameplate of DC Motor (Table 68.1)

Table 68.1 DC motor

DC motor			
Model	Z2-32	Excitation mode	Shunt
Power	2.2kV	Excitation voltage	220V
Voltage	220V	Working mode	Continuous
Current	12.4 amp.	Insulation grade	Stator B Armature E
Speed	1500 r/m	Weight	76 kg
Product No.		Ex-factory date	

Knowledge extention

General assembly of DC Motor is as follows (Fig.68.2).

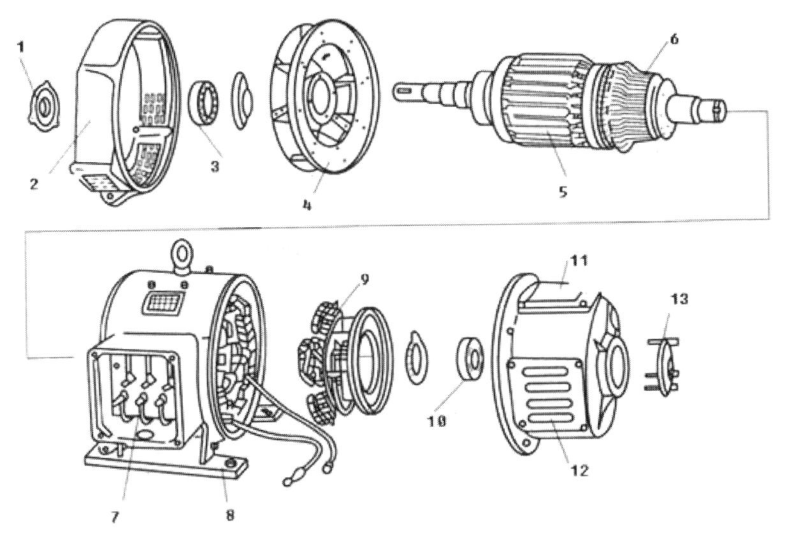

Fig.68.2 General assembly of DC Motor

1—rear bearing cover; 2—rear end cover; 3—bearing; 4—fans; 5—armature; 6—commutator; 7—junction box; 8—footing; 9—brush device; 10—breaing; 11—watch window; 12—vent hole; 13—front bearing cover

Exercises

1. Match the English with the Chinese. Draw lines.

(1) excitation mode	a. 励磁电压
(2) excitation voltage	b. 连续
(3) working mode	c. 励磁方式
(4) continuous	d. 工作方式
(5) insulation grade	e. 重量
(6) stator	f. 绝缘等级
(7) speed	g. 出厂日期
(8) product number	h. 定子
(9) weight	i. 转速
(10) ex-factory date	j. 产品编号

2. Understand the decomposition diagram of three-phase asynchronous motor (Fig.68.3).

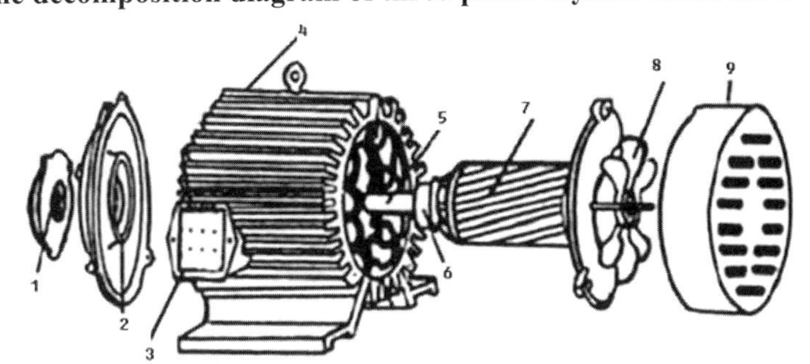

Fig.68.3 Exploded view of three-phase asynchronous motor

1—bearing cover; 2—end cover; 3—junction box; 4—stator; 5—rotating shaft; 6—bearing; 7—armature; 8—fans; 9—housing

Words and phrases

electrical current [i'lektrikəl 'kʌrənt] 电流
loop of wire 导线回路
excitation mode [ˌeksaɪ'teɪʃən] 励磁方式
excitation voltage 励磁电压
insulation grade [ˌɪnsju'leɪʃn] [greɪd] 绝缘等级
product No. ['prɒdʌkt] 产品编号
rear bearing cover [rɪə(r)] 后轴承盖
watch window [wɒtʃ] ['wɪndəʊ] 视察窗
commutator ['kɒmjuteɪtə(r)] 换向器
footing ['fʊtɪŋ] 底脚
vent hole [vent] [həʊl] 通风孔
asynchronous motor [eɪ'sɪŋkrənəs] 异步电机
rotating shaft [rəʊ'teɪtɪŋ] [ʃɑ:ft] 转轴
field magnet ['mægnət] 场磁铁
brush 电刷
shunt [ʃʌnt] 并励
working mode 工作方式
stator ['steɪtə] 定子
continuous [kən'tɪnjuəs] 连续的
rear end cover 后端盖
fan [fæn] 风扇
junction box ['dʒʌŋkʃn] 接线盒
electric brush device 电刷装置
front bearing cover 前轴承盖
exploded view [ɪk'spləʊdɪd] 分解图
housing ['haʊzɪŋ] 罩壳

Lesson 69 Bridge Circuit

Look and select

Look at the pictures and select the terms from the box.

clip	analog bridge	portable bridge	digital bridge

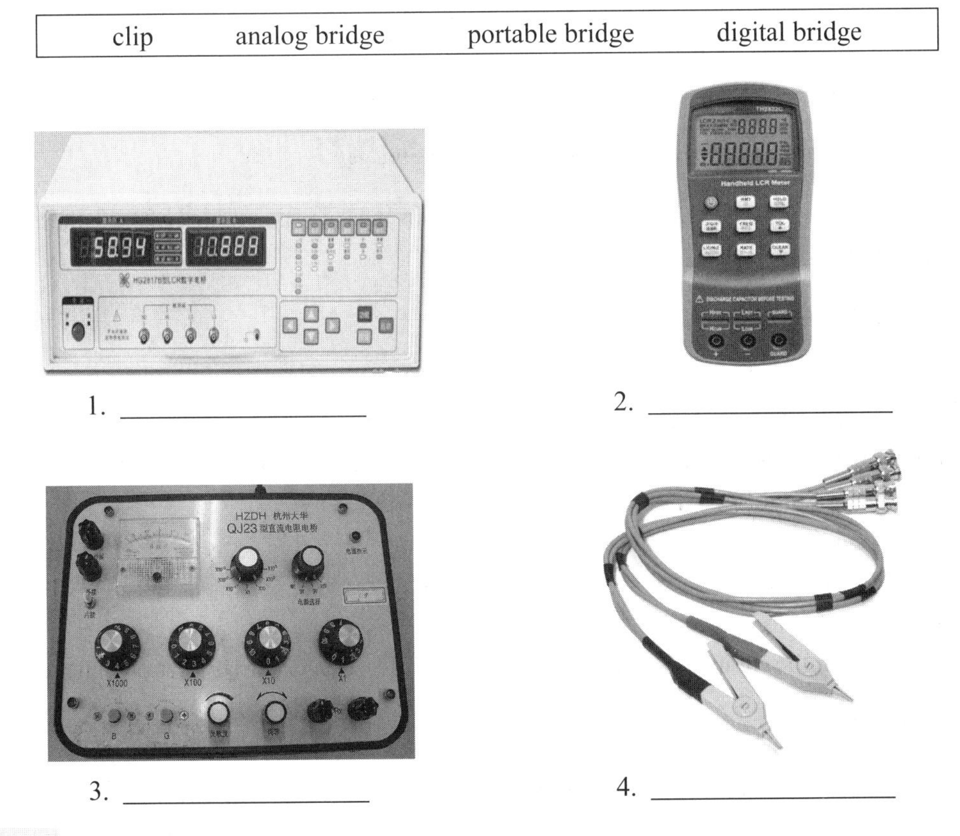

1. ____________

2. ____________

3. ____________

4. ____________

Text

Bridge circuits are special series-parallel circuits and are used to make measurements in electronic circuits.

Note zero volts appear between points A and B when the bridge is balanced (Fig.69.1). You will see this balance only when the ratios of the resistances in the left arm and in the right arm of the bridge are equal.

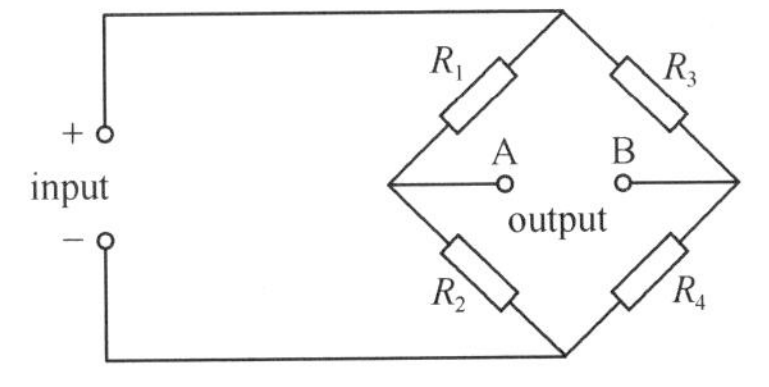

Fig.69.1 Balanced bridge

This balance condition (zero volts between points A and B) exists whenever the resistance ratios of the top resistor to bottom resistor in the left and right arm of the bridge are the same. That is:

Bridge is balanced when $R_1/R_2=R_3/R_4$

One of the common application bridges is named the Wheatstone bridge (Fig.69.2), which is used to measure the value of an unknown resistance.

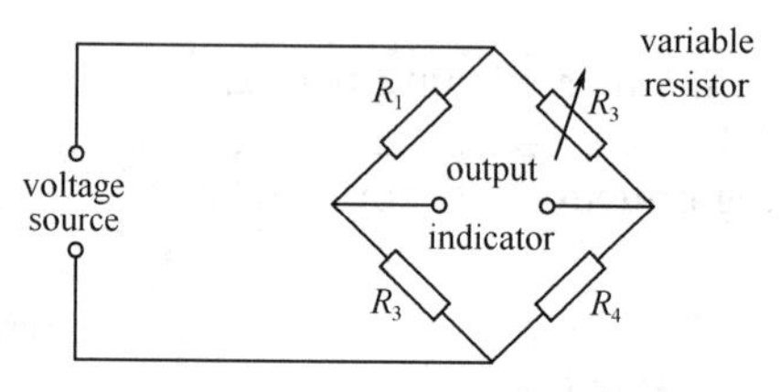

Fig.69.2 Wheatstone bridge

Variations of the bridge circuits can also be used to measure the values of the other types of components, such as capacitors.

Knowledge extention

1. Measurement

Measurement is made to measure a certain quantity. The quantity such as a current measured with instruments or meters is not always accurate. The difference is shown as an error.

2. Errors are made in different cases.

(1) Man-made error, for example, an incorrect reading.

(2) Instrument error, for example, digital instruments make mistakes during A/D or D/A conversion.

(3) Influence of the environment, for example, errors caused by external electromagnetic fields.

Exercises

1. Fill in the blanks with the words in the text.

(1) Bridge circuits are special ________ circuits and are used to make measurements in electronic circuits.

(2) Note ________ appear between points A and B when the bridge is ________. You will see this balance only when the ratios of the ________ in the left arm and in the right arm of the bridge are ________.

(3) One of the common application bridge is named the ___________, which is used to measure the ________ of an unknown resistance.

2. Match the English with the Chinese. Draw lines.

(1) analog bridge	a. 夹子
(2) digital bridge	b. 便携式电桥
(3) portable bridge	c. 数字电桥
(4) clip	d. 模拟电桥
(5) measurement	e. 人为错误

(6) man-made error f. 环境
(7) environment g. 影响
(8) influence h. 测量

3. Translate the following sentences into English.

a. 平衡状态 b. 电阻比相等 c. 一种常见的应用电桥

4. True or false.

(1) Actually, bridge circuits are special series-parallel circuits used for measuring electrical components. (　　)

(2) The Wheatstone bridge is used to measure the value of an unknown capacitor. (　　)

(3) Variations of the bridge circuits can also be used to measure the capacitors. (　　)

Words and phrases

bridge circuit [brɪdʒ] ['sɜːkɪt] 电桥
note [nəʊt] 注意，留心
series-parallel circuits ['sɪəriːz] ['pærəlel] 串并联电路
appear [ə'pɪə(r)] 出现
make measurement ['meʒəmənt] 测量
zero volt ['zɪərəʊ] [vəult] 零电压
point [ə'pɪə(r)] 点
balance ['bæləns] 平衡
ratio ['reɪʃɪəʊ] 比
left arm [left] [ɑːm] 左臂
right arm [raɪt] 右臂
be equal ['iːkwəl] 相等
condition [kən'dɪʃn] 情况
exist [ɪg'zɪst] 存在
top resistor [rɪ'zɪstə(r)] 上电阻
bottom resistor ['bɒtəm] 下电阻
common ['kɒmən] 常见
application [ˌæplɪ'keɪʃn] 应用
Wheatstone bridge [brɪdʒ] 惠斯通电桥
be named 叫作，称作
value ['væljuː] 值
unknown [ˌʌn'nəʊn] 未知
variation [ˌveəri'eɪʃn] 变化，变异
type [taɪp] 类型
component [kəm'pəʊnənt] 元件
certain ['səːtn] 某一，某种
quantity ['kwɒntəti] 量，数量
be shown as [ʃəʊn] 显示为
error ['erə(r)] 错误
man-made 人造的
incorrect [ˌɪnkə'rekt] 不正确的
reading ['riːdɪŋ] 读数
conversion [kən'vɜːʃn] 变换，转换
environment [ɪn'vaɪrənmənt] 环境
influence ['ɪnflʊəns] 影响
external [eks'təːnl] 外部的
caused by [kɔːzd] 由……引起

Lesson 70 Logic Gate

Look and select

Look at the pictures and select the correct terms from the box.

digital instrument	control	communication	computer

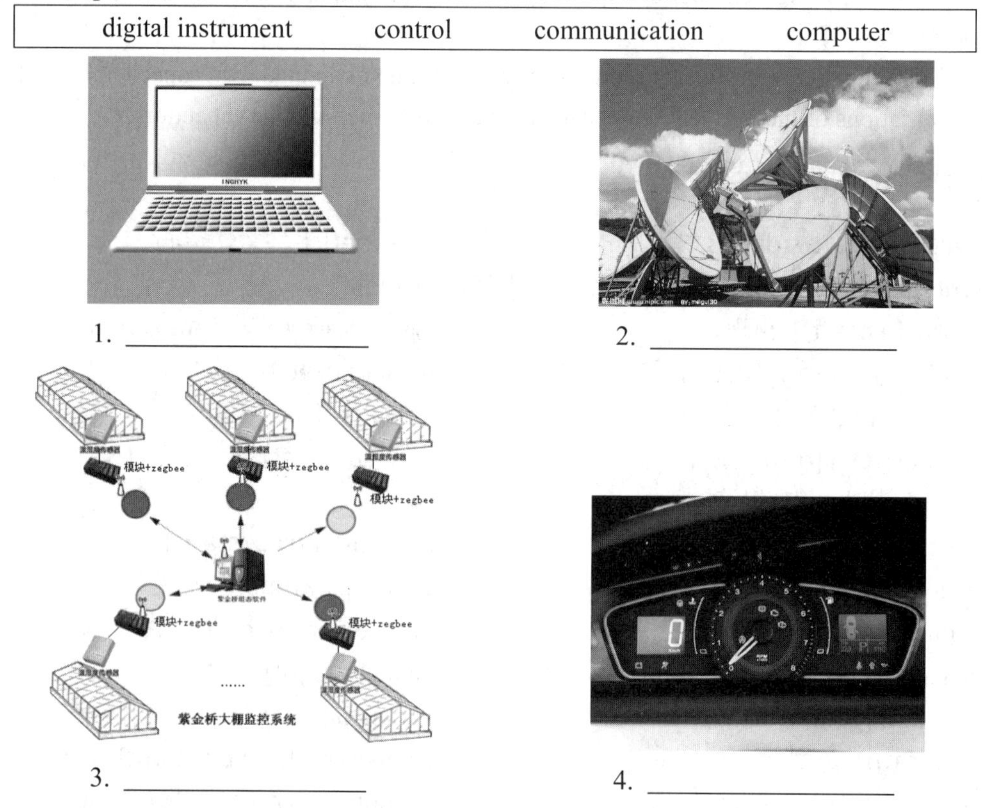

Text

The processors of most electronic systems contain an arrangement of switches called logic gate. These switches process the input signals by only allowing them to pass through if they fulfill certain conditions.

There are several different types of gates: AND, OR, NOT, NAND and NOR. Truth tables are used to explain the results of all the possible combinations of inputs.

The following (Fig.70.1) is an example of the use of gates.

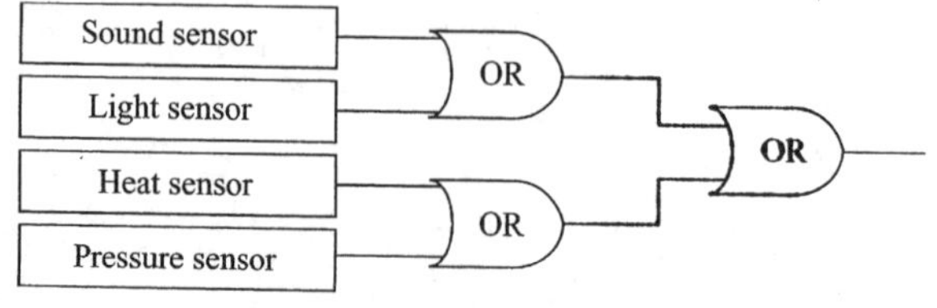

Fig.70.1 Example of the use of gates

Burglar alarm: connecting several gates together can produce quite sophisticated systems. Using this system, there are several different methods to detect a burglar (Fig.70.2).

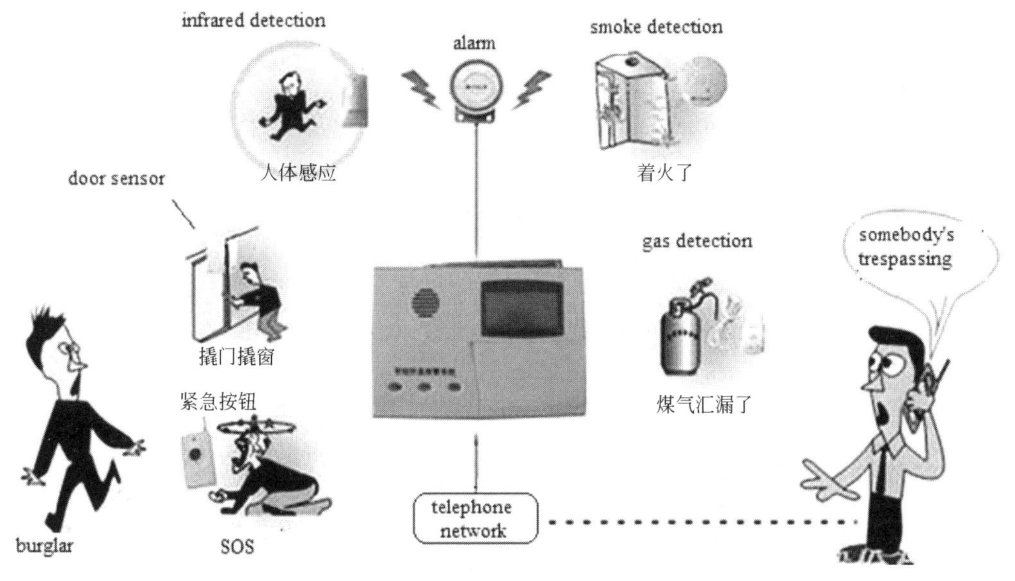

Fig.70.2 Security system of burglar alarm

Knowledge extention

Electronic system and control

Processors of some electronic system manipulate input signals in a continuous manner. These are called analog system. Some electronic systems work in a non-continuous way. These are called digital systems.

Digital processors such as computers use a binary method to handle information and perform calculations. The processors use just two alternatives: 1 and 0. These numbers can be represented by switches which have two possible positions: ON or OFF.

Exercises

1. Fill in the blanks with the words in the text.

(1) The processors of most electronic systems contain an ________________of switches called ________________.

(2) Truth tables are used to explain the ______________ of all the possible combinations of inputs.

(3) Burglar alarm: connecting several ____________ together can produce quite sophisticated systems.

2. Match the English with the Chinese. Draw lines.

(1) AND gate　　　　a. 或门

(2) OR gate　　　　b. 与门

(3) NOT gate　　　　c. 与非门

(4) NAND gate　　d. 非门

(5) NOR gate　　e. 或非门

3. True or False.

(1) Logic gates are electronic switches. (　　)

(2) Logic gates allow any signal to explain the output. (　　)

(3) Truth tables are used to explain the output. (　　)

(4) NAND is one instance out of many types of gates. (　　)

4. Translate the following passages into Chinese.

(1) These switches process the input signals by only allowing them to pass through if they fulfill certain conditions.

(2) Using this system, there are several different methods to detect a burglar.

5. Translate the following expressions into English.

a. 开关阵列　　b. 满足一定条件　　c. 将几个逻辑门连在一起

Words and phrases

processor ['prəʊsesə(r)] 处理器

communication [kəˌmju:nɪ'keɪʃn] 通讯，交际/流

arrangement [ə'reɪndʒmənt] 阵列，安排

process [prə'ses] 处理，过程

signal ['sɪgnəl] 信号

pass through 通过，经过

condition [kən'dɪʃn] 情况，条件

type [taɪp] 类型

truth table [tru:θ] table 真值表

result [rɪ'zʌlt] 结果

combination [ˌkɒmbɪ'neɪʃn] 组合，组成

burglar alarm ['bɜ:glə(r)] [ə'lɑ:m] 报警器

connect...together [kə'nekt] 把……连接起来

sophisticated [sə'fɪstɪkeɪtɪd] 复杂的

burglar ['bɜ:glə(r)] 强盗，盗贼

manipulate [mə'nɪpjʊleɪt] 操作

analog system ['ænəlɔ:g] ['sɪstəm] 模拟系统

perform [pə'fɔ:m] 实施，完成

be represented by [ˌrepri'zentid] 用……表示

continuous manner [kən'tɪnjuəs] ['mænə(r)] 连续的方式

digital ['dɪdʒɪtl] 数字的

control [kən'trəʊl] 控制

contain [kən'teɪn] 包含

input ['ɪnpʊt] 输入

allow [ə'laʊ] 允许

fulfill [fʊl'fɪl] 满足，完成

certain ['sə:tn] 某一

gate [geɪt] 门

explain [ɪk'spleɪn] 解释，说明

possible ['pɒsəbl] 可能的

position [pə'zɪʃn] 位置

handle ['hændl] 处理

produce [prə'dju:s] 产生，生成

detect [dɪ'tekt] 监测

sensor ['sensə(r)] 感应器

binary ['baɪnəri] 二进制的

information [ˌɪnfə'meɪʃn] 信息

calculation [□kælkju□le□□n] 计算

alternative [ɔ:l'tɜ:nətɪv] 交替的

Lesson 71 PLC

Look and select

Look at the pictures and select the correct terms from the box.

COM	PLC	control panel	certificate

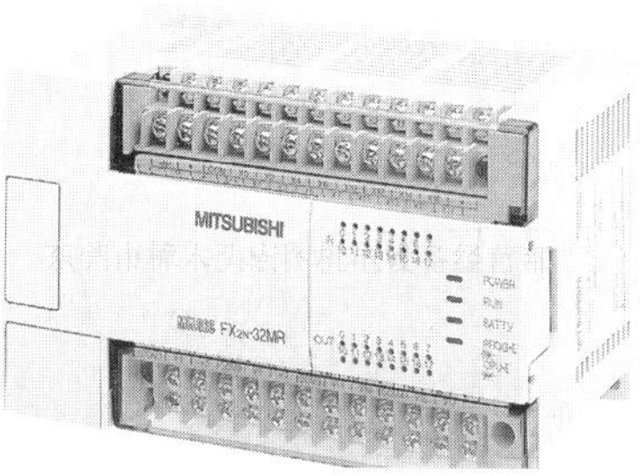

1. ____________________

2. ____________________

3. ____________________

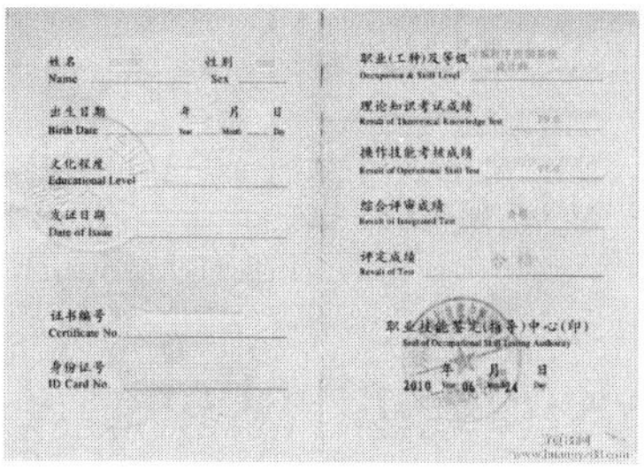

4. ____________________

Text

A PLC is a replacement for relay devices. It is the heart of modern automatic factories. It is programmed using a ladder diagram, which is standard electric wiring diagram.

PLCs vary in size and power. A large PLC can have up to 10000 I/O points and support many functions. There are also expansion slots to accommodate PC and other communication devices. For many applications, a small PLC is sufficient. Fig.71.1 shows a small PLC. It has 16 I/O points and a standard Siemens S7-224 serial communication port. The speed of PLC is constantly improving, even the low-end PLCs perform at high speed. One to two mocrosec/kbyte of memory speed is very common.

Like a general-purpose computer, a programmable controller consists of five major parts: CPU (processor), memory, input / output (I/O), power supply, and peripherals (Fig.71.2).

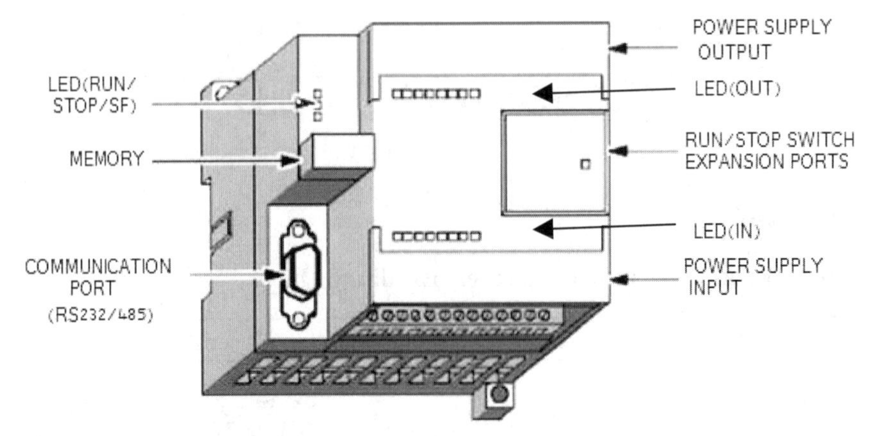

Fig.71.1 Siemens S7-224 PLC

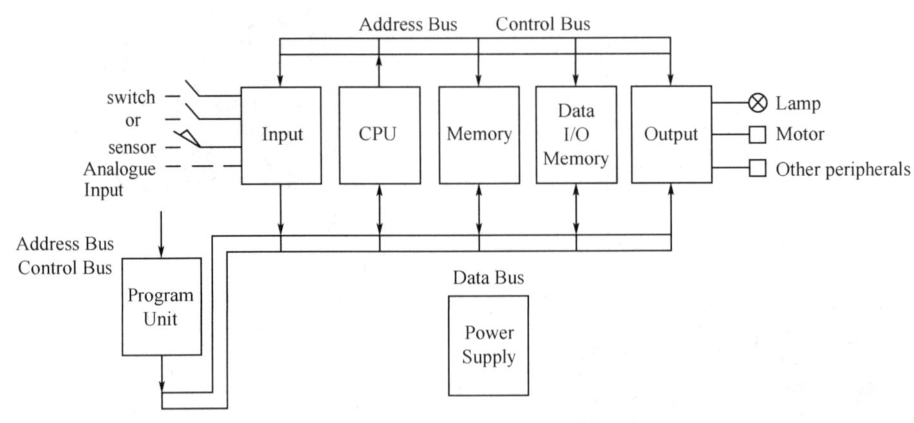

Fig.71.2 Structure of programmable logic controller

Knowledge extention

A PLC works by continually scanning a program. We can think of this scan cycle as consisting of 3 important steps (Fig.71.3). There are typically more than 3 but we can focus on the important parts and do not worry about the others. Typically the others are checking the system and updating the current internal counter and timer values.

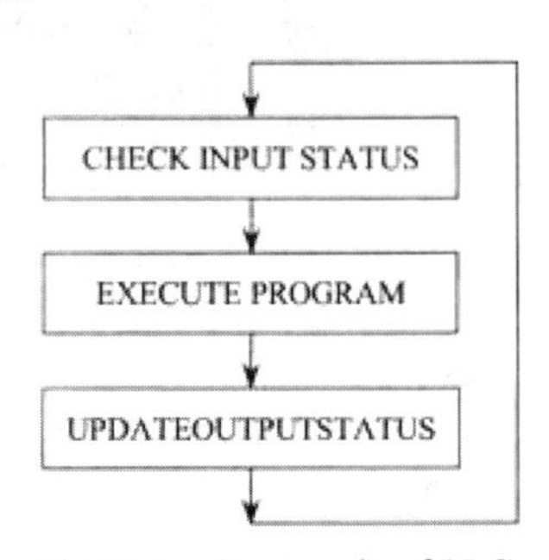

Fig.71.3 Scan cycle of PLC

Exercises

1. Fill in the blanks with the word in the text.

(1) A PLC is a replacement for __________ devices. It is the heart of modern automatic factories. It is programmed using a _________, which is standard electric __________ diagram.

(2) PLCs vary in _________ and _________. A large PLC can have up to 10000 I/O points and support many functions.

(3) A PLC works by continually scanning a program. We can think of this scan cycle as consisting of 3 important steps: ___________, execute program and ___________.

2. Match the English with the Chinese. Draw lines.

(1) CPU (processor)	a. 输入/输出
(2) memory	b. 电源
(3) I/O	c. 存储器
(4) power supply	d. 外设
(5) peripheral	e. 中央处理器

3. Answer the following questions.

(1) What is PLC?

(2) What are the PLC's main parts?

4. Understand the structure of PLC and translate them into Chinese:

Input Circuit
CPU
Memory
Output Circuit

⇨

Input Relays	Counters	Output Relays
Internal Utility Relays	Timers	Data Storage

Input Relays ________; Output Relays ________; Counters ________;
Timers ________; Internal Utility Relays ________; Data Storage ________.

Words and phrases

PLC (Programmable Logic Controller) ['prəʊgræməbl] ['lɒdʒɪk] [kən'trəʊlə(r)] 可编程逻辑控制器

modern automatic ['mɔdən] [ˌɔ:tə'mætɪk] 现代自动化

control panel [kən'trəʊl] ['pænl] 控制面板/柜

replacement [rɪ'pleɪsmənt] 替代物

major part ['meɪdʒə(r)] [pɑ:t] 主要部件/部分

power supply ['paʊə(r)] [sə'plaɪ] 电源

COM [kɑm] 接口，端口

differ from ['difəz] 与……不同

condition [kən'dɪʃn] 条件，状况

sensor ['sensə(r)] 传感器，感应器

symbolic [sɪm'bɒlɪk] 象征的，符号的

certificate [sə'tɪfɪkət] 资格证书

short [ʃɔ:t] 短的，缩略的

form [fɔ:m] 形式，格式

memory ['meməri] 存储器，内存

peripheral [pə'rɪfərəl] 外设

a large number of 许多，大量

extremely [ɪk'stri:mli] 极大/度地

millisecond ['mɪlisekən] 毫秒

understand [ˌʌndə'stænd] 理解

Lesson 72 Electrical System

Look and select

Look at the pictures and select the correct terms from the box.

lightning and grounding	lighting	substation	fireproof

1.  ____________ 2. ____________

3.  ____________ 4. ____________

Text

Electrical system: substation, power work, lighting, communication system, fireproof, lightning and grounding.

Substation: transformer, HV/LV switch cabinet, control/ emergency power supply equipment, bus duct, miscellaneous panel, cable installation and connection.

Power work: motor, local equipment, cable ladders / trays and supports (Fig.72.1), cabling and wiring, cable termination splicing.

Lighting: lighting panelboards, lighting fixtures, lighting in plant, street light, floodlight, receptacles.

Fire alarm system: fire alarm panel, smoke/heat detector, breakglass point.

Lightning and grounding: electrode, inspection box for electrode, copper line.

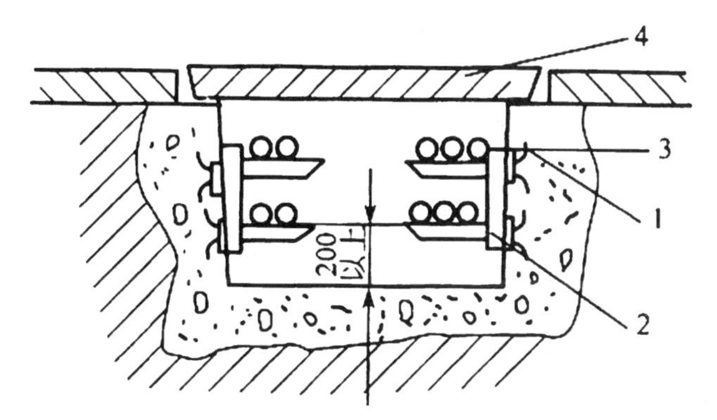

Fig.72.1 Laying of cable trench

1—embedded iron parts; 2—cable support; 3—cable; 4—cover plate

Knowledge extention

If an electric iron is not use, you must pull out the plug connected to 220V voltage source. If not, there is a danger of fire (Fig.72.2).

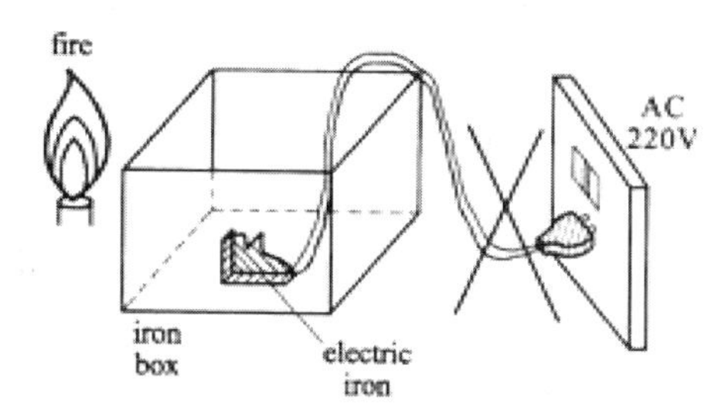

Fig.72.2 Safety alarm

Exercises

1. Match the English with the Chinese. Draw lines.

(1) electrical system	a. 变电站
(2) substation	b. 照明
(3) power work	c. 电气系统
(4) lighting	d. 电力工作
(5) fire alarm system	e. 避雷与接地
(6) lightning and grounding	f. 防火报警系统
(7) cable trench	g. 电缆沟
(8) embedded	h. 支架
(9) support	i. 预埋
(10) cover plate	j. 盖板

2. Fill in the blanks with the words in the text.

(1) Substation: transformer, HV/LV ____________, control/ ____________ power supply equipment, bus duct, miscellaneous panel, ____________ and connection.

(2) Power work: motor, local equipment, _________ / trays and supports, cabling and wiring, cable _________ splicing.

(3) Lighting: lighting panelboards, ___________________, lighting in plant, street light, floodlight, and _________.

(4) Fire alarm system: _________, smoke/heat _________, breakglass point.

(5) Lightning and grounding: ___________, inspection box for electrode, ___________.

3. Answer the following questions.

(1) What are included in electrical system?

(2) What are included in substation?

(3) Please brief the details of power work.

Words and phrases

communication system [kəˌmjuːnɪ'keɪʃn] ['sɪstəm] 通信系统

control/emergency power supply equipment [kən'trəʊl] [ɪ'mɜːdʒənsɪ] [ɪ'kwɪpmənt] 控制/紧急供电设备

miscellaneous panel [ˌmɪsə'leɪnɪəs] ['pænl] 零星配电盘

local equipment ['ləʊkl] [ɪ'kwɪpmənt] 地方设备

smoke/heat detector [sməʊk] [hiːt] [dɪ'tektə(r)] 感烟/热探测器

lightning ['laɪtnɪŋ] 避雷，闪电

lighting ['laɪtɪŋ] 照明

power work ['paʊə(r)] [wɜːk] 电力工作

transformer [træns'fɔːmə(r)] 变压器

switch cabinet [swɪtʃ] cabinet 开关柜

installation [ˌɪnstə'leɪʃn] 安装

termination [ˌtɜːmɪ'neɪʃn] 终端

panelboard [peɪ'nelbɔːd] 配电盘

lighting fixture ['laɪtɪŋ] ['fɪkstʃə(r)] 照明灯具

fire alarm ['faɪə(r)] [ə'lɑːm] 火灾报警

breakglass point 易碎玻璃手动按钮

inspection box [ɪn'spekʃn] 检查箱

grounding ['graʊndɪŋ] 接地

substation ['sʌbsteɪʃn] 变电站

fireproof ['faɪəpruːf] 防火

bus duct [dʌkt] 护埂

HV/LV 高压/低压

connection [ˌɪnstə'leɪʃn] 连接

splicing ['splaɪsɪŋ] 接头

plant [plɑːnt] 工厂

receptacle [rɪ'septəkl] 插座

floodlight ['flʌdlaɪt] 泛光灯

electrode [ɪ'lektrəʊd] 电极

copper line ['kɒpə(r)] [laɪn] 铜线

Lesson 73 Instrument System

Look and select

Look at the pictures and select the correct terms from the box.

pressure meter	level meter	cable tray	flow transmitter

1. ______________

2. ______________

3. ______________

4. ______________

Text

Instrument system: control center, local instruments.

Control center: DCS (Fig.73.1) control cabinets, power supply cabinets, MOV control system, emergency control system, CCTV system, communication system, UPS etc.

Local instruments: FAR, analysers and analyser houses, flow instruments, level instruments, pressure instruments, temperature instruments, control valves, fire and gas instruments, cable trays and supports, cabling and wiring, optical fiber cables, power supply and earthing etc.

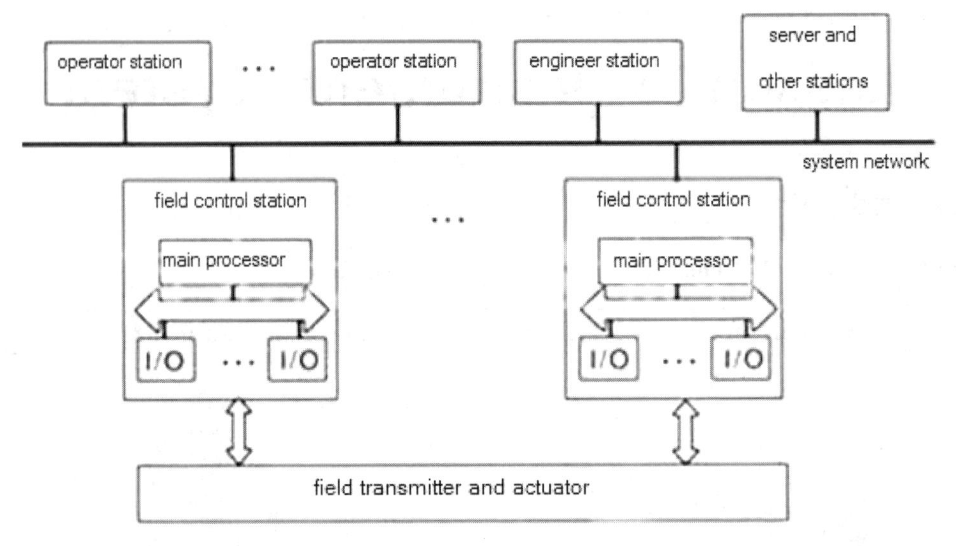

Fig.73.1 Typical system of DCS

Knowledge extention

Usually a microcontroller (Fig.73.2) is mainly made up of a universal CPU, memory, timer, counter, interruption, and many kinds of interface circuits, and has high price performance ratio.

Fig.73.2 Microcontroller

A microcontroller system is a typical embedded system which has such features as low cost, high performance, small size, and simple structure, etc.

Exercises

1. Match the English with the Chinese. Draw lines.

A.

(1) instrument system	a. 压力仪
(2) pressure meter	b. 电缆桥架
(3) level meter	c. 仪表系统
(4) cable tray	d. 液位计
(5) flow transmitter	e. 流量变送器

B.

(1) DCS	a. 闭路电视系统
(2) MOV	b. 集散控制系统
(3) CCTV	c. 电动阀
(4) UPS	d. 现场辅助房
(5) FAR	e. 无间断供电
(6) CPU	f. 中央处理器

2. Fill in the blanks with the words in the text.

(1) Control center include DCS control cabinets, ________________, MOV control system, ____________, CCTV system, communication system, UPS etc.

(2) Usually a microcontroller is mainly made up of a universal CPU, ____________, timer, _____________, interruption, and many kinds of ________________, and has high price performance ratio.

3. Answer the following questions.

(1) What are included in control center?

(2) What are included in local instruments?

(3) What are the features of microcontroller?

Words and phrases

DCS (distributed control system) control cabinet [dɪs'trɪbju:tɪd] ['kæbɪnət] 集散控制系统控制柜

flow transmitter [fləʊ] [træns'mɪtə(r)] 流量变送器

control center [kən'trəʊl] ['sentə] 控制中心

power supply cabinet ['paʊə(r)] [sə'plaɪ] 电力供应柜

emergency control system [ɪ'mɜ:dʒənsɪ] 紧急控制系统

CCTV (closed circuit television) system 闭路电视系统

communication system [kəˌmju:nɪ'keɪʃn] ['sɪstəm] 通信系统

UPS (uninterrupted power supply) [ˌʌnˌɪntə'rʌptɪd] 无间断供电

FAR (field auxiliary room) [fi:ld] [ɔ:g'zɪliəri] 现场辅助房

analysers and analyser houses ['ænəlaizəz] 分析器及分析器室

cable trays and supports [sə'pɔ:t] 电缆桥架及支架

power supply and earthing ['ɜ:θɪŋ] 电力供应及接地

CPU (central processing unit) [prəʊ'sesɪŋ] 中央处理器

instrument ['ɪnstrəmənt] 仪表

system ['sɪstəm] 系统

pressure meter ['preʃə(r)] ['mi:tə(r)] 压力表

level meter ['levl] 液/物位计

cable tray ['keɪbl] [treɪ] 电缆桥架

flow instrument 流量仪表

local instrument ['ləʊkl] 现场仪表

level instrument 液位/物仪表

temperature instrument 温度仪表

pressure instrument 压力仪表

fire and gas instrument 防火及气体仪表

control valve [vælv] 控制阀

optical fiber cable ['ɒptɪkl] ['faɪbə] 光缆

cabling and wiring 布线

be mainly made up of 主要由……组成

microcontroller 微控制器

memory 存储器

timer 定时器

counter 计数器

interruption [ˌɪntə'rʌpʃn] 中断

interface circuit ['ɪntəfeɪs] 接口电路

high price performance ratio [pə'fɔ:məns] ['reɪʃɪəʊ] 很高的性价比

Lesson 74 Flow Chart of Instrument Installation

Look and select

A. Look at the pictures and select the correct terms from the box.

wiring	installation	loop testing	DCS control center

1. ________________

2. ________________

3. ________________

4. ________________

B. Learn the terms about instrument acceptance regulation (Table 74.1).

Table 74.1 Instrument terms

Chinese	English	Chinese	English
自动化仪表	automation instrument	执行器	actuator
就地仪表	local instrument	取源部件	tap
现场	site	监测点	measuring point
检测仪表	detecting instrument	取压点	pressure measuring point
传感器	transducer	测温点	Temp. measuring point
转换器	converter	控制系统	control system
变送器	transmitter	仪表管道	instrument piping
显示仪表	display instrument	测量管道	measuring piping

Text

Understand the flow chart of instrument installation (Fig.74.1).

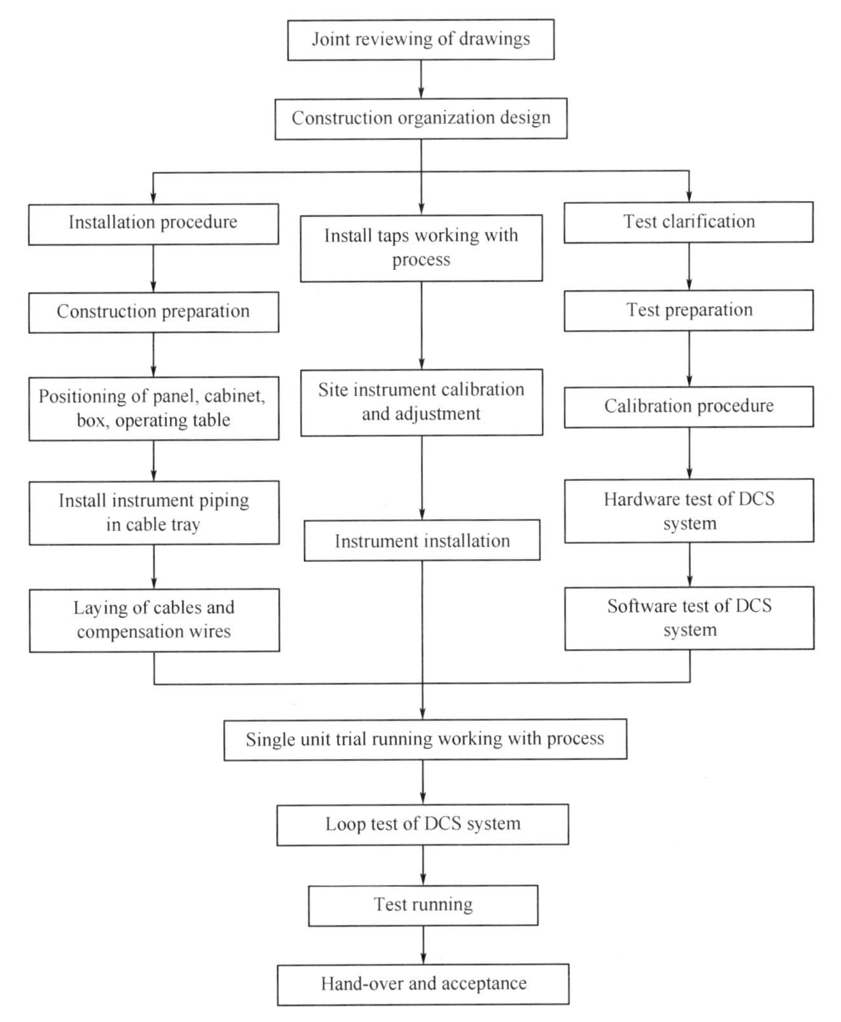

Fig.74.1 Flow chart of instrument installation

Knowledge extention

Understand the principle of auto-control system(closed loop) (Fig.74.2).

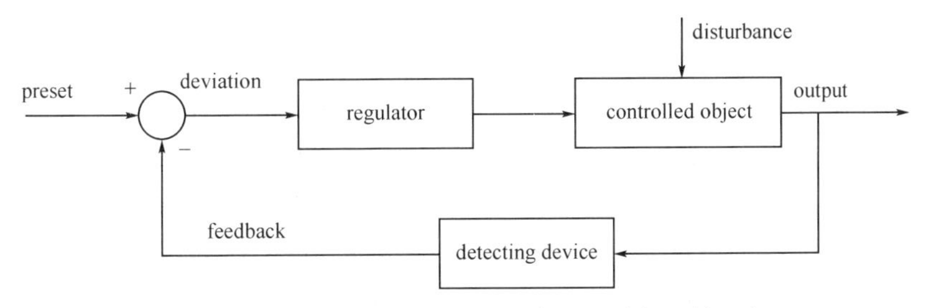

Fig.74.2 Principle of auto-control system(closed loop)

Exercises

1. Match the English with the Chinese. Draw lines.

A.

(1) wiring　　a. 回路测试
(2) installation　　b. 接线
(3) loop testing　　c. 安装
(4) transducer　　d. 变送器
(5) transmitter　　e. 传感器
(6) converter　　f. 转换器

B.

(1) local instrument　　a. 检测仪表
(2) detecting instrument　　b. 测量管道
(3) display instrument　　c. 就地仪表
(4) control instrument　　d. 显示仪表
(5) measuring piping　　e. 信号管道
(6) signal piping　　f. 控制仪表

2. Translate the following phrases into Chinese.

(1) joint reviewing of drawings
(2) construction organization design
(3) install taps working with process
(4) site instrument calibration and adjustment
(5) single unit trial running working with process
(6) loop test of DCS system
(7) test running
(8) hand-over and acceptance

3. Translate the following phrases into English.

(1) 仪表安装准备
(2) 盘、柜、箱、操作台就位
(3) 电缆槽中安装仪表配管
(4) 电缆及补偿导线敷设
(5) 试验交底
(6) 试验准备
(7) 校对程序
(8) DCS 硬件试验
(9) DCS 软件试验

Words and phrases

flow chart [fləu tʃɑ:t] 流程图
wiring ['waɪərɪŋ] 接线
installation [ˌɪnstə'leɪʃn] 安装
loop testing [lu:p 'testiŋ] 回路实验
DCS control center DCS 控制中心
automation [ˌɔ:tə'meɪʃn] 自动化
local ['ləʊkl] 地方的，就地，局部
detecting [dɪ'tektɪŋ] 检测
transducer [trænz'dju:sə(r)] 传感器
converter [kən'vɜ:tə(r)] 转换器

joint reviewing [dʒɔɪnt] [ri'vju:ɪŋ] 会审
process ['prəʊses] 工艺，过程
deviation [ˌdi:vi'eɪʃn] 偏差
regulator ['regjuleɪtə(r)] 调节器
actuator ['æktʃʊeɪtə] 执行器
preparation [ˌprepə'reɪʃn] 准备
feedback ['fi:dbæk] 反馈
disturbance [dɪ'stɜ:bəns] 扰动
detecting device [di'tektiŋ di'vais] 检测装置
hand-over and acceptance [ək'septəns] 交工验收
calibration and adjustment [ˌkælɪ'breɪʃn] [ə'dʒʌstmənt] 校验与调整
construction organization design [kən'strʌkʃn] [ˌɔ:gənaɪ'zeɪʃn] [dɪ'zaɪn] 施工组织设计
single unit trial running ['sɪŋgl] ['ju:nɪt] ['traɪəl] ['rʌnɪŋ] 单体试车
positioning of panel, cabinet, box, operating table 盘、柜、箱、操作台就位
laying of cables and compensation wires 电缆及补偿导线敷设

Lesson 75 Installation of Pressure Transmitters

Look and select

A. Look at the pictures and select the correct terms from the box.

level transmitter	pressure transmitter
temperature transmitter	flow transmitter

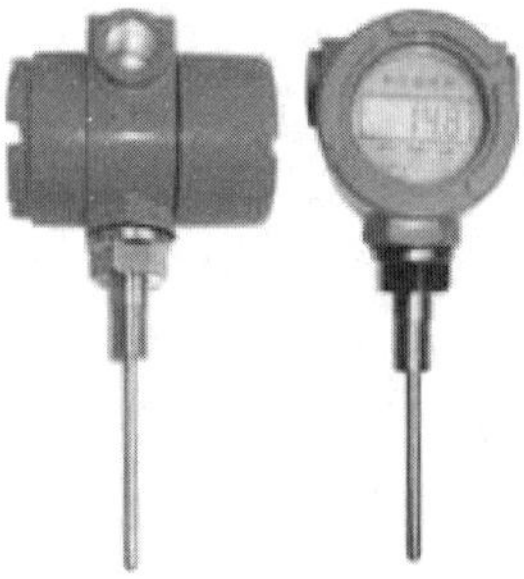

1. ____________________
2. ____________________ 
3. ____________________ 
4. ____________________ 

B. Learn the terms and symbols in Table 75.1 about instrument installation.

Table 75.1 Instrument terms and symbols

Chinese	English	Graphic symbol
就地仪表盘（柜子）	local instrument panel/ cabinet	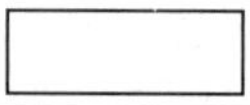
接线盒	junction box	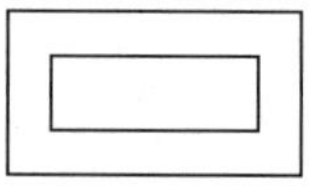
变送器	transmitter	
控制阀	control valve	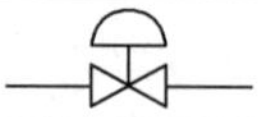

Text

Understand the drawing for installation of pressure transmitter (Fig.75.1).

GB	Hook-up dwg of liquid pressure measurement(transmitter below tap threaded gauge/root valve)		HG/T 21581-95 HK 02-138	
			Of sheet	Of total
Rating:6.3MPa	Conn type B.W	NO.	Tag NO.	Pipe(vessel)NO.
		1		

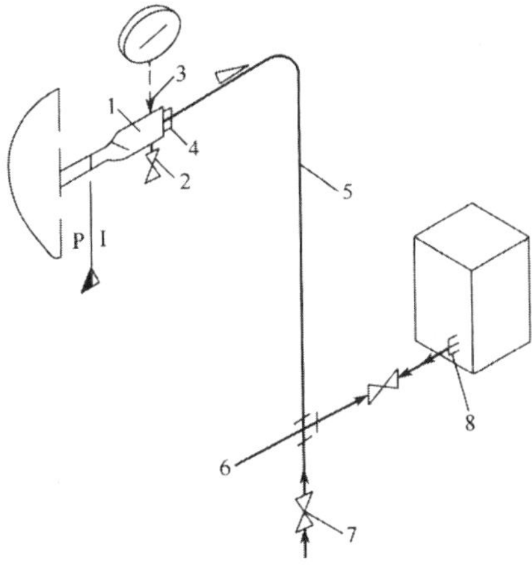

Note: 1. If the transmitter be mounted together with pressure switch or etc., NO.3 should be changed to suitable adapter.
2. Drain valve and tee can be deleted for clear liquid.

No.	Code	DWG& STD NO.	NAME&SIZE	MATERIAL	Q'TY	REMARKS
8	FB010 FB055		*PN* 6.3 1/2"NPT/ϕ14 B.W End Connector	CS 0Cr18Ni10Ti	1	
7	VB212 VB217	Q21F-64	*PN* 6.3 *DN*10ϕ14/ ϕ 14 Male threaded ball valve	CS 0Cr18Ni10Ti	2	
	VC210 VC211	J21W-64C P	*PN* 6.3 *DN*10ϕ14/ϕ14 Male threaded globe valve	CS 0Cr18Ni10Ti	2	
6	FB167 FB182		*PN* 6.3ϕ14 B.W Tee	CS 0Cr18Ni10Ti	1	
5	PL005 PL205	GB 8163-87 GB 2270-80	ϕ14×2 Seamless steel tube	20 0Cr18Ni10Ti		
4	FB009 FB054		*PN* 6.3 ZG1/2"/ϕ14 B.W End connector	CS 0Cr18Ni10Ti	1	
3			ZG1/2" Plug	CS 0Cr18Ni10Ti	1	With gauge/root valve
2			Bleeder valve	CS 0Cr18Ni10Ti	1	With gauge/root valve
1	VM102 VM107		*PN*16 *DN*15 ZG1/2"(M)/3xZG 1/2" (F) Gauge/root gate valve	CS 0Cr18Ni10Ti	1	
	VM122 VM127		*PN*16 *DN*15 ZG1/2"(M)/3xZG 1/2" (F) Gauge/root globe valve	CS 0Cr18Ni10Ti	1	
Installation Material List						

Fig.75.1 Installation of pressure transmitter

Knowledge extention

Try to know the expandor for thermometer (Fig.75.2).

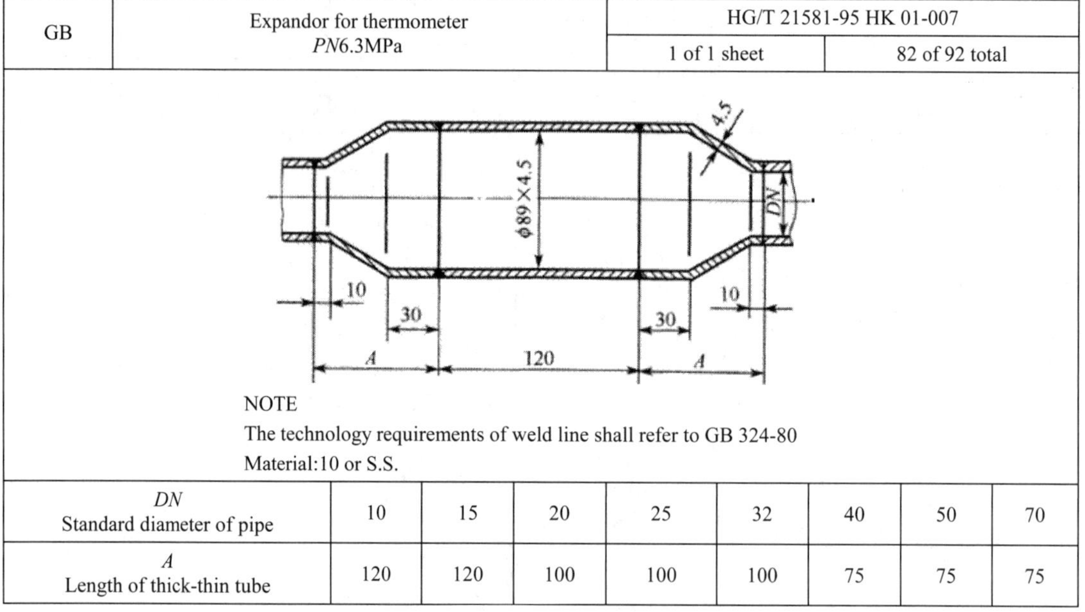

GB	Expandor for thermometer *PN*6.3MPa	HG/T 21581-95 HK 01-007	
		1 of 1 sheet	82 of 92 total

NOTE

The technology requirements of weld line shall refer to GB 324-80

Material:10 or S.S.

DN Standard diameter of pipe	10	15	20	25	32	40	50	70
A Length of thick-thin tube	120	120	100	100	100	75	75	75

Fig.75.2 Expandor for thermometer

Exercises

1. Match the English with the Chinese. Draw lines.

A.

(1) level transmitter	a. 压力变送器
(2) pressure transmitter	b. 温度变送器
(3) temperature transmitter	c. 液位变送器
(4) flow transmitter	d. 控制阀
(5) junction box	e. 流量变送器
(6) control valve	f. 接线盒

B.

(1) GB	a. 额定值
(2) Rating	b. 连接类型
(3) Conn type	c. 位号
(4) Tag NO.	d. 国标
(5) Pipe(vessel)NO.	e. 材料清单
(6) material list	f. 管道（容器）号

2. Translate the following sentences into Chinese.

(1) If the transmitter be mounted together with pressure switch or etc., NO.3 should be changed to suitable adapter.

(2) Drain valve and tee can be deleted for clear liquid.

(3) The technology requirements of weld line shall refer to GB 324-80.

3. Translate the following phrases into Chinese.

(1) Expandor for thermometer

(2) Standard diameter of pipe

(3) Length of thick-thin tube

Words and phrases

level transmitter ['levl] [træns'mɪtə(r)] 液位变送器

pressure transmitter ['preʃə(r)] 压力变送器

temperature transmitter ['temprətʃə(r)] 温度变送器

hook-up dwg of liquid pressure measurement 测量液体压力管路连接图

flow transmitter [fləʊ] 流量变送器

conn type(connection) 连接形式

... of ... sheet 第……张，共……张

... of ... total 总……张，第……张

mount [maʊnt] 安装

suitable adapter [ə'dæptə] 相应的转换接头

drain valve [dreɪn] [vælv] 排放阀

delete [dɪ'li:t] 删除，取消

male threaded [meɪl] 外螺纹

globe valve 截止阀

seamless steel tube ['si:mləs] [sti:l] 无缝钢管

bleeder valve ['bli:də(r)] 排放阀

code [kəʊd] 代码

material list [mə'tɪəriəl] [lɪst] 材料表

technology [tek'nɒlədʒi] 技术

weld line 焊缝

length of thick-thin tube 大小头长度

expandor for thermometer [ɪks'pændə] [θə'mɒmɪtə(r)] 温度计扩大管

DWG & STD NO.(drawing & standard No.) ['drɔ:ɪŋ] ['stændəd] 图纸与标准号

threaded ['θredɪd] 螺纹的

B.W (butt welding) 对焊

Tag NO. 位号

pipe (vessel) NO. 管道或设备号

pressure switch 压力开关

be changed to 改变为……

tee 三通

end connector [kə'nektə(r)] 终端接头

ball valve 球阀

B.W Tee 对焊式三通接头

plug [plʌg] 堵头

gauge/root gate valve 多路闸阀

remark [rɪ'mɑ:k] 备注

diameter 直径

requirement [rɪ'kwaɪəmənt] 要求

refer to 参照，按照

Appendix

Appendix A Vocabulary and phrases of Module Ⅰ Mechanical Equipments Part

abrasive [ə'bresɪv] 磨砂的
accessory [ək'sesəri] 附件
accomplish [ə'kʌmplɪʃ] 完成
actuating component ['æktʃuːˌeɪtɪŋ] [kəm'pəʊnənt] 执行元件
air compressor 空气压缩机
air tightness [eə(r)] [taɪtnəs] 气密性
air-tightness test 气密性试验
allowable [ə'laʊəbl] 允许的
allowance [ə'laʊəns] 允许度
alloy steel 合金钢
aluminum 铝
amiable ['eɪmɪəbl] 和蔼可亲
amicable ['æmɪkəbl] 和善，友好
analysis [ə'næləsɪs] 分析
angle ['æŋgl] 角
angle square ['æŋgl] [skweə(r)] 角尺
angling the boom 调节吊臂的角度
annealing [ə'niːlɪŋ] 退火
anti-corrosion and packing ['æntiːkər'əʊʒn] ['pækɪŋ] 防腐及包装
approve 审批
arbor ['ɑːbə] 刀杆
arc rays and noise [nɔɪz] 弧光及噪音
AS=American standard 美国标准
assemble [ə'sembl] 组装，拼装
assembly [ə'semblɪ] 装配
at high pressure ['preʃə(r)] 高压
atmosphere ['ætməsfɪə(r)] 大气
automated drafting 自动成图
automated factory 自动化工厂
automatic lathe [ˌɔːtə'mætɪk] 自动车床
automatically [ˌɔːtə'mætɪklɪ] 自动地
auxiliary [ɔːg'zɪliəri] 辅助
auxiliary view 辅助视图
axis ['æksɪs] 轴
backflow ['bækfləʊ] 回流
backward 向后地
ball valve 球阀
base [beɪs] 基座，底座
be perpendicular to [ˌpɜːpən'dɪkjələ(r)] 与/相对……垂直
bearing ['beərɪŋ] 轴承
bed way 床身导轨
bed 床身
bed-type 卧型

bench lathe [bentʃ] 台式车床
bench [bentʃ] 台子
bench-type drilling machine 台式钻床
bend [bend] 弯曲
between 在……之间
bismuth 铋
blade [bleɪd] 叶片
blower ['bləʊə(r)] 鼓风机
boring ['bɔ:rɪŋ] 钻孔
box-girder beam 箱形梁
brake [breɪk] 制动器
bridge crane 桥式起重机
brine [braɪn] 盐水
brittle ['brɪtl] 脆，淬
broken view 断面图
bulldozer ['bʊldəʊzə(r)] 推土机
butterfly valve ['bʌtəflaɪ] 蝶阀
CAD computer-aided design ['eɪdɪd] [dɪ'zaɪn] 计算机辅助设计
CAM computer-aided manufacture [ˌmænjʊ'fæktʃə(r)] 计算机辅助制造
cam [kæm] 凸轮
cantilever 悬臂
carbon steel 碳钢
carriage ['kærɪdʒ] 溜板，滑架
cast iron 铸铁
castability [kɑ:stə'bɪlɪtɪ] 可铸造性
catalyst particle ['kætəlɪst] 触媒颗粒
cavitation [ˌkævɪ'teɪʃən] 气穴
centerless grinder ['sentələs] 无心磨床
centrifugal blower [ˌsentrɪ'fju:gl] 离心式鼓风机
centrifugal pump 离心泵
check valve [tʃek] 止回阀
chiseling [t'ʃɪzlɪŋ] 錾削
chuck [tʃʌk] 卡盘，夹头
close safety belt 系安全带
clutch [klʌtʃ] 离合器
cock [kɒk] 旋塞
cold stream [stri:m] 冷流体
column ['kɔləm] 立柱，支柱
company ['kʌmpənɪ] 公司
comparison [kəm'pærɪsən] 比较
component [kəm'pəʊnənt] 构建，元件，零件
compress [kəm'pres] 压缩
compressible [kəm'presɪbl] 可压缩的
compression device [kəm'preʃn] [dɪ'vaɪs] 压缩装置
compression ratio [kəm'preʃn] ['reɪʃɪəʊ] 压缩比
compressor [kəm'presə(r)] 压缩机
computer system ['sɪstəm] 计算机系统
conductivity 传导性，传导率
conical ['kɒnɪkl] 圆锥型的
construction [kən'strʌkʃn] 施工
contain [kən'teɪn] 包含，包括
container [kən'teɪnə(r)] 容器，集装箱
control [kən'trol] 控制
controlling and regulating 控制及调节
cooler ['ku:lə(r)] 冷却器
corrosive [kə'rəʊsɪv] 腐蚀的
counterboring ['kaʊntəbərɪŋ]（平底）锪孔
countersink ['kaʊntəˌsɪŋk] 锪锥孔
coupling ['kʌplɪŋ] 联轴器
crack [kræk] 裂纹
crane [kre□n] 起重机
crawler crane ['krɔ:lə(r)] 履带式起重机
cross [krɒs] 交叉，十字
cross slide [kr□s] [slaɪd] 横向滑板/拖板
crowbar ['krəʊbɑ:(r)] 撬棍
cut [kʌt] 切割
cut hole in metal [həʊl] ['metl] 在金属中钻孔
cutter ['kʌtə(r)] 刀具
cylinder and motor ['sɪlɪndə(r)] ['məʊtə(r)] 缸与马达
cylinder block 气缸柱/座/体
cylindrical shell [sə'lɪndrɪkl] [ʃel] 圆柱壳体
damage ['dæmɪdʒ] 损坏
danger ['deɪndʒə(r)] 危险
data base 数据库

degree [dɪ'gri] 度数
density ['densəti] 密度
department [dɪ'pɑ:tmənt] 系部，部门
description [dɪ'skrɪpʃn] 明细，内容
design [dɪ'zaɪn] 设计
designer [dɪ'zaɪnə(r)] 设计师
develop very high pressure 产生高压
deviation [ˌdi:vi'eɪʃn] 偏差
device [dɪ'vaɪs] 仪器，装置
diameter [daɪ'æmɪtə(r)] 直径
dimension [dɪ'mɛnʃən, daɪ-] 尺寸
directional valve 方向阀
disassemble [ˌdisə'sembl] 拆装/卸
discipline ['dɪsəplɪn] 专业
double acting 双程运动
down milling 顺铣
draftsman ['drɑ:ftsmən] 制图人
drawing ['drɔ:ɪŋ] 图纸
dressing ['dresɪŋ] 着装
drill [drɪl] 钻，钻头
drilling machine 钻床
drilling ['drɪlɪŋ] 钻削
drinking ['drɪŋkɪŋ] 喝酒
ductility 展延性
elasticity 弹力，弹性
elbow ['elbəʊ] 弯管
electric shock [ɪ'lektrɪk] [ʃɒk] 触电，电击
electromechanical [ɪ'lektrəʊmɪ'kænɪkəl] 机电的
elevated track 高架轨道
elevation [ˌelɪ'veɪʃn] 标高
ell [el] 弯头
end 端部
engine lathe ['endʒɪn] [leɪð] 普通车床
engineer [ˌendʒɪ'nɪə(r)] 工程师
engineering [ˌendʒɪ'nɪərɪŋ] 工程
enhance heat transfer 加强热传递
enlarge [ɪn'lɑ:dʒ] 扩大，放大
equator zone [ɪ'kweɪtə(r)] 赤道带
equipment [ɪ'kwɪpmənt] 设备
exchanger [ɪks'tʃeɪndʒə] 交换器
expansion 膨胀
expensive [ɪk'spensɪv] 昂贵
experience [ɪk'spɪərɪəns] 经历，经验
exploding part [iks'pləudɪŋ] [pa:t] 爆炸物
external grinder [eks'tə:nl] 外圆磨床
factory management 工厂管理
fall [fɔ:l] 坠落
feed [fi:d] 进给
feed rod 进给杆
female/internal threading 攻螺纹
filing ['faɪlɪŋ] 锉削
filter ['fɪltə(r)] 过滤器
fin [fɪn] 翅，翅膀
finish ['fɪnɪʃ] 粗糙度
fit [fɪt] 配合
fitter ['fɪtə(r)] 钳工
fitting up 拼装
fixed bed reactor [fɪkst] 固定床反应器
flammable ['flæməbl] 可燃的
flange [flændʒ] 法兰
flow meter 流量计
flow valve 流量阀
fluidized bed reactor 流化床反应器
formability [fɔ:mə'bɪlɪtɪ] 可成形性
formation [fɔ:'meɪʃn] 形成
formed head 封头
forward 向前地
foundation [faun'deiʃən] 基础
frame [freɪm] 框，框架
function ['fʌŋkʃn] 功能
furnace ['fɜ:nɪs] 炉子
fusion ['fju:ʒn] 熔合
gantry beam [bi:m] 行车梁
gas [gæs] 气体
gas-liquid reactor 气液反应器
gate valve [geɪt] 闸阀
gear pump [gɪr] 齿轮泵
gear [gɪə(r)] 齿轮

gearbox ['giəbɔks] 齿轮箱，变速箱
gender ['dʒendə(r)] 性别
general-purpose ['dʒɛnərəl] ['pɜ:pəs] 通用型
geometric modeling [ˌdʒi:ə'metrɪk] ['mɒdlɪŋ] 几何建模
gland [glænd] 压盖
globe valve [gləʊb] 截止阀
grain [greɪn] 颗粒，晶粒，谷粒
granular ['grænjələ(r)] 颗粒状的
grinding machine ['graɪndɪŋ] [mə'ʃi:n] 磨床
hand abrasive wheel 手砂轮
hand wheel 手轮
handle 手柄
hardening ['hɑ:dnɪŋ] 淬火
headstock ['hedstɒk] 床头箱，主轴箱
health/ safety/environment [helθ] [ɪn'vaɪrənmənt] 健康/安全/环境
heat [hi:t] 热量，加热
treatment [hi:t] ['tri:tmənt] 热处理
heavy bob ['hevi] [bɒb] 重锤
high-carbon steel 高碳钢
hoisting 起重，升起
home pipe 家用管道
horizontally [ˌhɑrɪ'zɑntəlɪ] 水平地
hot [hɒt] 热的
hydraulic [haɪ'drɔ:lɪk] 液压/水压的
hydraulic press 液压机
hydraulic test 水压试验
imagine [ɪ'mædʒɪn] 想象
immerse [ɪ'mɜ:s] 浸没
in a workshop 在车间里
in aircraft ['ɛəkrɑ:ft] 在航空器中
in storage tank ['stɔ:rɪdʒ] [tæŋk] 在储罐里
in the air 在高空
in the deep ocean 在深海里
in the ground 在地里/下
inch 英寸
inclined view 斜视图
incorrect [ˌɪnkə'rekt] 不正确的
indirect [ˌɪndə'rekt] 间接的
industrial [ɪn'dʌstriəl] 工业的
industry ['ɪndəstrɪ] 工业
ingredient 成分
innovation [ˌɪnə'veɪʃn] 发明创造，创新
inside scaffolding 内脚手架
inspection [ɪn'spekʃn] 检查
inspector [ɪn'spektə(r)] 检查员
installation [ˌɪnstə'leɪʃn] 安装
interact [ˌɪntər'ækt] 相互作用/影响
interactive terminal [ˌɪntər'æktɪv] ['tɜ:mɪnl] 交互终端
interface ['ɪntəfeɪs] 计算机接口/界面
internal grinder [in'tə:nəl] 内圆磨床
internal surface 内表面
internal thread cock 内螺纹式旋塞
jack [dʒæk] 千斤顶
jacket wall ['dʒækɪt] 夹套/套筒壁
jet pump [dʒet] 喷射泵
jib crane [dʒɪb] [kren] 悬臂式起重机
joint [dʒɔɪnt] 接头
key [ki:] 钥匙，键
kiln and hearth furnace [kɪln] [hɑ: θ] ['fɜ:nɪs] 窑炉及床炉式反应器
kinematics [ˌkɪnə'mætɪks] 动态
knee [ni:] 升降台
large-diameter [daɪ'æmɪtə(r)] 大直径
lathe [leɪð] 车床
lead 铅
leadscrew. 导杆，丝杠
leak [li:k] 泄漏
leakproof ['li:kpru:f] 防泄漏
level gauge ['levl] [geɪdʒ] 水准仪
level ruler 水平尺
lie in [laɪ] 在于
lift pump [lɪft] 提升泵，抽扬泵，抽水机
lifting hook 吊钩
lining ['laɪnɪŋ] 划线
liquid ['lɪkwɪd] 液体

load [ləʊd] 负载，重物
locating by marking ['mɑ:kɪŋ] 标记定位
locating by sample plate [ləʊ'kəɪtɪŋ] ['sɑ:mpl] [pleɪt] 用样板定位
long boom [bu:m] 长臂
low-carbon steel 低碳钢
lower crown [kraʊn] 下极板
lower side ['lə□ə(r)] [saɪd] 向下，下部
lower temperate zone ['ləʊə(r)] ['tɛmpərɪt, 'tɛmprɪt] 下温带
luff [lʌf] 转舵
lumpy ['lʌmpɪ] 块料，粒状材
machinability [məʃi:nə'bɪlɪtɪ] 切削加工性
machine [mə'ʃi:n] 机器
machining flat surface 加工平面
machining [mə'ʃi:nɪŋ] 加工，切削
magnesium 镁
main motion [meɪn] ['məʊʃn] 主运动
main part [meɪn] [pa:t] 主要部件
maintain homogeneity [ˌhɒmədʒə'ni:əti] 保持均匀，同质
maintenance ['meɪntənəns] 维修保养
major ['meɪdʒə(r)] 专业
male/external threading 套螺纹
male [meɪl] 男，雄性
manufacturing plant 制造工厂
marking of material ['mɑ:kɪŋ] 号料
mass production [mæs] [prə'dʌkʃn] 大批量/大规模生产
material [mə'tɪəriəl] 材料
measure ['meʒə(r)] 测量
measured value ['meʒəd] ['vælju:] 实测值
mechanical [mɪ'kænɪkəl] 机械的
medium ['mi:diəm] 媒介，介质
medium outlet ['mi:diəm] ['aʊtlet] 介质排出口
medium-carbon steel 中碳钢
metal ['metl] 金属
method ['meθəd] 方法
milling ['mɪlɪŋ] 铣削
milling machine ['mɪlɪŋ] [mə'ʃi:n] 铣床
NC miller 数控铣床
modeling ['mɒdlɪŋ] 建模
molecule ['mɒlɪkju:l] 分子，微粒
motor ['məʊtə(r)] 电机，马达
moving back and forth 前后移动
MT 磁粉探伤
multispindle drilling machine 多轴钻床
nailing gun ['neɪlɪŋ] 射钉枪
nationality [ˌnæʃə'næləti] 民族
NC lathe 数控车床
NDT inspection [ɪn'spekʃn] 无损探伤/检测
nonreacting molecule ['mɒlɪkju:l] 未发生反应的分子
normalizing ['nɔ:məlaɪzɪŋ] 正火
numerically controll [nju:'merɪklɪ] 数字控制
object ['ɒbdʒɪkt] 物体
occur in the liquid phase 以液态出现
oil mist 油雾器
oil tank 油箱
oil tubing 油管
opening and nozzle ['nɒzlz] 开口/孔及接管
operation [ˌɒpə'reɪʃn] 运作，操作
orientation [ˌɔ:riən'teɪʃn] 方位，方向
orthographic [ˌɔ:θə'ɡræfɪk] 正交的
outside scaffolding [ˌaʊt'saɪd] ['skæfəldɪŋ] 外脚手架
overarm ['ɑ:bə] 横梁
owner ['əʊnə(r)] 所有者，业主
painting ['peɪntɪŋ] 油漆
partial view 局部视图
passage ['pæsɪdʒ] 通道
peel the chip from 从……剥离/削皮
perseverance [ˌpɜ:sə'vɪərəns] 恒心
petrochemical [ˌpetrəʊ'kemɪkl] 石化
piecing together [pɪsɪŋ] 拼装
pin [pɪn] 销
pipe [paɪp] 管，管道
pipe in process plant 工厂管道

pipeline ['paɪplaɪn] 管道
piper ['paɪpə(r)] 管工
piston ['pɪstən] 活塞
plain carbon steel 普通碳钢
plane surface [ple□n] ['sɜ:fɪs] 平面
planer ['pleɪnə(r)] 龙门刨床
planing miller ['pleɪnə(r)] 龙门型铣床
plumbing ['plʌmɪŋ] 水暖管道
plumbness [p'lʌmnəs] 铅垂度
pneumatic [nju:'mætɪk] 气动的
polythene manufacture ['pɒlɪθi:n] 聚乙烯制造/生产
positioning by jig frame [dʒɪg] [freɪm] 框/胎架定位
positioning by setting element ['setɪŋ] ['elɪmənt] 设置元件定位
positive-displacement pump ['pɒzətɪv] [dɪs'pleɪsmənt] 容积泵
power unit 动力装置
powered drum 有动力装置的滚筒
pressure ['preʃə(r)] 压力
pressure gauge [geɪdʒ] 压力计
pressure test 试压
pressure valve 压力阀
prevent [prɪ'vent] 阻止
principle ['prɪnsəpl] 原理
process [prə'ses] 工艺，过程
process planning 工艺规划，生产计划
produce [prə'dju:s] 产生，加工
product ['prɒdʌkt] 成品
project ['prɔdʒekt] 工程，项目
projection [prə'dʒekʃn] 投影
proper shape ['prɒpə(r)] [ʃeɪp] 适当的形状
property ['prɒpəti] 性能
protective goggles [prə'tektɪv] ['gɔglz] 护目镜
PT 渗透探伤
pump [pʌmp] 泵
punch [pʌntʃ] 冲切
quality ['kwɑlɪti] 质量
radiation [ˌreɪdi'eɪʃn] 辐射
raise [reɪz] 提升
reactor [ri'æktə(r)] 反应器
reaming ['ri:mɪŋ] 铰孔
reciprocating piston compressor [rɪ'sɪprəkeɪtɪŋ] ['pɪstən] 往复式活塞泵
reciprocating pump [rɪ'sɪprəkeɪtɪŋ] 往复式泵
recommendation [ˌrekəmen'deɪʃn] 推荐
record ['rekɔ:d] 记录
reduce [rɪ'dju:s] 减少
reducer [rɪ'dju:sə] 大小头
reference line ['refrəns] [laɪn] 参考/基准线
refine [rɪ'faɪn] 细化
regulate ['regjʊleɪt] 调节
remove [rɪ'mu:v] 去除
replacement [rɪ'pleɪsmənt] 替换，代替
require [rɪ'kwaɪə(r)] 要求
residence ['rezɪdəns] 滞留
resume [rezju:mei] 简历
revolved view 旋转视图
revolving [rɪ'vɒlvɪŋ] 旋转的
ribbon ['rɪbən] 带，带状物
risk of injury from burning [rɪsk] ['ɪndʒərɪ] ['bɜ:nɪŋ] 烧伤
risk of injury from hot coolant ['ku:lənt] 冷却液灼伤
riveter ['rɪvɪtə] 铆工
robot ['robət] 机器人
robotics [rəʊ'bɒtɪks] 机器人技术
roll [rəʊl] 旋转， 滚动
rotary blower and compressor ['rəʊtərɪ] 旋转式鼓风机及压缩机
rotating impeller [ɪm'pelə] 旋转的叶轮
RT 射线探伤
rubbing ['rʌbɪŋ] 研磨
run 运行
saddle ['sædl] 马鞍，鞍座，床鞍
safety belt 安全带
safety cap 安全帽
safety shoes 安全鞋，工作皮鞋
safety valve ['seɪfti] [vælv] 安全阀

sawing ['sɔːɪŋ] 锯削
scaffolding ['skæfəldɪŋ] 脚手架
scale [skeɪl] 比例
schedule ['ʃedjuːl] 计划
scraping ['skreɪpɪŋ] 刮削
screw [skruː] 螺纹，螺钉
screw pump 螺杆泵，螺旋泵
sealing ring ['siːlɪŋ] [rɪŋ] 密封圈
section view 剖视图
set up 安装，固定
shaft [ʃɑːft] 轴
shape [ʃeɪp] 形状
shaper ['ʃeɪpə] 牛头刨床
shaping ['ʃeɪpɪŋ] 牛头刨
shearing machine ['ʃɪərɪŋ] 剪切机
shell [ʃel] 壳，贝壳，壳牌
silencer ['saɪlənsə(r)] 消声器
single spindle 单轴
site [saɪt] 现场，工地
skill competition 技能竞赛
skilled 有技术的
skirt [skəːt] 裙子，裙座
smoking ['sməʊkɪŋ] 抽烟
solid ['sɒlɪd] 实体，实物
specification [ˌspesɪfɪ'keɪʃn] 规范
spherical tank [tæŋk] 球罐
spindle ['spɪndl] 主轴
spiral ['spaɪrəl] 螺旋式，盘管式
spline [splaɪn] 花键
spot facing [spɒt] ['feɪsɪŋ] 锪端面
spring [sprɪŋ] 弹簧
stage 级，阶段
stamping [stæmpɪŋ] 冲压
standard ['stændəd] 标准
stationary ['steɪʃənrɪ] 静止的
steel beam 钢梁
steel distortion [dɪ'stɔːʃn] 钢材变形
steel factory 钢铁厂
steel tape 钢卷尺
stirred tank [stəːd] [tæŋk] 湍流槽
structure ['strʌktʃə(r)] 结构
sturdy construction ['stɜːdɪ] [kən'strʌkʃn] 坚固结实的构造
supervision [ˌsjuːpə'vɪʒn] 监理，监督
support [sə'pɔːt] 支架，支撑
surface finish 表面粗糙度
surface grinder 平面磨床
suspension [sə'spenʃn] 悬浮
sweep [swiːp] 吹扫，清扫
swing [swɪŋ] 摆动
system ['sɪstəm] 系统
table ['teɪbl] 工作台
Tag No. [tæg] ['nʌmbə] 位号，工号
tailstock ['teɪlstɒk] 尾座
tap [tæp] 水龙头
team leader ['liːdə(r)] 队/组长
technical ['teknɪkl] 技术的
technology [tek'nɒlədʒɪ] 技术
teepipe 三通管
temperature ['temprətʃə(r)] 温度
tempering ['tempərɪŋ] 回火
tensile strength 抗拉强度
theodolite [θi'ɒdəlaɪt] 经纬仪
thermal 热的，热量的
thick-walled 厚壁
thin-walled vessel 薄壁容器
thread machining [θred] [mə'ʃiːnɪŋ] 螺纹加工
three-dimensional view 3D 视图
tolerance ['tɒlərəns] 公差，允许度
tool [tuːl] 工具，刀具
tool and cutter grinder 工具磨床
tool head [tuːl] [hed] 刀具主轴箱
tool post 刀架，刀座
touch [tʌtʃ] 触摸
toughness 韧性，刚性
tower crane 塔吊
toxic ['tɒksɪk] 有毒的
transfer [træns'fɜː(r)] 传送

transmission [træns'mɪʃn] 传输，传送
traveling ['trævəlɪŋ] 移动的
tree trunk [trʌŋk] 木料堆
tridimensional [ˌtraɪdɪ'menʃənəl] 立体的
trim [trɪm] 修边
trolley 滑车
truck crane [trʌk] 卡车起重机
tube [tju:b] 管，小管
tubular flow reactor ['tju:bjələ(r)] 管道式流动反应器
turbine ['tɚbɪn] 涡轮
turbocompressor [tɜ:bəʊ'kɒmpresər] 涡轮压缩机
turning ['tə:nɪŋ] 车削
turret lathe ['tɜ:rɪt, 'tʌr-] 六角头车床
two-dimensional drawing 2D 图纸
type [taɪp] 类型，型号
typical example 典型例子
understanding drawing and lofting 识图及放样
uniform ['ju:nɪfɔ:m] 均匀的，统一的
unit 单元，单位，装置
up milling 逆铣
upper crown 上极板
upper temperate zone ['ʌpə(r)] 上温带
upright ['ʌpraɪt] 竖立的，直立的
upright drilling machine 立式钻床
UT 超声波探伤
vacuum ['vækjʊəm] 真空
valve [vælv] 阀
valve body 阀体
valve stem [stem] 阀杆
vertical lathe ['vɜ:tɪkl] 立式车床
vessel ['vesl] 容器
view [vju:] 视图
vocational college [vəʊ'keɪʃənl] 职业院校
volatile liquid ['vɒlətaɪl] 挥发性液体
Walmart Supermarket 沃尔玛超市
wear safety helmet ['seɪftɪ] ['helmɪt] 戴安全帽
wearing ['weərɪŋ] 穿戴
weld 焊接，焊缝
welder ['weldə(r)] 焊工
welding [weldɪŋ] 焊接
wheel [wi:l] 轮，砂轮
wheel head [wi:l] [hed] 磨头
windlass ['wɪndləs] 卷扬机，绞盘
wire rope 绳
working area 工作区域
workpiece ['wɜ:kˌpi:s] 工件
workshop ['wɜ:kʃɒp] 工作车间
work 工件
wrap 缠绕
wrench [rentʃ] 扳手
zinc 锌
zone [zəʊn] 地带，地域

Appendix B Vocabulary and phrases of Module Ⅱ Welding Part

AC supply [sə'plaɪ] 交流电源
active gas ['æktɪv] 活性的
adjusting dial [ə'd□□st□ŋ] ['daɪəl] 调谐钮
air valve [vælv] 空气阀门
alternating current ['ɔːltəneɪtɪŋ] ['kʌrənt] 交流电
American Welding Society [ə'merɪkən] [sə'saɪətɪ] 美国焊接协会
analysis [ə'næləsɪs] 分析
angle ['æŋgl] 角度，角
angle bar ['æŋgl] [bɑː(r)] 角钢
anode ['ænəʊd] 阳极
aperture ['æpətʃə(r)] 孔
application [ˌæplɪ'keɪʃn] 应用，适用
arc [ɑːk] 电弧
arc ending 收弧
arc ignition end [ɪg'nɪʃn] 引弧端
arc path [pɑːθ] 电弧空间
arc voltage ['vəʊltɪdʒ] 电弧电压
arc welding process 弧焊工艺
argon shielded arc welding ['ɑːgɒn] 氩弧焊
assembly [ə'sembli] 组装
at the output terminal ['tɜːmɪnlz] 输出端
atmosphere ['ætməsfɪə(r)] 大气
automatic welding [ˌɔːtə'mætɪk] 自动焊
automatically [ˌɔːtə'mætɪklɪ] 自动地
axis [□æks□s] 轴
back bead [bæk [biːd]] 封底焊缝
back surface [bæk 'sɜːfɪs] 背面
backhand ['bækhænd] 反手
backing bead ['bækɪŋ] 打底焊缝
bare metal electrode [beə(r)] ['metl] 裸焊条，焊丝
barrel ['bærəl] 桶，筒体
base metal [beɪs] 母材
beam [biːm] 束
bend [bend] 弯曲
bevel ['bevl] 斜边/斜面
brazing [breɪzɪŋ] 钎焊
breaking arc 断弧
burn [bɜːn] 燃烧
butt welding [bʌt] 对焊
cabinet ['kæbɪnət] 柜
cable ['keɪbl] 电缆
carbon arc air gouging ['kɑːbən] [ɑːk] [eə(r)] ['gaʊdʒɪŋ] 碳弧气刨
casting ['kɑːstɪŋ] 铸造
cathode ['kæθəʊd] 阴极
cathode-ray [□kæθə□d] [reɪ] 阴极射线
certificate of welding inspection 焊接检验资格证
channel steel ['tʃænl] 槽钢
characteristic [ˌkærəktə'rɪstɪk] 特性
check [tʃek] 检查
circular magnetization ['sɜːkjələ(r)] [ˌmægnətɪ'zeɪʃən] 周向磁化
circulating electric current ['sɜːkjʊleɪtɪŋ] 环形电流
CO_2 (carbon dioxide) ['kɑːbən] [daɪ'ɒksaɪd] 二氧化碳
coalescence [ˌkəʊə'lesns] 连接，结合
concentrate ['kɒnsntreɪt] 集中
condition [kən'dɪʃn] 条件
confined area [kən'faɪnd] ['ɛriə] 有限的区域

contact tube ['kɒntækt] [tju:b] 导电管/导电嘴
contact ['kɒntækt] 接触
control cable [kən'trəʊl] ['keɪbl] 控制电缆
control monitor ['mɒnɪtə(r)] 控制显示器
controller [kən'trəʊlə(r)] 控制器
cooling unit ['ku:lɪŋ] ['ju:nɪt] 冷却装置
corrosion [kə'rəʊʒn] 腐蚀
covered/coated [□k□vəd] ['kəʊtɪd] 覆盖的，有涂层的
covering ['kʌvərɪŋ] 覆盖物，药皮
crack [kræk] 裂纹
crater ['kreɪtə(r)] 弧坑
crescent ['kresnt] 月牙形
cut [kʌt] 切割
cylinder ['sɪlɪndə(r)] 气瓶
defect ['di:fekt] 缺陷
defective [dɪ'fektɪv] 有瑕疵/缺陷的
density variation ['densətɪ] 密度变异
deposition [ˌdepə'zɪʃn] 熔敷
detect [dɪ'tekt] 检测
developing agent [dɪ'veləpɪŋ] ['eɪdʒənt] 显像剂
diagram ['daɪəgræm] 图表
diameter button [daɪ'æmɪtə(r)] 直径调节键
diameter [daɪ'æmɪtə(r)] 直径
diffusion welding [dɪ'fju:ʒn] 热剂焊
digit ['dɪdʒɪt] 数字
direction of cut 切割方向
downward ['daʊnwəd] 向下的
drag [dræg] 后拖量
ductility [dʌk'tɪlɪtɪ] 展延性
eddy current ['edɪ] 涡流
edge [edʒ] 边缘， 端部
efficiency [ɪ'fɪʃnsɪ] 效率
electric arc [ɪ'lektrɪk] [ɑ:k] 电弧
electric conductor [ɪ'lektrɪk] [kən'dʌktə(r)] 导电嘴
electric switch [ɪ'lektrɪk] [swɪtʃ] 电开关
electrically conductive material [kən'dʌktɪv] 导电性材料
electrode [ɪ'lektrəʊd] 焊条
electrode cable [ɪ'lektrəʊd] 焊条电缆
electrode torch [ɪ'lektrəʊd] [tɔ:tʃ] 焊钳，焊枪
electron beam welding [ɪ'lektrɒn] [bi:m] 电子束焊
enough shielding gas [ɪ'nʌf] [ʃi:ldɪŋ] 足够的保护气体
equipment [ɪ'kwɪpmənt] 设备
ET (eddy-curent testing) 涡流探伤（检测）
excessive undercut [ɪk'sesɪv] [ˌʌndə'kʌt] 过咬边
exhaust [ɪg'zɔ:st] 排出
explosion [ɪk'spləʊʒn] 爆炸
fastening device 紧固装置
feed roll [fi:d] [rəʊl] 送丝辊
feed [fi:d] 进给
ferromagnetic material [ˌferəʊmæg'netɪk] [mə'tɪərɪəlz] 铁磁性的材料
ferrous metal ['ferəs] ['metl] 铁类（黑色）金属
figure eight shape 8 字形
filler ['fɪlə(r)] 填充物
fillet weld size ['fɪlɪt] 焊脚
fillet welding ['fɪlɪt] 角焊
filling bead ['fɪlɪŋ] 填充焊道
film viewer (illuminator) [fɪlm] ['vju:ə(r)] [ɪ'lju:mɪneɪtə] 观片灯
film [fɪlm] 底片，胶片
fixed [fɪkst] 固定的
flame cutting [fleɪm] 火焰切割
flanged edge weld [flændʒd] [edʒ] 卷边焊
flash [flæʃ] 闪光
flat [flæt] 平的， 平面
horizontal [ˌhɔri'zɔntəl] 水平的
flow meter [flə□] ['mi:tə(r)] 流量计
flow [fləʊ] 流动
flux cored electrode [flʌks] [kɔ:d] 药芯焊丝
flux hopper ['hɒpə(r)] 焊剂漏斗
flux [flʌks] 焊剂
forearm ['fɔ:rɑ:m] 前臂
forehand ['fɔ:hænd] 正手
friction ['frɪkʃn] 摩擦
fume [fju:m] 烟雾

function ['fʌŋkʃn] 功能
fusion ['fju:ʒn] 熔化
electroslag [ɪ'lektrəʊslæg] 电渣
gamma radiation ['gæmə] [ˌreɪdi'eɪʃn] 伽玛辐射
gamma ray 伽玛射线
gas flow [fləʊ] 气流
gas hose [həʊz] 气体软管
gas passage 气体通道
gas pressure regulator ['preʃə(r)] ['regjuleɪtə(r)] 气压校准器
gas solenoid valve ['sɒlənɔɪd] [vælv] 燃气电磁阀
gas supply [sə'plaɪ] 气源
gas test 气体试验/检测
Gas Tungsten Arc Welding ['tʌŋstən] 钨极气体保护电弧焊
generator ['dʒenəreɪtə(r)] 发电机，焊机
geometry variation [dʒɪ'ɒmətrɪ] [ˌveəri'eɪʃn] 几何变异
give a permanent film record ['pɜ:mənənt] [fɪlm] ['rekɔ:d] 留下永久性胶片记录
GMAW (gas metal arc welding) 熔化极气体保护电弧焊
groove [gru:v] 坡口
ground cable [graʊnd] ['keɪbl] 接地电缆
grounding/earthing ['graʊndɪŋ] ['ɜ:θɪŋ] 地线，接地
half round steel [hɑ:f] [raʊnd] 半圆钢
hazardous ['hæzədəs] 有害的
heat radiation [hi:t] [ˌreɪdi'eɪʃn] 热辐射
heating effect ['daɪəgræm] [ɪ'fekt] 热影响
helmet ['helmits] 面罩
high vacuum ['vækjʊəm] 高度真空
high voltage ['vəʊltɪdʒ] 高压
higher frequency ['haɪə(r)] ['fri:kwənsɪ] 高频
highly sensitive ['haɪli] ['sensətɪv] 高灵敏度
high-temperature steel ['temprətʃə(r)] 高温钢
holding end ['həʊldɪŋ] [end] 夹持端
horn [hɔ:n] 角，喇叭
host material [həʊst] [mə'tɪəriəl] 基材
hot crack [kræk] 热裂纹
ignition temperature [ɪg'nɪʃn] ['temprətʃə(r)] 点燃/燃烧温度
immerse [ɪ'mɜ:s] 浸入
inclined [ɪn'klaɪnd] 倾斜的
inclusion [ɪn'klu:ʒn] 夹,包含
incomplete joint penetration [ˌɪnkəm'pli:t] [ˌpenɪ'treɪʃn] 未焊透
industrial [ɪn'dʌstriəl] 工业的
inert gas [ɪn'ɜ:t] 惰性气体
inspection [ɪn'spekʃn] 检验
insufficient throat [ˌɪnsə'fɪʃnt] [θrəʊt] 焊缝厚度不足
insulated floor ['ɪnsjuleɪtɪd] [flɔ:(r)] 绝缘地面
insulating sheath ['insjuleitɪŋ] [ʃi:θ] 绝缘套
intelligent [ɪn'telɪdʒənt] 智能的
intensity of labour [ɪn'tensətɪ] ['leɪbə(r)] 劳动强度
interior [ɪn'tɪərɪə(r)] 内部
internal flaw [in'tə:nəl] [flɔ:] 内部缺陷
interpass [ɪntɜ:'pɑ:s] 道间
join [dʒɔɪn] 连接
joint [dʒɔɪnt] 接头，关节，结合处
kerf [kɜ:f] 割口
key lock switch [swɪtʃ] 开关锁
lack of fusion [læk] ['fju:ʒn] 未熔合
lap [læp] 搭接，部分重叠
laser beam ['leɪzə(r)] [bi:m] 激光束
lead gate [li:d] [geɪt] 铅门
leak [li:k] 泄漏
lens [lenz] 透镜
liquid-penetrant examination ['lɪkwɪd] ['penətrənt] [ɪgˌzæmɪ'neɪʃn]（液体）渗透检测
low-alloy steel 低合金钢
low-carbon steel 低碳钢
machine flame cutting 机械火焰切割
machine [mə'ʃi:n] 机器
magnet ['mægnət] 磁铁
magnetic field [mæg'netɪk] [fi:ld] 磁场
magnetic lines of force [mæg'netɪk] [laɪnz] [fɔ:s] 磁力线
magnetic particle [mæg'netɪk] ['pa:tɪklz] 磁粉

magnetic permeability [ˌpɜːmɪə'bɪlətɪ] 磁导率
magnetizing coil [kɔɪl] 励磁线圈
main bang [meɪn] [bæŋ] 主脉冲
manipulation [məˌnɪpjʊ'leɪʃn] 运作，操作
mass [mæs] 质量
material button [mə'tɪəriəl] 焊接材料选择键
material [mə'tɪəriəl] 材料
max [mæks] 最大
measure ['meʒə(r)] 测量
mechanical [mɪ'kænɪkəl] 机械的
metal rod ['metl] [rɒd] 金属棒
meter ['miːtə(r)] 仪表/器
method ['meθəd] 方法
min/mm 分/毫米
misalignment [ˌmɪsə'laɪnmənt] 错边
mixture ['mɪkstʃə(r)] 混合
mm/min 毫米/分
mode button [məʊd] 焊枪操作模式键
molten flux ['məʊltən] [flʌks] 熔渣
molten pool ['məʊltən] [puːl] 熔池
molten weld metal 熔化金属
MT (magnetic particle testing) 磁粉探伤
multi-layer/pass ['mʌlti] ['leiə] 多层/多道
NASA 美国航空航天局
NDT staff [stɑːf] 无损检测/探伤人员
negative ['negətɪv] 否定的，阴性/极
neutron ['nuːˌtrɔn, 'njuː-] 中子
non-consumable [kən'sjuːməbl] 非熔化的
nondestructive ['nɒndɪs'trʌktɪv] 非破坏性的
nonferrous [nɒn'ferəs] 非铁的
north pole [nɔːθ] [pəʊl] 北极
nozzle ['nɔzəl] 喷嘴
nugget ['nʌgɪt] 熔核
oblique [ə'bliːk] 斜的
oil couplant ['kuːplɑːnt] 耦合剂
opening ['əʊpnɪŋ] 开口
operation box [ˌɒpə'reɪʃn] 操作盒
operation skill [ˌɒpə'reɪʃn] [skɪl] 操作技能
oscilloscope [ə'sɪləskəʊp] 示波器
overhead ['əuvəhed] 上面的，头顶上的
overlap [ˌəʊvə'læp] 焊瘤
oxidation [ˌɒksɪ'deɪʃn] 氧化
oxygen jet [dʒet] 氧气流
parameter [pə'ræmɪtəz] 参数
parent metal ['peərənt] 母材
penetrameter ['penɪtrəmɪtər] 像质计
penetrant ['penətrənt] 渗透剂
penetration [ˌpenɪ'treɪʃn] 渗透，熔透
PETS (Public English Test System) ['pʌblɪk] ['sɪstəm] 全国英语等级考试
pipe [paɪp] 管道
plasma arc cutting ['plæzmə] 等离子弧切割
plasma arc welding ['plæzmə] 等离子弧焊
plastic ['plæstɪk] 塑料
plug [plʌg] 塞子
polarity [pə'lærəti] 极性
poor condition [pɔː(r)] [kən'dɪʃn] 恶劣条件
poor grounding connection [pʊə(r)] ['graʊndɪŋ] [kə'nekʃn] 不良接地
pore ['pɔːz] 孔
porosity [pɔː'rɒsətɪ] 气孔
positional welding [pə'zɪʃənəl] 全位置焊接
positive ['pɒzətɪv] 明确的，肯定的，阳性/极
power setting ['paʊə(r)] ['setɪŋ] 电源设置
power source ['paʊə(r)] [sɔːs] 电源
PQR (procedure qualification record) [ˌkwɒlɪfɪ'keɪʃn] ['rekɔːd] 工艺评定报告
preheat [ˌpriː'hiːt] 预热
preheating flame ['priːhiːtɪŋ] 预热火焰
pre-set [p'riːs'et] 预设置
press together [pres] [tə'geðə(r)] 挤压在一起
press [pres] 按，压
pressure ['preʃə(r)] 压力
oxyacetylene [ˌɒksiə'setəliːn] 氧乙炔
pressure vessel ['preʃə(r)] ['vesl] 压力容器
prevent [prɪ'vent] 阻止/防止
principle ['prɪnsəpl] 原理
problem of spatter ['prɒbləm] ['spætə(r)] 飞溅问题

procedure [prə'si:dʒə(r)] 程序，工艺
process button [prə'ses] 焊接方式选择键
process ['prəʊses] 工艺，方法，过程
prod-type contact ['kɒntækt] 触头
property ['prɒpətɪ] 性能
provide the arc [prə'vaɪd] 提供电弧
purge button [pɜ:dʒ] 气体检测键
PWHT 焊后热处理
quality ['kwɒlətɪ] 质量
radiography testing [ˌreɪdɪ'ɒgrəfɪ] ['testɪŋ] 射线测试/探伤
rate [reɪt] 率
reinforcement [ˌri:ɪn'fɔ:smənt] 加强，增援，余高
remote control unit [rɪ'məʊt] [kən'trəʊl] ['ju:nɪt] 遥控器
repetitive [rɪ'petətɪv] 重复的
resistance [rɪ'zɪstəns] 电阻
reversed [rɪ'vɜ:st] 反，颠倒，反接
robot body 机器人本体
roll [rəʊl] 旋转， 辊子
root [ru:t] 根部
root face [ru:t] [feɪs] 坡口钝边
rotating [rəʊ'teɪtɪŋ] 旋转的
RT (Radiographic examination) ['reɪdɪəʊ 'græfɪk] [ɪgˌzæmɪ'neɪʃn] 射线检测/探伤
rust [rʌst] 锈
sawtooth ['sɔ:tu:θ] 锯齿形
scratch [skrætʃ] 擦伤
seam [si:m] 焊缝
section of uncut metal ['sekʃn] 未切割的金属
seep into [si:p] 渗进
semi-automatic ['semi] 半自动
sensing system 感应系统
shallow subsurface discontinuities ['ʃæləʊ] ['sʌb'sɜ:fɪs] 近表面的缺陷（不连续）
sheet [ʃi:t]（纸）张，薄板
shield [ʃi:ld] 屏蔽，保护
shielded medium ['ʃi:ldɪd] ['mi:dɪəm] 屏蔽介质
shielding gas [ʃi:ldɪŋ] [gæs] 保护气体
single bevel groove ['bevl] 单边 V 形坡口
slag [slæg] 焊渣
slag and molten metal [slæg] ['məʊltən] 渣和熔化金属
slot [slɒt] 槽
smog [smɒg] 烟雾
solid wire ['sɒlɪd] ['waɪə(r)] 实心焊丝
sound wave [saʊnd] [weɪv] 声波
south pole [saʊθ] 南极
spark [spɑ:k] 火花，火星
spatter ['spætə(r)] 飞溅
specification [ˌspesɪfɪ'keɪʃn] 规范，规格
specimen 试件，样品，标本
specimen to be examined ['spesɪmən] [ɪg'zæmɪnd] 要检测的的试件
speed [spi:d] 速度
spot weld [spɒt] 点焊
spray [spreɪ] 喷射
square groove [skweə(r)] 平头坡口
square [skweə(r)] 方形，直角
stainless steel 不锈钢
standard ['stændəd] 标准
steel plate [sti:l] [pleɪt] 钢板
steel [sti:l] 钢材
step down transformer [træns'fɔ:mə(r)] 降压变压器
stiffness [stɪfnəs] 强度
store button [stɔ:(r)] ['bʌtn] 存储键
straight [streɪt] 直接的，正接
strike [straɪk] 直击
submerged [səb'mɜ:dʒd] 埋伏的，潜伏的
surface 表面
surfacing welding ['sɜ:fɪsɪŋ] 堆焊
swing [swɪŋ] 摆动
symbol ['sɪmbl] 符号
system ['sɪstəm] 系统
teach device [dɪ'vaɪs] 示教仪
technician [tɛk'nɪʃən] 技师
temperature ['temprətʃə(r)] 温度
the material being inspected 待检材料

the part to be inspected [in'spektid] 待检部件

the size and shape [saɪz] [ʃeɪp] 尺寸和形状

theoretical throat [ˌθɪə'retɪkl] [θrəʊt] 焊缝计算厚度

thermit ['θɜːmɪt] 热剂

thickness range ['θɪknəs] [reɪndʒ] 厚度范围

torch tip for cutting [tɔːtʃ] [tɪp] 割炬喷嘴

transducer [trænz'djuːsə(r)] 探头

triangle ['traɪæŋgl] 三角形

trolley ['trɒlɪ] 移动小车

trunk [trʌŋk] 躯干

T-steel [stiːl] T 型钢

tungsten electrode 钨极焊条

tungsten ['tʌŋstən] 钨

twist [twɪst] 扭曲

ultrasonic examination (UT) [ˌʌltrə'sɒnɪk] [ɪgˌzæmɪ'neɪʃn] 超声波检测

ultrasonic [ˌʌltrə'sɒnɪk] 超声的

undercut 咬边

undimensioned 未注明尺寸的

upward ['ʌpwəd] 向上的

vacuum ['vækjuəm] 真空

valve [vælv] 阀

variation [ˌveəri'eɪʃn] 变异

ventilation [ˌventɪ'leɪʃn] 通风

vertical ['vɜːtɪkl] 竖立的

vibration [vaɪ'breɪʃn] 震动，摆动

visually ['vɪʒʊəlɪ] 视觉地，直观地

void [vɔɪdz] 空隙，孔洞

water in/out 进水/出水

water-washable system ['wɒʃəbl] ['sɪstəm] 水洗型渗透探伤

weld length [lɛŋkθ] 焊缝长度

weld metal ['metl] 焊缝金属

weld puddle 熔池

weld spacing ['speɪsɪŋ] 焊缝间距

weldability [weldə'bɪlɪtɪ] 可焊性

welder ['weldə(r)] 焊工，焊机

welding [weldɪŋ] 焊接

welding area ['eərɪə] 焊接区

welding direction [də'rekʃn] 焊接方向

welding machine 焊机

welding material [mə'tɪərɪəlz] 焊接材料

welding operation [ˌɒpə'reɪʃn] 焊接操作

welding procedure specification (WPS) 焊接工艺规程

welding robot ['rəʊbɒt] 焊接机器人

welding rod [rɒd] 焊丝

welding wire ['waɪə(r)] 焊丝

weld 焊缝

width [wɪdθ] 宽度

wire core ['waɪə(r)] [kɔː(r)] 焊芯

wire feed ['waɪə(r)] [fiːd] 送丝

wire feeder ['waɪə(r)] ['fiːdə(r)] 送丝机

wire reel 焊丝盘

with a stream of oxygen [striːm] ['ɒksɪdʒən] 氧气流

work (workpiece) ['wɜːkˌpiːs] 工件

worn out [wɔːn] 破旧，磨损

wrist [rɪst] 手腕

wrong contact tube [rɒŋ] ['kɒntækt] [tjuːb] 导电铜管/导电嘴损坏

X-ray [reɪ] X 射线

X-ray source [sɔːs] X 射线源

Appendix C Vocabulary and phrases of Module Ⅲ Electrical and Instrument Part

AC circuit 交流电路
accuracy ['ækjərəsi] 精度，准确度
actuator ['æktʃʊeɪtə] 执行器
air core 空芯
alternating voltage ['ɔːltəneɪtɪŋ] 交流电压
alternative [ɔːl'tɜːnətɪv] 交替的
ammeter ['æmiːtə(r)] 安培表
ampere ['æmpeə(r)] 安培
amplifier ['æmplɪfaɪə(r)] 放大器
analog ['ænəlɔːg] 模拟
analog style ['ænəlɔːg] [staɪl] 模拟式
analog system ['ænəlɔːg] ['sɪstəm] 模拟系统
analyser and analyser house ['ænəlaizəz] 分析器及分析器室
anode ['ænˌəʊd] 阳极
anvil ['ænvɪl] 测砧
armature ['ɑːmətʃə(r)] 电枢，转子
arrangement [ə'reɪndʒmənt] 阵列，安排
array [ə'reɪ] 排列
asynchronous motor [eɪ'sɪŋkrənəs] 异步电机
at random ['rændəm] 随意，随机
automation [ˌɔːtə'meɪʃn] 自动化
auxiliary ruler [ɔːg'zɪliəri] 辅尺
B.W (butt welding) 对焊
B.W Tee 对焊式三通接头
back [bæk] 背，后面
balance ['bæləns] 平衡
ball valve 球阀
base [beɪs] 基极
basic part 基本部分
battery ['bætəri] 电池
bearing ['beərɪŋ] 轴承
bezel ['bezl] 表圈
binary ['baɪnəri] 二进制的
bleeder valve ['bliːdə(r)] 排放阀
bottom resistor ['bɒtəm] 下电阻
braking screw 制动螺钉
branch [brɑːntʃ] 支路
breakglass point 易碎玻璃手动按钮
bridge [brɪdʒ] 桥，电桥
brush circuit ['sɜːkɪt] 电刷线路
brush [brʌʃ] 电刷
bulb [bʌlb] 灯泡
burglar alarm ['bɜːglə(r)] [ə'lɑːm] 报警器
burglar ['bɜːglə(r)] 强盗，盗贼
burn out 烧
bus duct [dʌkt] 护埂
cable tray ['keɪbl] [treɪ] 电缆桥架
cable tray and support [sə'pɔːt] 电缆桥架及支架
cabling and wiring 布线
calculation [ˌkælkju'leɪʃn] 计算
calibration and adjustment [ˌkælɪ'breɪʃn] [ə'dʒʌstmənt] 校验与调整
capacitance [kə'pæsɪtəns] 电容
capacity [kə'pæsəti] 容量
cathode ['kæθəʊd] 阴极
cathode-ray ['kæθəʊd] [reɪ] 阴极射线
cathode-ray tube [tjuːb] 阴极射线管
CCTV (closed circuit television) system 闭路电视系统

cell [sel] 电池
ceramic capacitor [sə'ræmɪk] 陶瓷电容器
characteristic [ˌkærəktə'rɪstɪk] 特性，特征
charge flow 电荷流动
charge [tʃɑ:dʒ] 电荷
charger ['tʃɑ:dʒə(r)] 充电器
check [tʃek] 检查
circuit ['sɜ:kɪt] 电路
closed circuit 闭合电路
code [kəʊd] 代码
coil [kɔɪl] 圈，线圈
collector [kə'lektə(r)] 集电极
COM [kɑm] 接口，端口
common-base current gain ['kɒmən] [geɪn] 共基极放大系数/增益
common-emitter current gain 共发射极放大系数/增益
communication system [kəˌmju:nɪ'keɪʃn] ['sɪstəm] 通讯系统
commutator ['kɒmjuteɪtə(r)] 换向器
complete 完整的，闭合的
completed circuit 闭合电路
component [kəm'pəʊnənt] 元件，组件
conductor [kən'dʌktə(r)] 导体
conn type (connection) 连接形式
connecting wire 连接导线
connection group [kə'nekʃn] [gru:p] 连接组
construction organization design [kən'strʌkʃn] [ˌɔ:gənaɪ'zeɪʃn] [dɪ'zaɪn] 施工组织设计
consumer [kən'sju:mə(r)] 用户
continuous manner [kən'tɪnjuəs] ['mænə(r)] 连续的方式
control [kən'trəʊl] 控制
control center [kən'trəʊl] ['sentə] 控制中心
control device 控制仪器/设备
control grid [kən'trəʊl] [grɪd] 控制栅
control panel [kən'trəʊl] ['pænl] 控制面板
control valve [vælv] 控制阀
conversion [kən'vɜ:ʃn] 变换，转换
converter [kən'vɜ:tə(r)] 转换器
copper line ['kɒpə(r)] [laɪn] 铜线
core [kɔ:(r)] 芯，核
core type 芯型
counter 计数器
CPU (central processing unit) [prəʊ'sesɪŋ] 中央处理器
cross-section area 横截面
current ['kʌrənt] 电流
damage ['dæmɪdʒ] 损坏
DC circuit 直流电路
DCS (distributed control system) control cabinet [dɪs'trɪbju:tɪd] ['kæbɪnəts] 集散控制系统控制柜
DCS control center DCS 控制中心
deflection plate [dɪ'flekʃn] [pleɪt] 偏转板
delete [dɪ'li:t] 删除，取消
design [dɪ'zaɪn] 设计
desoldering tool 吸锡器
detect [dɪ'tekt] 监测
detecting device [di'tektiŋ di'vais] 检测装置
development [dɪ'veləpmənt] 开发
deviation [ˌdi:vi'eɪʃn] 偏差
device [dɪ'vaɪs] 仪器，设备
dial gauge ['daɪəl] [geɪdʒ] 百分表
dialplate ['daiəl pleit] 表盘
diameter 直径
digital style 数字式
digital ['dɪdʒɪtl] 数字的
diode ['daɪəʊd] 二极管
discrete component 分散的独立元件
distribution network [ˌdɪstrɪ'bju:ʃn] ['netwɜ:k] 配电网络
distribution power line [ˌdɪstrɪ'bju:ʃn] 配电线路
disturbance [dɪ'stɜ:bəns] 扰动
DMM (digital multimeter) ['dɪdʒɪtl] ['mʌltɪmi:tə] 数字万用表
drain valve [dreɪn] [vælv] 排放阀
drive shaft [draɪv] [ʃɑ:ft] 驱动轴

dual in-line package (DIP) 双列式封装
DWG & STD NO.(drawing & standard No.) ['drɔ:ɪŋ] ['stændəd] 图纸与标准号
electric appliance 电气设备
electric brush device 电刷装置
electric supply [sə'plaɪ] 电源/力供应
electric [ɪ'lektrɪk] 电的
electrical [ɪ'lektrɪkl] 电气的
electrical pen 电工笔
electricity [ɪˌlek'trɪsətɪ] 电
electrode [ɪ'lektrəʊd] 极，电极，焊条
electrolytic capacitor [ɪˌlektrə'lɪtɪk] 电解电容器
electromagnetic-force law 电磁定律
electromagnetism [ɪˌlektrəʊ'mægnətɪzəm] 电磁学
electron [ɪ'lektrɒn] 电子
electron flow 电子流动
electron gun [ɪ'lɛkˌtran] [gʌn] 电子枪
electronic industry 电子工业
electronic measuring instrument 电子测量仪器
electronics [ɪˌlek'trɒnɪks] 电子学
electrotechnics [ɪ'lektrə'teknɪks] 电工学
element ['elɪmənt] 元件，元素
emergency control system [ɪ'mɜ:dʒənsɪ] 紧急控制系统
emitter [ɪ'mɪtə] 发射极
end connector [kə'nektə(r)] 终端接头
engineer [ˌendʒɪ'nɪə(r)] 工程师
engineering [ˌendʒɪ'nɪərɪŋ] 工程
environment [ɪn'vaɪrənmənt] 环境
equation [ɪ'kweɪʒn] 公式，等式
error ['erə(r)] 错误
etc. [ˌet 'setərə] 等等
excitation mode [ˌeksaɪ'teɪʃən] 励磁线圈
excitation voltage 励磁电压
ex-factory 出厂
exist [ɪg'zɪst] 存在
expander for thermometer [ɪks'pændə] [θə'mɒmɪtə(r)] 温度计扩大管
exploded view [ɪk'spləʊdɪd] 分解图
face-plate 面板
FAR (field auxiliary room) [fi:ld] [ɔ:g'zɪliəri] 现场辅助房
farad 法拉
faucet ['fɔ:sɪt] 插孔
feedback ['fi:dbæk] 反馈
field magnet ['mægnət] 场磁铁
field winding [fi:ld] ['waɪndɪŋz] 励磁绕组
fire alarm ['faɪə(r)] [ə'lɑ:m] 火灾报警
fire and gas instrument 防火及气体仪表
fireproof ['faɪəpru:f] 防火
fixed resistor 不变/固定电阻
fixed sleeve [f□kst] [sli:v] 固定套管
floodlight ['flʌdlaɪt] 泛光灯
flow chart [fləu tʃɑ:t] 流程图
flow instrument 流量仪表
flow transmitter [fləʊ] [træns'mɪtə(r)] 流量变送器
footing ['fʊtɪŋ] 底脚
force [fɔ:s] 力，力量
force measuring device 测力装置
form [fɔ:m] 形式，格式
formula ['fɔ:mjələ] 公式
forward biased ['fɔ:wəd] ['baɪəst] 正向偏置
foundation [faun'deiʃən] 基础
frequency ['fri:kwənsi] 频率
front bearing cover 前轴承盖
fuel tank ['fju:əl] [tæŋk] 油箱
fuel-burning station ['fju:əl] ['bɜ:nɪŋ] ['steɪʃn] 热电站
function ['fʌŋkʃn] 功能
fuse [fju:z] 保险丝，熔丝
gas relay ['ri:leɪ] 气体继电器
gate [geɪt] 门
gauge/root gate valve 多路闸阀
generator ['dʒenəreɪtə(r)] 发电机
give power ['paʊə(r)] 提供动力/电源
globe valve 截止阀
grid [gr□d] 高压输电
grounding ['graʊndɪŋ] 接地

hand-over and acceptance [ək'septəns] 交工验收

high price performance ratio [pə'fɔ:məns] ['reɪʃɪəʊ] 很高的性价比

hook-up dwg of liquid pressure measurement 测量液体压力管路连接图

housing ['haʊzɪŋ] 罩壳

HV bushing ['bʊʃɪŋ] 高压套管

HV/LV 高压/低压

i.e. [ˌaɪ 'i:] 即

impedance voltage [ɪm'pi:dns] 阻抗电压

impede [ɪm'pi:d] 阻止，阻抗

inductance [ɪn'dʌktəns] 电感

inductance loop 电感线圈

inductor [ɪn'dʌktə] 电感

industrial [ɪn'dʌstriəl] 工业

information [ˌɪnfə'meɪʃn] 信息

input ['ɪnpʊt] 输入

input loop circuit 输入线圈电路

input power 输入功率/电源

input ['ɪnpʊt] 输入

inspection box [ɪn'spekʃn] 检查箱

installation [ˌɪnstə'leɪʃn] 安装

instrument ['ɪnstrəmənt] 仪表

insulation grade [ˌɪnsju'leɪʃn] [greɪd] 绝缘等级

integrated ['ɪntɪgreɪtɪd] 集成的

interface circuit ['ɪntəfeɪs] 接口电路

interruption [ˌɪntə'rʌpʃn] 中断

iron core ['aɪən] 铁芯

joint reviewing [dʒɔɪnt] [ri'vju:ɪŋ] 会审

junction box ['dʒʌŋkʃn] 接线盒

junction ['dʒʌŋkʃn] 结

kilo ['ki:ləʊ] 千，千克

knife [naɪf] 电工刀

knob switch [nɒb] 旋钮开关

layout ['leɪaʊt] 布置图，布局，草图

LCD (liquid crystal display) 液晶显示

lead [lid] 导线

LED (light-emitting diode) [ɪ'mɪt] 发光二极管

left arm [left] [ɑ:m] 左臂

length [lɛŋkθ] 长度

length of thick-thin tube 大小头长度

level meter ['levl] 液/物位计

level transmitter ['levl] [træns'mɪtə(r)] 液位变送器

light-dependent resistor (LDR) 光敏电阻

light-emitting diode [ɪ'mɪtɪŋ] 发光二极管

lighting fixture ['laɪtɪŋ] ['fɪkstʃə(r)] 照明灯具

lightning ['laɪtnɪŋ] 避雷，闪电

line resistance 线路电阻

load [ləʊd] 负荷，负载

local equipment ['ləʊkl] [ɪ'kwɪpmənt] 地方设备

local instrument ['ləʊkl] 就地仪表

locking device 锁紧装置

logic ['lɒdʒɪk] 逻辑

loop of wire 导线回路

loop testing [lu:p 'testiŋ] 回路实验

magnetic field [mæg'netɪk] [fi:ld] 磁场

magnetic [mæg'netɪk] 磁性的

magnitude ['mægnɪtju:d] 幅度，大小

mains [meɪnz] 电源/电力/电网

major part ['meɪdʒə(r)] [pɑ:t] 主要部件/部分

make measurement ['meʒəmənt] 测量

male threaded [meɪl] 外螺纹

manipulate [mə'nɪpjʊleɪt] 操作

man-made 人造的

manufacturer [ˌmænju'fæktʃərə(r)] 制造商，厂家

material [mə'tɪəriəl] 材料

material list [mə'tɪəriəl] [lɪst] 材料表

MCB (miniature circuit breaker) 小型断路器

measure ['meʒə(r)] 测量

measuring meter 测量仪器

measuring range 测量范围，量程

mega ['megə] 兆

mega ohmmeter ['megə] ['əʊmmi:tə(r)] 兆欧表

megger ['megə] 兆欧表

memory ['memərɪ] 存储器，内存

metal ['metl] 金属

metering head [hed] 测量头

metering rod [rɒd] 量杆

micro ['maɪkrəʊ] 微
microcontroller 微控制器
microdrum [maɪk'rɒdrʌm] 微分筒
micrometer screw [skru:] 测微螺杆
microprocessor [ˌmaɪkrəʊ'prəʊsesə(r)] 微处理器
microprocessor chip 微处理器芯片
milli 毫
millisecond ['mɪlisekən] 毫秒
miniaturization [ˌmɪnətʃəraɪ'zeɪʃn] 小型化
minus ['maɪnəs] 减，负的
miscellaneous panel [ˌmɪsə'leɪnɪəs] ['pænl] 零星配电盘
mode [məʊd] 方式
modern automatic ['mɔdən] [ˌɔ:tə'mætɪk] 现代自动化
moisture meter ['mɔɪstʃə(r)] 吸湿器
MOS (metal oxide semiconductor) 金属氧化物半导体
motor frame [freɪm] 电动机架
motor ['məʊtə(r)] 电机，马达
mount [maʊnt] 安装
multi-functions ['fʌŋkʃn] 多功能
multimeter ['mʌltɪmi:tə] 万用表
mV meter 毫伏表
nameplate 铭牌
neck [nek] 颈，管颈
negative ['negətɪv] 负的，阴极
obstacle ['ɒbstəkl] 障碍，阻碍
oil conservator [kən'sɜ:vətə(r)] 储油柜
oil discharge valve [dɪs'tʃɑ:dʒ] [vælv] 放油阀门
oil immersed [ɪ'mɜ:st] 油浸式
oil meter 油表
optical fiber cable ['ɒptɪkl] ['faɪbə] 光缆
oscilloscope [ə'sɪləskəʊp] 示波器
outer surface ['sɜ:fɪs] 外表面
output loop circuit 输出线圈电路
output power 输出功率/电源
output voltage 输出电压
outside micrometer [maɪ'krɒmɪtə(r)] 外径千分尺
overhead line 高架线
oxide ['ɒksaɪd] 氧化物
package ['pækɪdʒ] 包裹，封装
palm [pɑ:m] 手掌
panelboard [peɪ'nelbɔ:d] 配电盘
paper capacitor 纸质电容器
parallel ['pærəlel] 并联
parallel circuit 并联电路
PCB (printed circuit board) [p'rɪntɪd] [bɔ:d] 印刷电路板
photo-diode 光电二极管
physicist ['fɪzɪsɪst] 物理学家
pin grid array package (PGA) 阵列式引脚封装
pin [pɪn] 管脚
pipe (vessel) NO. 管道或设备号
plant [plɑ:nt] 工厂
plastic film capacitor（塑料）薄膜电容器
plier ['plaɪə] 虎钳
plug [plʌg] 堵头
plus [plʌs] 加，正的
pointer ['pɔɪntə(r)] 指针
positive ['pɑzɪtɪv] 正的，阳极
power plant 发电厂
power station 发电站
power supply ['paʊə(r)] [sə'plaɪ] 电源
power supply and earthing ['ɜ:θɪŋ] 电力供应及接地
power supply cabinet ['paʊə(r)] [sə'plaɪ] 电力供应柜
power switch 电源开关
power work ['paʊə(r)] [wɜ:k] 电力工作
preparation [ˌprepə'reɪʃn] 准备
pressure instrument 压力仪表
pressure meter ['preʃə(r)] ['mi:tə(r)] 压力表
pressure switch 压力开关
pressure transmitter ['preʃə(r)] 压力变送器
principle ['prɪnsəpl] 原理
probe [prəʊb] 探针
process ['prəʊses] 工艺，过程
processor ['prəʊsesə(r)] 处理器
product No. ['prɒdʌkt] 产品编号

quantity ['kwɒntəti] 量，数量

radiator ['reɪdieɪtə(r)] 散热器

radio ['reɪdɪəʊ] 无线电，收音机

range of measurement 量程

rated voltage ['reɪtɪd] 额定电压

ratio ['reɪʃɪəʊ] 比

RCD (residual current device) 漏电保护器

reading ['ri:dɪŋ] 读数

rear bearing cover [rɪə(r)] 后轴承盖

rear end cover 后端盖

receptacle [rɪ'septəkl] 插座

rectification [ˌrektɪfɪ'keɪʃn] 调校，修整

rectifier ['rektɪfaɪə] 整流器

refrigerator [rɪ'frɪdʒəreɪtə(r)] 冰箱

regulator ['regjuleɪtə(r)] 调节器

relationship [rɪ'leɪʃnʃɪp] 关系

replacement [rɪ'pleɪsmənt] 替代物

requirement [rɪ'kwaɪəmənt] 要求

resist [rɪ'zɪst] 阻止

resistor [rɪ'zɪstə(r)] 电阻（器）

result [rɪ'zʌlt] 结果

reversed biased [rɪ'vɜ:st] 反向偏置

right arm [raɪt] 右臂

right measuring range 正确的量程

rotate [rəʊ'teɪt] 旋转

rotating shaft [rəʊ'teɪtɪŋ] [ʃɑ:ft] 转轴

ruler frame [freɪm] 尺架

safety airway ['eəweɪ] 安全气道

safety sign 安全标志

threaded sleeve 螺纹轴套

screwdriver 旋具，螺丝刀

seamless steel tube ['si:mləs] [sti:l] 无缝钢管

semiconductor [ˌsemikən'dʌktə(r)] 半导体

sensor ['sensə(r)] 传感器，感应器

separate unit 单独元件/装置

series ['sɪəri:z] 串联

series circuit 串联电路

series-parallel circuits ['sɪəri:z] ['pærəlel] 串并联电路

servo motor ['sɜ:vəʊ] 伺服电动机

shell type 壳型

short circuit 短路

shunt [ʃʌnt] 并励

signal generator ['sɪgnəl] ['dʒenəreɪtə(r)] 信号发生器

signal monitor ['s□gnəl] ['mɒnɪtə(r)] 信号监测器

silicon ['sɪlɪkən] 硅

single in-line package (SIP) 单列式封装

single unit trial running ['sɪŋgl] ['ju:nɪt] ['traɪəl] ['rʌnɪŋ] 单体试车

single-phase [□s□ŋgl] [feɪz] 单相

smoke/heat detector [sməʊk] [hi:t] [dɪ'tektə(r)] 感烟/热探测器

soldering iron 烙铁

splicing ['splaɪsɪŋ] 接头

stator ['steɪtə] 定子

step motor [step] 步进电动机

step-down substation 降压变电站

step-up substation ['sʌbsteɪʃn] 增压变电站

stereo ['sterɪəʊ] 立体声

substation ['sʌbsteɪʃn] 变电站

suitable adapter [ə'dæptə] 相应的转换接头

switch cabinet [swɪtʃ] 开关柜

symbol ['sɪmbəl] 符号，象征

system ['sɪstəm] 系统

Tag NO.位号

take a measurement 测量

take down 拆卸

technology [tek'nɒlədʒi] 技术

tee 三通

telecommunication system [ˌtelɪkəˌmju:nɪ'keɪʃn] ['sɪstəm] 电讯系统

temperature ['temprətʃə(r)] 温度

temperature instrument 温度仪表

temperature transmitter ['temprətʃə(r)] 温度变送器

terminal ['tɜ:mɪnl] 端子，端部

test equipment [ɪ'kwɪpmənt] 测试设备

tester ['testə(r)] 测试仪

thermistor [θɜ:'mɪstə] 热敏电阻
thermometer [θə'mɒmɪtə(r)] 温度计
threaded ['θredɪd] 螺纹的
thumb [θʌm] 大拇指
timer 定时器
top resistor [rɪ'zɪstə(r)] 上电阻
transducer [trænz'dju:sə(r)] 传感器
transformer [træns'fɔ:mə(r)] 变压器
transistor [træn'zɪstə(r)] 晶体管
transmission line [træns'mɪʃn] 输电线
transmit [træns'mɪt] 传输，传送
triode ['traɪəʊd] 三极管
trouble shooting ['trʌbl] ['ʃu:tɪŋ] 故障排斥
truth table [tru:θ] table 真值表
TTL (transistor-transistor logic) 晶体管-晶体管逻辑
type [taɪp] 类型
ULSIC (ultra large scale integrated circuit) 超大规模集成电路
UPS (uninterrupted power supply) [ˌʌnˌɪntə'rʌptɪd] 无间断供电
value ['vælju] 数值，价值
variation [ˌveəri'eɪʃn] 变化，变异
vent hole [vent] [həʊl] 通风孔
vernier caliper ['vɜ:nɪə] ['kælɪpə] 游标卡尺
voltage ['vəʊltɪdʒ] 电
voltage measurement 电压测量
voltmeter ['vəʊltmi:tə(r)] 电压表
VOM (volt-ohm-milliammeter) [mɪlɪ'æmɪtə] 伏特，欧姆，毫安表
washing machine ['wɒʃɪŋ] [mə'ʃi:n] 洗衣机
watch window [wɒtʃ] ['wɪndəʊ] 视察窗
weight [weɪt] 重量
Wheatstone bridge [brɪdʒ] 惠斯通电桥
winding insulation ['waɪndɪŋ] [ˌɪnsju'leɪʃn] 绕组绝缘
wire stripper 剥线钳
wire ['waɪə(r)] 电线，导线
wiring ['waɪərɪŋ] 接线
working mode 工作方式
Zener diode 稳压/齐纳二极管
zero volt ['zɪərəʊ] [vəult] 零电压

References

[1] 杨春生. 机电专业英语. 北京：电子工业出版社，2007.

[2] 徐鸿，董其伍. 过程装备与控制工程专业英语. 北京：化学工业出版社，2002.

[3] 全国十五所高等院校联合编写. 焊接专业英语文选. 南宁：广西人民出版社，1983.

[4] 赵丽玲. 焊接专业英语. 北京：机械工业出版社，2008.

[5] 赵丽玲. 焊接英语. 北京：中国劳动社会保障出版社，2011.

[6] 王晓江，李学哲. 热加工专业英语. 北京：机械工业出版社，2008.

[7] 王晓敏. 电工英语. 北京：中国劳动社会保障出版社，2004.

[8] 周虹. 电子英语. 北京：中国劳动社会保障出版社，2003.

[9] 陈国祥. 机械设备安装工. 北京：中国劳动社会保障出版社，2011.

[10] 杨育红. 化工设备与维护. 北京：化学工业出版社，2008.

[11] 徐慧波. 化工机械维修铆焊工艺. 北京：化学工业出版社，2005.

[12] 伍广. 焊接工艺. 北京：化学工业出版社，2009.

[13] 汪兴云. 过程仪表安装与维护. 北京：化学工业出版社，2006.